Gulls and Plovers

Studies in Behavioural Adaptation

Series Editor: John Lazarus, Department of Psychology,
The University of Newcastle-upon-Tyne

Gulls and Plovers: The Ecology and Behaviour of Mixed Species Feeding Groups
C.J. Barnard and D.B.A. Thompson

In preparation:
Modelling in Behavioural Ecology
Dennis Lendrem
Motivation and Decision-making in Animals
Robin McCleery

GULLS AND PLOVERS

The Ecology and Behaviour of Mixed-Species Feeding Groups

C.J. BARNARD, Department of Zoology, University of Nottingham and D.B.A. THOMPSON, Department of Zoology, University of Liverpool

Springer-Science+Business Media, B.V.

British Library Cataloguing in Publication Data

Barnard, C.J.
Gulls and plovers: the ecology and behaviour of mixed species feeding groups.—(Croom Helm studies in behavioural adaptation)
1. Gulls—Food 2. Plovers—Food 3. Birds — Food
I. Title II. Thompson, D.B.A.
598'.3 QL696.C46

ISBN 978-94-010-8652-3 ISBN 978-94-009-4864-8 (eBook)
DOI 10.1007/978-94-009-4864-8

Typeset in Plantin by Leaper & Gard Ltd, Bristol, England

Contents

Series Editor's Foreword		vii
Preface		x
Chapter 1.	Why Feed in Flocks?	1
Chapter 2.	Gulls and Plovers	46
Chapter 3.	Choosing Where to Feed. Choice of Field	66
Chapter 4.	Choosing Where to Feed: Choice within Fields	93
Chapter 5.	Choosing What to Eat	116
Chapter 6.	Time Budgeting and Feeding Efficiency	143
Chapter 7.	Flock Dynamics: Patterns of Arrival and Departure	192
Chapter 8.	Kleptoparasitism: Host and Prey Selection by Gulls	217
Chapter 9.	Vigilance, Alarm Responses and an Early Warning System	255
Chapter 10.	Gulls and Plovers: an Overview	275
References		280
Index		296

For Dawn and Anna, Lucy and Matthew

Series Editor's Foreword

In the early years of this century a Scottish doctor speculated on the evolutionary origin of human tears. It seemed to him that with the increase in brain size and cognitive powers of our early ancestors many events in the struggle for existence would be just too distressing to observe. How comforting then, for the mother, distraught by the sight of her child being devoured by a lion, to cloud her vision with a flood of tears!

Just so, though if the good doctor had pondered further, the following picture might have occurred to him, comfortable in his speculative armchair, and given him some pause for thought.

These stories do not, of course, get us very far in understanding the evolution of tears or anything else, but they do remind us how far the study of behavioural adaptation has come this century. This is, in fact, an exciting time for students of behaviour. The last twenty years have seen a great advance in the theoretical armoury for tackling questions of behavioural evolution and adaptation, and a parallel expansion in empirical studies, particularly in the field. The concepts of inclusive fitness and the evolutionarily stable strategy, for example, have helped to explain major features of social behaviour and have generated entirely new questions and predictions for the field worker to examine. Cost-benefit analysis and optimisation theory have done the same for behaviour in general, and links with population biology and population genetics are becoming stronger.

The heady days which saw the birth of behavioural ecology and sociobiology are now over, the new concepts have been refined and consolidated, and field data and comparative studies have accumulated at an impressive rate. Now seems a good time to take stock, to review the state of the art, and to point some directions for future work. These are the aims of the present series, which will examine questions of behavioural adaptation and evolution in depth. As for our intended readership, we hope that all those interested in such problems, from advanced undergraduate to research worker and lecturer, will find these books of value. Some contributions to the series will encompass particular areas of study, reviewing theory and data and presenting fresh ideas. Others will report the findings of new empirical studies of an extensive nature, which make a significant contribution by examining a range of interrelated questions. The richness, but also the difficulty, of functional enquiry results from the multiple effects of selection pressures and the complex causal relationships between the behavioural responses to evolutionary forces. Studies which measure a comprehensive set of related behavioural attributes, and their ecological correlates, are therefore particularly valuable.

The present book describes just such a study. Chris Barnard and Desmond Thompson show how the feeding requirements of three interacting species of birds influence their social lives in a variety of complex and subtle ways. Chapter by chapter they build a story in which foraging demands are seen to determine not only where the birds feed and what they feed on, but also their time budgets and flock dynamics, while flock structure, in turn, is shown to have important effects on the foraging strategy. They achieve this by integrating two areas of behavioural ecology that have until now remained largely independent: foraging theory and group living. In addition, they describe a number of intra- and interspecific tactics for feeding competition and the early warning of predatory approach; particularly striking are the ingenious manoeuvres adopted by two of the species to thwart the food stealing attempts of the third.

Barnard and Thompson's method has been to start with the basic questions of foraging theory and flock dynamics, and then to develop and test a series of new hypotheses in order to gain a more complete understanding of the system. In this way they generate a quite extraordinary amount of data and reveal an unsuspected subtlety and sensitivity in the individual's behavioural solutions to the moment-to-moment problems of feeding in a social context. Their work demonstrates at once both the intellectual allure of evolutionary studies and the daunting complexity in nature that remains to be explained.

John Lazarus
Department of Psychology
University of Newcastle-upon-Tyne

Preface

Since its advent in the 1960s, behavioural ecology (the synthesis of ethology, population ecology and evolutionary theory) has revolutionised our understanding of the evolution of behaviour. Until recently, however, models specific to certain aspects of behavioural ecology have been proposed and tested largely in isolation. Models of foraging behaviour, for instance, have been developed mainly without regard for the complex social environment in which many predators feed. Conversely, models of social behaviour have not catered for the changes in foraging criteria which may be imposed on predators as a consequence of their social organisation. With increasingly detailed laboratory and field studies, however, it is now possible to integrate different aspects of the behavioural ecology of certain species, and assess their role in determining individual survivorship and reproductive potential. This book examines the relationship between individual foraging behaviour and group dynamics in a mixed-species feeding association. It shows how consideration of adaptive behaviour at the level of the individual can lead to testable predictions about the formation of feeding groups. A major point emerging from the book is that the consequences of mixed association for the individual may not only be very different from those of associating with conspecifics, but that, even within particular associations, they may depend on the size and species composition of the group.

The book focuses on a winter feeding association between three charadriiform bird species, lapwings (*Vanellus vanellus*), golden plovers (*Pluvialis apricaria*) and black-headed gulls (*Larus ridibundus*), which we have studied on agricultural land for the past six years. The association is based on feeding flocks of lapwings which are exploited by golden plovers as indicators of the best places to start feeding. Lapwings also provide local information within the flock on the whereabouts and profitability of prey (in this case earthworms). Birds of both species are exploited by black-headed gulls, which steal food and cause changes in individual feeding efficiency and the size and species composition of the flock.

In Chapter 1, we look at the way single- and mixed-species flocking might affect individual survivorship and discuss the relationship between this and flock dynamics. Chapter 2 then introduces the species and study area and speculates about the evolution of pasture use by plovers. Chapters 3 and 4 move on to consider the distribution of birds around the study area, looking not only at preferences for particular types of field but also at the relationship between the concentration and movement of birds within fields and the density and distribution of their earthworm prey. Chapters 5 and 6 examine the effects of flock size and composition on individual feeding efficiency, taking into account the effect of climatic factors on food requirement and availability. We develop a simple optimal-diet model which incorporates a number of time and probability costs of prey selection and show how optimal diets and prey selection shift with flock composition. Chapter 6 also shows how changes in the allocation of time to different activities affects individual feeding efficiency and energy balance.

Chapter 7 uses the results of the previous chapters to make some predictions about the build-up and stability of flocks. Rates of arrival and departure at feeding sites are examined in relation to expected feeding efficiency and we discuss the apparent 'assembly point' function of pre-foraging flocks. Chapter 8 looks in detail at kleptoparasitism by gulls. Optimal-host and prey-selection models are developed and tested and we look at the ability of gulls to meet their daily energy demand by stealing from plovers. In Chapter 9, we investigate the relationship between flock composition and responsiveness to alarm and show how gulls may provide plovers with an early-warning system. Finally, Chapter 10 provides an overview of the book's major findings.

Throughout the book, the association between gulls and plovers is discussed in the context of recent ideas and investigations in behavioural ecology and comparisons are drawn with other single- and mixed-species associations. Discussion of data is therefore integrated with review material so that conclusions can be generalised to other species. Each chapter finishes with a résumé and summary to develop and emphasise the main points.

Naturally, a lot of people have helped in the development of the study, from assisting in data collection and analysis to reading manuscripts and offering criticism and advice. In particular, we should like to thank Hilary Stead for assistance in the field and with data-handling and John Lazarus for his, as usual, helpful and detailed editorial comments on the manuscript. We should also like to thank John Goss-Custard, John Lazarus, Bill Sutherland, Tim Reed, Hans Källander, Derek Ratcliffe, Dick Waite, Chris Brown, Peter Evans, Mike Pienkowski, Andy Wood, Nick Davidson, Neil Metcalfe, Hector Galbraith, Pat Monaghan, Bob Furness, John Speakman, Dennis Lendrem, Bill Hale, Dave Curtis, John Smyth, Pat Thompson, Alisdair Houston, Chris Spray, Colin

Galbraith, Rob Fuller, Philip Burton, Stefan Ulfstrand, Nils Stenseth, Ingvar Byrkjedal, Sven-Axel Bengtson, Robert Moss, Dave Pons, Terry Burke, Dave Parkin, Ray Parr and Desmond and Maimie Nethersole-Thompson for helpful discussion, advice or correspondence. Dick Waite, Hans Källander, John McLennan, Mike Pienkowski, John Speakman, H.M.S. Blair, A. Dobbs, Keith Bannister, Rob Kirkwood and Keith Futter kindly allowed us to quote from their unpublished data. We are heavily indebted to Rob Fuller, Richard Vaughan and Tony Holley for their generosity in providing us with photographs, to Sam Grainger, Brian Case and Martin Hutchings who helped in the preparation of some of the figures and plates and Rebecca Torrance for preparing the chapter vignettes. Finally we must thank David and John Allsopp for allowing us to work on and dig up their land, the Natural Environment and Science and Engineering Research Councils and Professors P.N.R. Usherwood and C.I. Howarth for financial and other support, and Dawn Thompson and Siân, Anna, Lucy and Matthew Barnard for support, assistance and patience.

C.J. Barnard, D.B.A. Thompson

Chapter 1

Why Feed in Flocks?

In one way, this book is about three species of bird feeding on some fields in Nottinghamshire. In another, it is about the interaction between ecology, social behaviour and predation. That a study of the former can lead to conclusions about the latter is a reflection of the recent, dramatic increase in our understanding of the evolution of animal behaviour, an increase which stems from an integration of ideas from population biology, evolutionary theory and ethology. Behavioural ecology, as the new synthetic approach has become known, seeks to interpret the evolution of behaviour in terms of the constraints imposed and opportunities provided by an animal's ecology. It assumes that variation in behaviour has a genetic basis and that natural selection is the principal agent governing the differential survival of variants. This simple assumption allows precise, quantitative questions to be asked about the evolution of behaviour in any particular context: How many types of prey should a predator take? How many conspecifics should it associate with? How long should a male court a female before attempting to copulate? The answers are couched in terms of changes in the animal's reproductive potential as a consequence of its choice from a range of feasible alternatives. The strength of this approach lies in its broad applicability. The same assumptions and methods of analysis underlie the study of parent-offspring relationships in baboons and of choice of prey by foraging water boatmen. This book is concerned with one particular feeding association, but its conclusions are aimed at the behaviour of socially feeding predators in general. We begin, however, by considering the reasons why predators might form groups in the first place.

In most animals, the potential for survival and reproduction is determined by success in acquiring limited resources. Food, water, shelter and other essential commodities are often scarce or otherwise difficult to find. Competition may therefore be severe, and diverse adaptations have evolved to reduce it. In addition, animals may themselves be a limited resource for others higher up the food chain. Any solution to their problems of resource acquisition must therefore take into account the risk of predation. One way to get around both problems may be to join a group. Mounting evidence from phylogenetically diverse species shows that grouping behaviour may simultaneously reduce individual risk of predation and enhance feeding efficiency. In the first part of this chapter, we review some of the evidence. In doing so, however, we are concerned mainly with those properties of grouping which are relevant to bird flocks and therefore to the study discussed in this book (although not all the examples come from birds). A summary of the costs and benefits of flocking in a range of bird species is provided in Tables 1.1 and 1.2. The tables are not exhaustive taxonomic surveys. For more general reviews of social behaviour see Bertram (1978), Wittenberger (1981), Pulliam and Millikan (1982), Gosling and Petrie (1981) and Pulliam and Caraco (1984).

While some of the effects of flocking apply equally to single- and mixed-species aggregations, associations with heterospecifics can bring advantages (and disadvantages) which do not accrue in single-species flocks. In the second section of this chapter, we discuss some of the ways in which this may occur. We then consider the interaction between individual costs and benefits and feeding-flock dynamics and examine critically the concept of optimal flock size and species composition.

To Eat or Be Eaten: Costs and Benefits of Forming Flocks

Avoiding predation

There is an obvious truth in the maxim 'safety in numbers'. Safety, however, can arise in several different ways (see also Table 1.1).

Avoiding detection by predators. While we might expect a group of animals to be more conspicuous than a solitary individual, groups are likely to be scarcer because more of the population is concentrated at fewer points in space. They may therefore be missed more often by predators wandering randomly across their home range. The success rate of a predator which is likely to capture only one individual per group at best may thus be considerably reduced by the tendency for prey to form groups.

Counteracting the scarcity effect, however, is the fact that large numbers of closely spaced individuals are usually more conspicuous. Birds in flocks sometimes use 'contact' calls which make them audibly conspicuous (Odum 1942) and large roosts may produce a strong odour which renders them more vulnerable to olfactory detection. Andersson and Wicklund (1978) found that colonially nesting fieldfares (*Turdus pilaris*) attracted more predators than solitary nesting birds, although, in this case, greater attraction was offset by the fieldfares' effective mobbing response. Treisman (1975) and Vine (1971, 1973) have put forward detailed mathematical models of the advantages of grouping in relation to detectability by predators.

Early detection of predators. Many predators depend on surprise for making a successful attack (e.g. Schaller 1972, Brown 1976, Ratcliffe 1980). The sooner a potential prey individual can spot an approaching predator, the more likely it is to avoid capture. In a group, more sets of sense organs are likely to be scanning the environment at any one time. We might therefore expect predators to be detected earlier. This appears to be the case. Kenward (1978), for instance, found that a trained goshawk (*Accipiter gentilis*) was less successful when it attacked large flocks of woodpigeons (*Columba palumbus*). This was because the pigeons took off while the hawk was still some distance away, implying that the latter had been spotted earlier than when attacking small flocks. Careful laboratory experiments by Lazarus (1979a), however, have shown that detection may occur even earlier than indicated by flight responses and that flocking may be a means of delaying and modifying responses to alarm. These experiments are discussed more fully in Chapter 9.

Confusing predators. When a group of potential prey is alarmed, it may break up and individuals scatter in different directions. Scattering behaviour is well known in a variety of species. Impala (*Aepyceros melampus*) and other bovid herds disperse with erratic jumping movements when a predator appears. To the human observer at least, individuals are lost in the explosion of activity and it is impossible to track any one of them. The same can be said of coastal flocks of waders (Charadriidae) responding to the appearance of a peregrine falcon (*Falco peregrinus*). Several studies have investigated the confusion effect. Neill and Cullen (1974) showed that pike (*Esox lucius*) and cephalopod predators had lower success rates against large shoals of minnows and guppies than against single fish or small shoals. Similarly, Milinski (1977) found that sticklebacks (*Gasterosteus aculeatus*) hesitated and appeared to equivocate when attacking large swarms of independently moving *Daphnia*. The marine insects (*Halobates robustus*) aggregate into flotillas and scatter when fish approach from beneath. This causes the sun to reflect from their bodies at different angles, producing a confusing flashing effect (Treherne and Foster 1981). Treherne and Foster (1981) also showed that the acceleration of movement at the approach of fish facilitates the rapid communication of alarm to other members of the flotilla (the so-called Trafalgar effect).

Deterrence. The aggregation of prey into groups may actually deter predators from attacking. In some cases this is because prey groups pose a physical threat to the predator. Adult muskoxen (*Ovibos moschatus*), for instance,

Table 1.1a: Benefits of feeding in single-species flocks

Species	Early detection of predators	Deterring predators	Reduced individual risk	More time to feed	Exploiting good food supplies	Decreased risk of low feeding efficiency	Local information about food	References
House sparrow (*Passer domesticus*)	—	—	Bigger flocks in more vulnerable areas	Less time scanning, more time feeding	Flock size positively correlated with food density	—	Birds copy and steal food from one another	Barnard (1980a,b,c). Barnard and Sibly (1981)
Starling (*Sturnus vulgaris*)	—	—	Birds in centre of flock scan less	Less time scanning, more time feeding	—	—	Birds imitate feeding neighbours	Powell (1974), Jennings and Evans (1980), Feare (1984)
Yellow-eyed juncos (*Junco phaeonotus*)	—	—	Bigger flocks when attacks likely	Less time scanning, more time feeding	—	Subordinates benefit through reduction in foraging-time variance	—	Caraco (1979b, 1981), Caraco *et al.* (1980a,b,c)
Weaver bird (*Quelea quelea*)	Birds in big flocks detect stimuli sooner	—	Birds in big flocks less likely to take off when alarmed	—	Birds attracted by others feeding	—	Birds copy each other. Roosts may act as 'information centres'	Crook (1964), Lazarus (1979a,b), de Groot (1980)
Harris sparrow (*Zonotrichia querula*)	—	—	—	—	Subordinates use dominants to defend good food supplies	—	Dominants copy subordinates	Rohwer and Ewald (1981)
Great tit (*Parus major*)	—	—	—	—	—	—	Birds copy successful companions	Krebs *et al.* (1972)
Whinchat (*Saxicola rubetra*)	—	—	—	Birds feed more intensively in large flocks	Birds flush insects which are then eaten by others	—	—	Draulans and Vessem (1982)
Fieldfare (*Turdus pilaris*)	—	—	—	Less time scanning, more time looking for prey (crouching)	—	—	—	Barnard and Stephens (1983)

Downy woodpecker (*Picoides pubescens*)	—	—	—	Less time scanning, more time feeding	—	—	—	Sullivan (1984)
Great blue heron (*Ardea herodias*)	—	—	—	Less time scanning, more time feeding	Birds prefer to feed near others	Variation in feeding rate smaller in large flocks	Roosts may act as 'information centres'	Krebs (1974)
Barnacle goose (*Branta leucopsis*)	—	—	—	—	Birds prefer to feed near others	—	—	Drent and Swierstra (1977), Drent and van Eerden (1980)
Brent goose (*Branta bernicla*)	—	—	—	—	Flocking synchronises visits to renewing food supplies	—	—	Prins *et al.* (1980)
White-fronted goose (*Anser albifrons*)	—	—	Birds scan less when nearer neighbours	—	—	—	—	Lazarus (1978)
Shelduck (*Tadorna tadorna*)	—	—	—	More time feeding when near neighbours	Most flocking where food is most abundant	—	Rapid foragers quickly surrounded by others	Buxton (1981), Thompson (1981, 1982), Patterson (1982)
Ostrich (*Struthio camelus*)	Overall flock vigilance increases with flock size	—	—	Less time scanning in large flocks	—	—	—	Bertram (1980)
Barbary dove (*Streptopelia risoria*)	—	—	—	Less time scanning, more time feeding	—	—	—	Lendrem (1984)
Ground dove (*Geopelia striata*)	Birds respond soonest in medium-sized flocks	—	—	—	—	—	—	Siegfried and Underhill (1975), Krebs and Barnard (1980)
Woodpigeon (*Columba palumbus*)	Take off sooner in big flocks when attacked	—	Hawks less successful against big flocks	Birds prefer to feed near others	—	—	Birds copy each other	Murton (1971b), Kenward (1978)

Table 1.1a: continued.

Species	Early detection of predators	Deterring predators	Reduced individual risk	More time to feed	Exploiting good food supplies	Decreased risk of low feeding efficiency	Local information about food	References
Sandpipers (*Calidris* spp.)	—	Falcons less likely to attack medium-sized flocks	Falcons less successful against medium-sized flocks	—	—	—	—	Page and Whitacre (1975)
Curlew (*Numenius arquata*)	—	—	—	Less time scanning, more time feeding	Birds most concentrated where food is abundant	—	—	Abramson (1979), Rands and Barkham (1981)
Black-headed gull (*Larus ridibundus*)	—	—	—	—	Flock density positively correlated with food density	—	Birds copy and steal food from one another	Curtis and Thompson (1985)
Rook (*Corvus frugilegus*)	—	—	—	Less time scanning, more time feeding	Large flocks indicate good food supplies	—	—	Lockie (1956), Waite (1981)
Merganser (*Mergus serrator*)	—	—	—	—	Flocks drive fish into shallow water	—	—	Parks and Bressler (1963)
Coot (*Fulica atra*)	—	Collective splashing may distract or confuse predator	—	—	—	—	—	Lack (1954), D.B.A.T. pers. obs.
Black vulture (*Coragyps atratus*)	—	—	—	—	Individuals descending to carcase attract others	Birds co-operate in catching live prey	—	Brent (1937), McIlhenny (1937)

Table 1.1b: Costs of feeding in single-species flocks

Species	Attracting predators	Impaired vigilance	Aggression and food stealing	Prey depletion	Prey disturbance and reduction of availability	Area copying	Reduction in searching/ feeding efficiency	Increased incidence of disease	References
House sparrow (*Passer domesticus*)	—	—	More fighting in big flocks. Some 'copiers' snatch food from searchers. Dominants exclude subordinates from food and force them to edge	—	—	Area copying on patchy supply. 'Copiers' spend little time looking for their own food	—	—	Barnard (1980a, b, c), Barnard and Sibly (1981)
Starling (*Sturnus vulgaris*)	Large winter flocks attract predators	—	Reduction in feeding space results in aggression	—	Birds feeding close together disturb prey, making them unavailable	—	—	Several bacterial diseases transmitted most rapidly amongst dense flocks of birds	Feare and Inglis (1979), Feare (1984)
Yellow-eyed junco (*Junco phaeomotus*)	—	—	More aggression in large flocks	—	—	—	Disruption of the search path in large flocks	—	Caraco (1979b, 1980), Caraco *et al.* (1980a)

Table 1.1b: continued.

Species	Attracting predators	Impaired vigilance	Aggression and food stealing	Prey depletion	Prey disturbance and reduction of availability	Area copying	Reduction in searching/ feeding efficiency	Increased incidence of disease	References
Boat-tailed grackle (*Cassidix mexicanus*)	—	—	—	—	—	—	Reduced searching efficiency in dense flocks	—	Smith (1977)
Weaverbird (*Quelea quelea*)	Large gatherings attract predatory raptors	—	—	—	—	Individuals which do badly follow successful foragers to good feeding sites	—	—	Ward and Zahavi (1973), Newton (1979), de Groot (1980)
Harris sparrow (*Zonotrichia querula*)	—	—	—	—	—	Subordinates area-copied by dominants when food is concealed and patchy	—	—	Rohwer and Ewald (1981)
Willow-tit (*Parus montanus*)	Subordinates more vulnerable than dominants to predation	—	Subordinates excluded from preferred habitat by dominants	—	—	—	—	—	Ekman *et al.* (1981), Ekman and Askenmo (1984)
Downy woodpecker (*Picoides pubescens*)	More easily located in large flocks	—	—	—	—	—	—	—	Sullivan (1984)
Brent goose (*Branta bernicla*)	—	—	—	Prey depletion at high feeding densities results in some birds moving elsewhere	—	—	Birds at front obtain more, but less nutritious, food than birds at rear	—	Prins *et al.* (1980), Prater (1981)

Shelduck (*Tadorna tadorna*)	—	—	Aggression increases with flock density and flock size	—	—	Rap d foragers used by others to locate food	—	—	Patterson (1977, 1982), Thompson (1981), Evans and Pienkowski (1982)
Perching ducks (*Cairinini*)	—	—	—	—	—	—	—	Close proximity of birds results in build-up of faeces and can lead to tuberculosis	Hillgarth and Kear (1981)
Ground dove (*Geopelia striata*)	—	Impaired vigilance at high feeding density	—	—	—	—	—	—	Siegfried and Underhill (1975)
Sandpipers (*Calidris* spp.)	Largest flocks attacked more frequently and successfully than small to medium-sized flocks	—	Birds most aggressive at high feeding density	—	—	—	—	—	Page and Whitacre (1975), Goss-Custard (1977c)
Ringed plover (*Charadrius hiaticula*)	—	—	—	—	Prey disturbance at high feeding density	—	Birds feed more efficiently in small loosely-spaced flocks	—	Pienkowski (1983)
Oystercatcher (*Haematopus ostralequs*)	—	—	More food stealing at high feeding densities	—	—	—	Food intake (especially in subordinates) declines at high feeding density	—	Sutherland and Koene (1982), Ens and Goss-Custard (1984)
Redshank (*Tringa totanus*)	—	—	—	—	Shrimps disappear from surface in response to many birds	—	Visual foraging birds capture fewer prey at high feeding density because of search path disruption	—	Goss-Custard (1970b, 1976, 1980)

Table 1.1b: continued

Species	Attracting predators	Impaired vigilance	Aggression and food stealing	Prey depletion	Prey disturbance and reduction of availability	Area copying	Reduction in searching/ feeding efficiency	Increasing incidence of disease	References
Bar-tailed godwit (*Limosa lapponica*)	—	—	—	Large flocks deplete food supply when most prey are available	The presence of many birds causes prey to burrow deep down	—	—	—	Smith and Evans (1973), Bryant (1979)
Curlew (*Numenius arquata*)	—	—	—	—	—	—	Food intake lowest at high feeding density	—	Zwarts (1978), Bryant (1979)
Black-headed gull (*Larus ridibundus*)	—	—	More fighting amongst dense flocks	Greater depletion of prey where birds are concentrated	Shrimps driven down by presence of large number of gulls	—	Capture rate of shrimps lowest where birds feed close together	—	Curtis and Thompson (1985)
Rook (*Corvus frugilegus*)	—	—	Most aggression in dense flocks	—	Prey burrow deep when large number of birds present on soil surface	—	Prey intake lowest at highest feeding density	—	Patterson (1975), Waite (1981, 1983)
Raptors (Falconiformes)	—	—	—	—	—	—	—	Diseases including botulism, tuberculosis and aspergillosis affect birds breeding/ feeding in close proximity	Keymer (1972), Newton (1979)

Table 1.2: Some costs and benefits of mixed-species flocking (associations between bird species only)

Species	Early detection of predators	Reduced individual risk	More time to feed	Extension of foraging niche and/or change in feeding strategy	More food made available	Increasing success rate of predators	Rapid depletion of food	Increase in food fighting, kleptoparasitism and competition	Disturbance of prey	References
Chickadees and titmice (*Parus* spp.), woodpeckers (*Picoides* spp.), nuthatches (*Sitta* spp.), creepers (*Certhia* spp.), kinglets (*Regulus satrapa*) and warblers (*Dendroica* spp.)	Mixed flocks respond rapidly to predators through calls made by wary species	Low frequency of predator attacks. Mixed flocks mob predators	Dominant species probably achieve a greater energy intake than subdominant species	—	Chickadees and titmice lead mixed flocks to rich food supplies	—	—	Interspecific dominance hierarchy exists through supplanting attacks, chases and fights. Commonest species overlap greatly in terms of feeding space	—	Morse (1967, 1969 1970), Moynihan (1967), Brockmann and Barnard (1979)
Crested tits (*Parus cristatus*), willow tits (*P. montanus*) and coal tits (*P. ater*)	—	Mixed flocking may reduce success rate of predators	—	Coal tits feed in more areas in mixed- than in single-species flocks	—	—	—	Willow tits spread out when crested tits are absent, coal tits spread out when willow and crested tits are absent	—	Morse (1968), Högsted (1978), Perrins (1979), Ekman *et al.* (1981)
Black-capped chickadee (*Parus atricapillus*) and chestnut-backed chickadee (*P. rufescens*)	—	—	—	Birds feed in new areas when in mixed flocks	—	—	—	—	—	Krebs (1973)
Yellow-faced grassquit (*Tiaris olivacea*), black seedeater (*Sporophila aurita corvina*) and white-collared seedeater (*S. torquilla*)	—	—	Birds surrounded by heterospecifics spend longer feeding	Grassquit feeds in additional areas when seedeaters present	—	—	—	—	—	Rubenstein *et al.* (1977)

Table 1.2: continued

Species	Early detection of predators	Reduced individual risk	More time to feed	Extension of foraging niche and/or change in feeding strategy	More food made available	Increasing success rate of predators	Rapid depletion of food	Increase in food fighting, kleptoparasitism and competition	Disturbance of prey	References
Pine siskins (*Carduelis pinus*) and evening grosbeaks (*Hesperiphona vespertina*)	—	—	—	—	Pine siskins use evening grosbeaks to break up large seeds	—	—	—	—	Balph and Balph (1979)
Downy woodpecker (*Picoides pubescens*), black-capped chickadees and tufted titmice (*Parus bicolor*)	Alarm calls given by wary species warn others of danger	Woodpeckers may reduce their individual risk of predation by joining mixed flocks. Some mixed flocks mob predators	Birds spend more time feeding and less time scanning in mixed flocks	Chickadees attract others to good feeding sites. Woodpeckers copy feeding methods of chickadees and titmice	—	—	—	Chickadees steal food from downy woodpeckers	—	Sullivan (1984), Morse (1970), Ficken (1981)
Fieldfares (*Turdus pilaris*) and redwings (*T. iliacus*)	Redwings may use fieldfares to signal danger	—	Fieldfares spend more time feeding with redwings than when alone	—	Both species achieve higher rates of energy intake in mixed flocks	—	—	Redwings spend less time on prey assessment when fieldfares are present	—	Barnard and Stephens (1983), H. Stephens, unpubl.
Turnstones (*Arenaria interpres*), purple sandpipers (*Calidris maritima*) and other waders (Charadriiformes)	—	Increase in heterospecific density reduces time spent vigilant	Birds spend more time feeding and less time scanning in mixed flocks	—	—	Presence of large wader species may attract predators and reduce field of view of vigilant birds	—	Increase in interspecific aggression at high densities	—	Metcalfe (1984a, c), Recher and Recher (1969)

Dunlin (*Calidris alpina*) and golden plovers (*Pluvialis apricaria*)	Dunlins use golden plovers to signal danger	—	Dunlins spend more time feeding and less time scanning in mixed flocks	—	Dunlins use golden plovers to indicate good feeding sites	—	Food is depleted more rapidly by mixed flocks	More competition in mixed flocks, golden plovers attack dunlings but not conspecifics	Close proximity of dunlins to golden plovers implicated in reduced availability of prey at surface	Byrkjedal and Kålås (1983), Thompson (1986)
Gulls (Laridae), skuas (Stercorariidae), auks (Alcidae), and shearwaters (*Puffinus* spp.)	—	—	—	Mixed flocking results in species feeding outwith normal feeding sites	Some 'catalyst' species are used to indicate available food	—	Prey depletion results in mixed flock moving on	Gulls and skuas steal from auks	—	Hoffman *et al.* (1981), Nelson (1980)
Rolland's grebe (*Rollandia rolland*), silver grebe (*Podiceps occipitalis*) and brown-hooded gull (*Larus maculipennis*)	Gulls provide grebes with early warning of predation in nesting colonies	Grebes nesting with gulls lose fewer eggs to predators	—	—	—	—	—	—	—	Burger (1984)
Carrion crow (*Corvus corone*), rook (*C. frugilegus*), jackdaw (*C. monedula*), and magpie (*Pica pica*)	Crows may provide early warning of danger	—	—	—	Rooks indicate good feeding sites	—	Large mixed flocks deplete food supply more rapidly	Rooks and jackdaws attain a lower energy intake when crows are present. Crows steal food from heterospecifics	Crows attain less energy in large mixed flocks because of reduced activity of large prey at the surface	Waite (1981, 1983, 1984a, b, unpubl.), D.B.A.T., pers. obs.

tend to surround their calves and threaten when wolves (*Canis lupus*) attack. Similarly, the aggregation of some birds into tightly-knit flocks when attacked by raptors may act as a deterrent, because the predator may be injured if it collides with any but its target bird (Tinbergen 1951). Mobbing, too, may be a deterrent where attacks by prey groups pose a threat to the predator (e.g. Hoogland and Sherman 1976).

Predators may, however, also be deterred because of their reduced success against groups. The early-detection and confusion effects discussed above may mean that a predator would do better to ignore groups of prey and hunt lone individuals. There is evidence for this. Neill and Cullen (1974) showed that their pike and cephalopods tended to ignore or avoid the large shoals of minnows against which they were less successful. Similarly, Milinski (1977) found that sticklebacks, when presented with different densities of *Daphnia* in glass tubes, preferentially attacked tubes containing single individuals (singletons) rather than those with higher densities. Fish appeared to be confused by the different directions of movement by *Daphnia* in dense swarms. The degree of preference for singletons increased with the density of *Daphnia* next to them. In the field, Page and Whitacre (1975) found that coastal merlins (*Falco columbarius*) attacking flocks of sandpipers (*Calidris* spp.) were least successful against intermediate-sized flocks. Observations showed that these flocks were also least likely to be attacked.

Reducing individual risk. Potential prey may also be safer in a group through the 'dilution' effect. Because the predator has a number of different victims from which to choose, the probability of any given individual being selected is the reciprocal of the group size (assuming that all individuals are equally vulnerable). Furthermore, those individuals not caught can escape while the predator handles its victim. An example where reduced vulnerability through the dilution effect seems to hold is flotilla formation in *Halobates robustus* (Foster and Treherne 1981).

Individual risk is also likely to vary with group density, i.e. the distance between one animal and its nearest neighbours. Hamilton (1971) called this unoccupied area the individual's 'domain of danger'. The 'domain' encloses those points which are nearer to the individual in question than to any other. Hamilton suggested that, where a predator is likely to emerge at an unpredictable point within a prey group, individuals with the largest domains will have the highest risk of capture because they are likely to be nearest the predator. Selection will therefore favour individuals which minimise the size of their domain by moving towards their nearest neighbour (the 'selfish herd' effect). Lazarus (1978) tested the consequences of the 'domain of danger' idea for vigilant behaviour in flocks of white-fronted geese (*Anser albifrons*). He assumed that birds which were relatively isolated within the flock (i.e. had a large domain around them) would be more vulnerable to predators because of Hamilton's argument and also because they would be in a poor position to take advantage of alarm responses by neighbours. He predicted that they would therefore compensate by being more vigilant. This turned out to be the case. There was a significant negative relationship between the duration of vigilant 'head-up' bouts and the number of birds within nine goose-lengths of a focal bird. As Lazarus and Inglis (1978) point out, however, there is an alternative explanation for these results based on the number of birds at the edge of the flock (see Chapter 7).

In many species, particularly small birds, vulnerability to predators increases with distance to cover. Grouping is one way in which individuals can offset the increase in individual risk and capitalise on food supplies which otherwise would be unobtainable. House sparrows (*Passer domesticus*) in arable fields feed in larger flocks and for shorter periods of time the farther they go from perimeter hedges (Barnard 1980a). In this way they can feed on the higher seed densities away from the edge of the field where otherewise they would be too vulnerable. Similarly, Pulliam and Mills (1976) found that vesper sparrows (*Pooecetes gramineus*) form flocks only when they are far from cover in sparsely-vegetated habitats. One of the most convincing demonstrations that grouping has an anti-predator function is Caraco *et al.*'s (1980a) study of yellow-eyed juncos (*Junco phaeonotus*): Caraco *et al.* found that releasing a tame hawk near feeding birds resulted in increased flock size and increased individual scanning rates for any given flock size.

Finding food

Grouping behaviour can also help animals become more efficient predators, for instance by providing information about the location of food or reducing variability in food intake. Feeding benefits, however, are not necessarily alternative adaptive explanations to predator avoidance. They may arise as a consequence of it.

Time budgeting. One spin-off of the 'safety in numbers', early-detection and other anti-predator effects of grouping is that individuals can afford to spend less time looking out for predators. Time saved by not scanning can be used for other important activities, such as finding food. There is now a wealth of evidence from flock-feeding bird species that this change in time allocation (time budgeting) does occur. In flocks of white-fronted geese, for example, the number of birds vigilant at any one time increases with flock size at a rate lower than that expected simply by multiplying flock size by the vigilance level of a solitary bird (Lazarus 1978). As flock size increases, therefore, the proportion of birds which are vigilant goes down. In house sparrows, juncos (*Junco spp.*) and starlings (*Sturnus vulgaris*), birds scan less and peck more as flock size increases (Barnard 1980a, Caraco 1979a, Powell 1974). This 'double benefit' of grouping, however, may be more complex than appears at first sight. Barnard's (1980a) study of time budgeting in house sparrows provides a good example.

The sparrows in Barnard's study fed in two different habitats: cattlesheds and open fields. In the cattlesheds, birds were taking barley seed from bedding straw. In the fields, the same birds were feeding on harvest debris and the newly-sown winter crop of barley. An important difference between the habitats lay in the apparent risk of predation to feeding birds. In the sheds, birds were almost completely sheltered from aerial detection (raptors were the most important predators) and predators were seldom seen there. In the fields, however, they were exposed except when sheltering in perimeter hedges. At first sight, flock size appeared to correlate positively with individual pecking rate and negatively with scanning rate. When food density and distance from cover were taken into account, however, the flock-size effect disappeared in the sheds but not in the fields. The allocation of time to feeding and scanning in the sheds was almost completely governed by local food availability. Flock size was positively related to food density, but did not itself influence time budgeting. In the fields, flock size influenced time budgeting independently of food availability and other environmental factors, although scanning rate increased with distance from cover. Similar relationships between time budgeting and risk of predation have been found in yellow-eyed juncos by Caraco *et al.* (1980b). Lendrem (1984) carried out a detailed laboratory analysis of the effects of predation risk on time budgeting. Working with barbary doves (*Streptopelia risoria*), he found that time spent feeding and *instantaneous feeding rate* (seeds taken per second of foraging) increased with flock size. When a restrained predator (a ferret, *Mustela furo*) was introduced, instantaneous feeding rate in any given flock size was reduced. Increased risk of predation thus caused birds to feed less efficiently even when they were actually feeding.

Finding good feeding places. A hungry animal may wander over a number of potential feeding sites before deciding which to exploit. One cue it could use in deciding where to feed is the number of individuals already at a site. Several studies have shown that predators tend to aggregate where food is most abundant (e.g. Goss-Custard 1970, Barnard 1980b, and see Begon *et al* 1985). Aggregations may build up at good feeding areas in at least two ways. Firstly, individuals may spend longer in areas where food availability is high (e.g. Murdie and Hassell 1973, Smith 1974a, b, Barnard 1980a, b) and, secondly, one or a few individuals may locate an area of high food availability and in some way attract others to the site (e.g. Crook 1964, Murton 1971a, Krebs 1974). While it may be difficult to recognise good feeding areas from the level of food availability alone (e.g. because food is often cryptic or concealed), the number of individuals present can provide a simple and reliable indicator. Supporting this, comparative studies have shown that species exploiting ephemeral, patchily-distributed food often tend to feed in groups.

The use of conspecific aggregations was shown clearly by Krebs (1974) in his study of great blue herons (*Ardea herodias*). Herons fly over large areas of mudflat searching for suitable pools in which to feed. They usually feed in flocks, and flock size correlates positively with individual rate of food intake. To see whether birds used flock size as a guide to the profitability of different pools, Krebs put out life-sized models of herons in different 'flock' sizes. He then recorded the number of birds passing over which alighted near the models. There was a strong positive relationship between the number of birds landing at a pool and the number of models present. Furthermore, birds landed close to the models and immediately started foraging.

If group size is to provide a reasonably accurate index of food availability, it is important that it changes in relation to food availability. Barnard (1980b) tested this in a field experiment with house sparrows. The experiment consisted of baiting areas near three different hedges in the fields where sparrows were feeding. Counts of the number of birds feeding at each hedge were made before baiting and then each area was baited with a different density of millet seed. The number of birds feeding at each hedge was then counted over the next two hours. The procedure was repeated twice more, with the seed densities being rotated around the three hedges until all had been tested at each density. If flock size was a good indicator of food availability, the largest flock should always be at the hedge with the highest seed density. The results showed very clearly that flock size was positively correlated with seed density and that most sparrows were always recorded on the highest density. While this appeared to be due to birds staying longer where seeds were more abundant, birds were also attracted to the hedges from the farm buildings. The total number of birds feeding at the three hedges therefore increased during the course of the experiment.

Local information about food. Once a group has built up at a site, further feeding benefits may arise as the result of individuals using information generated by other foragers. When animals feed in sensory contact with one another, information about their foraging success is incidentally transmitted between them. Depending on the type of food supply, individuals can greatly enhance their feeding efficiency by making use of this information. Several studies have shown that individuals in socially feeding species interact with one another while foraging (e.g. Krebs *et al.* 1972, Krebs 1973, Rubenstein *et al.* 1977, Barnard and Sibly 1981). In some cases, interactions are forms of social facilitation or local enhancement (Thorpe 1953), where individuals copy the behaviours or types of foraging area used by successful foragers. Krebs *et al.* (1972) showed that perch-copying by great tits (*Parus major*) resulted in birds doing better in flocks than singly or in pairs. Clearly, however, the profitability of different types of interaction is likely to vary with the pattern of food distribution. For instance, area-copying (Barnard and Sibly 1981), whereby individuals search in the area immediately surrounding the point where another forager appeared to find food, is likely to be successful only where food is patchily distributed. Krebs *et al.* (1972) found that great tits trained on uniform food distributions were less likely to copy each other than were those trained on patchy food. Likewise, interactions between the house sparrows studied by Barnard (1978) varied with different food distributions. Area-copying predominated on patchy food; following (whereby one bird followed close behind another and sometimes usurped food from it) on randomly-distributed food; and snatching on uniformly-distributed food. These changes meant that birds tended to use the most effective form of interaction on any given distribution. In a study of flock-feeding Harris' sparrows (*Zonotrichia querula*), Rohwer and Ewald (1981) found that dominant birds tended to area-copy subordinates only when seed was concealed and patchily distributed. Furthermore, they copied only those subordinates that were actively feeding. Rohwer and Ewald's study is also interesting because it showed that copying interactions were initiated by particular individuals — those high up the dominance hierarchy. Indeed, dominant birds defended their subordinate 'hosts' against other dominants and, in doing so, suffered reduced feeding efficiency as a result of spending time fighting. Similar individual specialisation was found in house sparrow flocks by Barnard and Sibly (1981). Here, 'copier' birds spent little time looking for food themselves, but instead usurped items from 'searchers' who foraged for their own food. Most interactions between sparrows involved area-copying, but copiers sometimes forcibly snatched food from successful searchers, thus behaving as intraspecific kleptoparasites (see later)

Many bird species form dense, communal roosts at night. In some cases, temporary roosting or resting associations may also develop during the day. Ward and Zahavi (1973) suggested that communal roosts and nesting colonies may function as 'information centres' in which birds find out about the location of good feeding areas. Birds which have fared badly during the day might be able to recognise those which have done well and follow them to their, presumably better-quality, feeding areas the next day. Successful foragers might be recognised because, for example, they are fat or they fly out faster the next morning. While field studies have lent some support to the 'information centre' idea (e.g. Krebs 1974, Emlen and Demong 1975, Gilbert 1975, but see Chapter 7), de Groot (1980) provides the most direct test in some laboratory experiments with queleas (weaver-birds). In the wild, queleas roost in colossal groups which sometimes exceed a million birds and are serious pests of grain crops in the areas they inhabit. De Groot housed two groups of birds together in an aviary. From the aviary, birds had access via funnel-shaped entrances to four compartments in which food or water could be found. De Groot trained one group of birds (A) to find water in one compartment and the other group (B) to find food in another compartment. The two groups were then deprived of food or water and allowed to roost together. It turned out that, when birds of group A were hungry (food-deprived), they tended to follow B birds to their food supply and when B birds were thirsty (water-deprived), they tended to follow A birds to their water supply. In other words, naive birds were somehow able to recognise 'knowledgeable' birds when they needed food or water, and to follow them to a supply which they themselves had never visited.

At first sight, information-centre gatherings appear to hinge on an exploitative relationship between efficient and less efficient foragers. Individuals who have done badly exploit the know-how or good fortune of those who have done well. Evans (1982a), however, has considered a number of ways in which information transfer might affect the two parties.

Parasitism. The use of information can be regarded as parasitic (see Barnard 1984a) where followers benefit to the detriment of leaders or without their gaining any advantage. Increased competition at feeding sites is an obvious potential cost to leaders. The fact that in some cases leaders advertise their departure by conspicuous calls or displays (Evans 1982a and see Chapter 7), however, makes it unlikely that parasitism is a general explanation for communal roosting. Parasite pressure would tend to select for birds which sneak away quietly and inconspicuously (Bertram 1978).

Reciprocal altruism. Information transfer might evolve through reciprocal altruism (Trivers 1971) if bird A leads bird B to food when it has done well, then B leads A to food when the tables are turned. As pointed out by several workers (e.g. Trivers 1971, Maynard Smith 1982, Davies and Krebs 1978), however, reciprocal relationships are vulnerable to exploitation by cheats, in this context birds that follow others to food but do not lead them to rich feeding sites themselves. In a system based on reciprocity, leaders could pay the cost of competition without the compensation of finding food when they are not doing so well.

Mutual benefit. There is a number of circumstances under which both leaders and followers might benefit from leaving the colony in groups. Leaving in groups may reduce individual risk of predation, both in transit to and on the feeding site. Alternatively, or in addition, there may be feeding benefits from associating in groups at the food source (through social facilitation and local enhancement, for example). Benefits of associating away from the colony may thus produce a strong selection pressure for aggregation on departure. Clearly, these independent effects of grouping are difficult to distinguish from the effects of information transfer *sensu* Ward and Zahavi (1973) and are an important caveat in interpreting group formation at colonies.

Kleptoparasitism. As we have seen in house sparrows, feeding interactions may involve the forcible theft of prey. If certain individuals obtain a substantial proportion of their food by theft, they are referred to as pirates or kleptoparasites (see Vollrath 1984 for a rigorous definition of terminology). Many interspecific feeding associations are based on kleptoparasitism, especially in birds. Brockmann and Barnard (1979) and Barnard (1984a) have discussed the evolutionary consequences of kleptoparasitism and the ecological circumstances in which it occurs.

Brockmann and Barnard's (1979) review of kleptoparasitism in birds showed a remarkably uneven distribution across taxonomic groups. Over 60% of species showing kleptoparasitism belong to the order Falconiformes (hawks, falcons, eagles, etc) and Charadriiformes (gulls, skuas, waders), although these contain only 7% of known bird species. There is a similar taxonomic bias in other groups (see Barnard 1984a). At least in birds, the preponderance of kleptoparasitism within particular orders appears to be a function of well-defined ecological factors which make kleptoparasitism more or less profitable. These are summarised in Table 1.3.

Kleptoparasitism is most likely where close association between individuals has evolved and where food items are costly to procure, but conspicuous and profitable once procured by a host. It is particularly common as one of several alternative feeding strategies in dietary opportunists such as gulls, egrets and house sparrows. While the term kleptoparasitism is used mainly in connection with interspecific relationships, it is clear that similar interactions occur within species, and Brockmann and Barnard (1979) point out how kleptoparasitism might have evolved through the escalation of more subtle feeding interactions such as area-copying.

Clearly, exploitative feeding strategies like kleptoparasitism are likely to have a serious impact on host individuals. Hosts spend time and energy procuring food only to lose it to kleptoparasites. The impact of kleptoparasites may thus set up a counteradaptive arms race between parasite and host, with the host lineage being selected to minimise the inroads of kleptoparasitism and the parasite lineage to counter the host and maximise its attack efficiency. Barnard (1984a) discusses different counteradaptive strategies which hosts and parasites appear to have adopted. Among hosts, the chief strategies seem to be (a) evasion (the avoidance of attack by e.g. escaping or shifting to a less vulnerable diet), (b) retaliation (reclaiming stolen food items), and (c) toleration/compensation (simply putting up with the loss or feeding at a higher rate to make up for it). While kleptoparasites show various counteradaptations to these strategies, they also face the problem of competition from other kleptoparasites. To get around this, they may (a) adopt alternative kleptoparasitic strategies (e.g. sitting and waiting instead of chasing), (b) adopt alternative non-kleptoparasitic strategies (searching for their own food), or (c) limit competition (defend hosts against other kleptoparasites). Since many kleptoparasites are dietary opportunists, (a) and (b) are common (see Barnard 1984a for a review). Many kleptoparasites are also territorial with respect to their hosts and, as we have seen in the Harris' sparrows, the defence of hosts occurs within as well as between species.

Table 1.3: Ecological and behavioural factors favouring kleptoparasitism (from Barnard, 1984a)

Food and host availability		Behavioural factors
1. *Host concentration.* Kleptoparasitism is common where potential hosts are aggregated as in breeding colonies (e.g. gulls, auks, digger wasps) and feeding groups (e.g. lapwing flocks, eider rafts). 2. *Food availability.* In some species, kleptoparasitism occurs only at times when hosts are making large quantities of food available, e.g. systematic piracy of auks by skuas, gulls and terns occurs only in the nesting season when hosts are feeding chicks. 3. *Food quality.* Many kleptoparasites appear to attack selectively host individuals which are carrying large or high-quality food items (e.g. gulls selectively attack swans bringing mussels rather than weed to the water surface).	4. *Food predictability.* Kleptoparasitism is more likely where the habits of the host make food availability temporally and spatially predictable (e.g. puffins repeatedly return to the same nest burrow, where skuas and gulls lie in wait). 5. *Food detectability.* Kleptoparasitism is more likely where food items are easily detectable, thus reducing searching-time costs. Long prey-handling times make predators particularly vulnerable to piracy. 6. *Food shortage.* Kleptoparasitism may be a 'best of a bad job' strategy adopted during times of food shortage. In birds, for instance, it is subordinate individuals that show kleptoparasitism.	7. *Behaviour of parasite.* Many kleptoparasites are dietary opportunists, and piracy occurs as one of a flexible range of feeding strategies when appropriate food items are made available. The ability to move quickly and with agility is also a common feature of kleptoparasites. 8. *Behaviour of host.* Kleptoparasitism is more likely where there is little defence or retaliation by hosts and where there is little likelihood of costly chases or host escape (see also 1).

Other feeding costs and benefits. As well as the above, group living may provide other feeding benefits. For instance, it may reduce the risk of individuals faring badly (e.g. Thompson *et al.* 1974). In some situations, minimising risk may be more important than maximising intake. In a recent review, Pulliam and Caraco (1984) also point out that grouping may be a means of minimising searching time between patches where these are distributed unpredictably in the environment. More individuals covering a greater area are more likely to discover patches. This effect will be particularly important where groups do not appreciably deplete food patches and interpatch search time is the main time cost of feeding (Pulliam and Millikan 1982).

Flocking may also increase the efficiency with which renewing food supplies are exploited. If visits to feeding sites are synchronised, individuals can regulate revisits so that food has time to replenish itself (Prins *et al.* 1980, see Chapters 4 and 7). In some cases, grouping may help individuals tackle large prey which would be impossible to tackle singly. This is well known in carnivorous mammals (e.g. Schaller 1972), but is also known in birds. Black vultures (*Coragyps atratus*) may join forces in catching live prey (e.g. McIlhenny 1939) which could not be tackled by a single bird. Also, mergansers (*Mergus* spp.) and terns (*Sterna* spp.) co-operate in driving large shoals of fish into shallow water (see Rand 1954).

An important factor which needs to be borne in mind is that individuals may vary in the costs and benefits accruing to them from grouping behaviour. Because individuals differ in their size, metabolism, competitive ability and so on, the costs of interference and competition incurred by joining a group will vary in their severity. In brent geese (*Branta bernicla*), variation in feeding efficiency seems to be mainly a consequence of position in the flock (Prins *et al.* 1980): birds at the front get to the sea plantain (*Plantago maritima*) food first and leave only the bases of plants for those at the rear. The bases, however, are the most nutritious part of the plant so, although front birds get a greater bulk of material, rear birds get better quality. All in all, therefore, individuals probably fare equally well. In house sparrow flocks, however, feeding efficiency depends on individual competitive ability. Barnard (1978) found that high-ability (dominant) birds had priority of access to localised food supplies (in this case a bowl of seed) and could therefore feed in large flocks, reaping the benefits of reduced predation risk and commitment to scanning. This meant that dominants tended to feed briefly but with a high rate of food intake, and therefore had plenty of time for other important maintenance activities. Subordinates, on the other hand, were aggressively excluded from the food supply and could approach it only when no, or very few, low-ranking birds were present. Consequently, they tended to make frequent visits between those of other birds, and spent a long time at each visit because most of their feeding time was taken up by scanning. In the field, sparrows which were pushed to the edge of feeding flocks often remained behind when companions were scared off. They then moved in to exploit the denser seed supplies from which they had previously been excluded (Barnard 1980c). As we shall see later, Caraco (1979a, b) predicted and found similar effects of competitive ability in flock-feeding yellow-eyed juncos.

Foraging in a group is likely to affect prey availability (through depletion, disturbance, etc) more than foraging solitarily. In some cases, as in so-called beating associations (Rand 1954), prey disturbance can be put to good use. Here the activity of some group members flushes out prey, which are then caught by others. This seems to be the reason why cattle egrets (*Bubulcus ibis*) and starlings associate with cattle (e.g. Heatwole 1965). In other cases, however, disturbance is deleterious and may feed back to influence group size and density, as in redshanks (*Tringa totanus*) foraging on mudflats (Goss-Custard 1976). Redshanks tend to forage in tight flocks at night, but in loose-knit flocks during the day. It turns out that, during the day, birds feed on small shrimps (*Corophium volutator*) by *visually* detecting their tails sticking out of mud. At night, visual detection is impossible and birds switch to feeding on tiny snails (*Hydrobia ulvae*), which are detected by probing with the bill (*touch* feeding). The differences in flock density arise because surface-dwelling shrimps are easily disturbed by the movements of birds which therefore space themselves out to prevent neighbours scaring their prey away. Snails, on the other hand, are more or less sessile and do not react quickly to the activity of birds. Interference is thus not so serious and birds can afford to bunch together. While birds are less at risk from predation in dense flocks, their ability to capitalise on flocking is compromised by the effects of feeding density on prey availability. Similar explanations apply to variation in flock density in other shorebirds (Goss-Custard 1970, 1980, Pienkowski 1981, Curtis and Thompson 1985). Sometimes, interference within flocks may be more direct. Kleptoparasitism (see above), or other aggressive forms of interaction, may result in increased inter-neighbour distance because interactors tend to do better when they are near victims (e.g. Barnard 1978, Burger *et al.* 1979, Vines 1980, Ens and Goss-Custard 1984), and it pays potential victims to move away from them.

While various feeding and anti-predator costs and benefits of grouping have been discussed under separate headings, this is purely for descriptive convenience. Historically, there has been a tendency to try to explain grouping in terms of one particular 'function' (e.g. avoiding predation or finding more food) (see Barnard 1978 for a review), but it is clear that many advantages and disadvantages accrue simultaneously in any given association. This is made clear in Table 1.1 and in the third section of this chapter, where we discuss the relationship between individual costs and benefits and group dynamics.

Costs and Benefits of Mixed-species Association

So far, we have dealt with the advantages and disadvantages of grouping mainly as they apply to single-species associations. In many cases, however, species associate together in mixed groups. As with single-species associations, there has been a tendency to explain mixed groups, especially among birds, in terms of performing a particular 'function' (e.g. Morse 1970, Miller 1922). Powell (1980), for instance, analysed data from neotropical mixed flocks to see whether flocking improved feeding efficiency or protection from predators. Even where single functions have not been explicitly sought, mixed associations have generally been regarded as conferring similar selective advantages to single-species associations. While many, if not all, the factors discussed above may still apply to mixed groups, it is reasonable to suppose that association between species might affect individuals in ways not found within species. Since even closely-related species are likely to differ in, for example, feeding requirements and foraging strategies, we might expect the way individuals benefit or suffer from joining a group to depend on the group's species composition (qualitative and quantitative). In this section we shall discuss some of the possible consequences of mixed association, again with particular reference to feeding birds, and then detail a worked example where single- and mixed-species groups have been compared directly.

Consequences of Mixed Association

There are at least four ways in which the feeding/anti-predator consequences of mixed association may differ quantitatively from those of single-species association (see also Table 1.2):

Changes in 'foraging niche'. Individuals may change the type of place in which they search for food (foraging niche), either by copying species which search in different places, or through competitive exclusion from their single-species foraging niche. Good evidence for a shift based on copying comes from Krebs' (1973) study of chickadees. In an aviary experiment, Krebs trained two species of chickadee in single-species flocks to search in different parts of artificial trees. Black-capped chickadees (*Parus atricapillus*) were trained to search the lower, inner branches of trees, while chestnut-backed chickadees (*P. rufescens*) were trained to search the upper, outer branches. When the two species were allowed to forage together, both showed a significant decrease in the amount of time spent in their own foraging niche; they spent more time in the upper, inner branches of the trees. Earlier experiments had shown that birds of both species copied successful con- and hetero-specifics in the same way as the great tits in Krebs *et al.*'s (1972) study (see above). In the field, however, Smith (1967) found no significant change in foraging niche when the two species associated in mixed flocks.

The 'attendant' species joining 'nucleus' species in neotropical bird flocks often show pronounced shifts in foraging height (Buskirk 1974). Birds tend on the whole, however, to join nucleus species with foraging heights similar to their own. In mixed flocks of tits (*Parus* spp.), Hogstad (1978) found that, in the presence of crested tits (*P. cristatus*), willow tits (*P. montanus*) tended to feed nearer to the trunks of trees. Conversely, they expanded their foraging niche when crested tits were absent. Similarly, coal tits (*P. ater*) had a broader foraging niche when both willow and crested tits were absent. In these flocks, therefore, changes in foraging niche appeared to be due to competitive exclusion. Morse (1970) provides an extensive review of mixed insectivorous bird flocks and cites several examples of foraging-niche shifts. For instance, in the presence of red-breasted nuthatches (*Sitta canadensis*), black-capped chickadees moved towards the inner parts of vegetation, although in the presence of kinglets (*Regulus calendula*) they showed no change. Similarly, in the presence of brown-headed nuthatches (*S. pusilla*), pine warblers (*Dendroica pinus*) foraged more often in the inner parts of trees. When pine warblers were present, the nuthatches showed a greater tendency to forage in the outer parts of trees.

Herrera (1979) distinguished 'flock-positive' from 'flock-negative' species among mixed passerine associations in southern Spain. The former were found in mixed flocks more often than expected from their occurrence in the mixed woodland habitat, the latter less often. Correlated with this, flock-positive species doubled their foraging success rate by joining flocks, whereas flock-negative species did no better than by foraging solitarily. This difference appeared to be due to flock-positive birds expanding their foraging niche in mixed flocks, both in height and type of substrate, but flock-negative birds *contracting* theirs. Herrera suggested this as evidence that flock-negative birds were joining flocks to reduce their risk of predation rather than to find more food.

Changes in food availability. The foraging behaviour of other species may also make new types of food available. Extreme examples can be found among kleptoparasitic associations. In puffin (*Fratercula arctica*) and guillemot (*Uria aalge*) colonies, jackdaws (*Corvus monedula*), carrion crows (*C. corone*) and ravens (*C. corax*) steal fish from the auks, feed on fish fragments dropped near the auks' nests and eat eggs broken open by gulls (e.g. Birkhead 1974, Cramp *et al.* 1974, Richford 1978). Emlen and Ambrose (1970) found that snowy egrets (*Egretta thula*), which are wading birds, used flocks of red-breasted mergansers (*Mergus serrator*) to obtain fish from deep water. The egrets made no attempt to feed prior to the arrival of the mergansers. Only as the mergansers approached the shore, driving fish in front of them, did the egrets begin fishing.

New food may also be made available by the so-called 'beating' effect (Rand 1954). Antcatchers (*Gymnopithys* spp.) and alethes (*Alethe* spp.) follow swarms of army or driver ants and feed on the arthropods they frighten out of hiding (e.g. Willis 1967). Carmine bee-eaters (*Merops nubicus*) ride on the backs of large birds such as bustards (*Ardeotis* spp.), storks (*Ciconia* spp.), ostriches and secretary birds (*Sagittarius serpentarius*), as well as large mammals, flying out to catch insects disturbed by the host (Boswall 1970). Gulls and terns associate with other aquatic bird species (including waders, ducks, mergansers, divers, grebes and auks) as they feed, and sometimes catch organisms frightened to the water surface by the birds' feeding activities (see references in Brockmann and Barnard 1979).

In other cases, one species may be able to procure or break up food items which are too large or difficult for another to obtain or handle. This seems to be the case in the mixed finch flocks studied by Balph and Balph (1979).

Balph and Balph found that pine siskins (*Carduelis pinus*) were unable to husk large sunflower seeds. Evening grosbeaks (*Hesperiphona vespertina*), however, could husk them. When pine siskins were fed large seeds, they tended to associate with grosbeaks which husked the seeds and dropped fragments, thus making them available to the siskins. When fed small seeds, the siskins did not associate with grosbeaks.

In a fascinating recent study, Rasa (1983) showed active association between dwarf mongooses (*Helogale undulata rufula*) and hornbills (*Tockus deckeni* and *T. flavirostris*). The association was based partly on mongooses making available to hornbills prey that they would otherwise not have been able to obtain. These included lizards, mice and centipedes, hidden under logs and therefore invisible to the birds, which were detected olfactorily by the mongooses and dug out. In addition, mongoose groups also flushed various insect species out of the grass as they moved through and these were caught by the hornbills. A more subtle change in food availability as a result of mixed association was noted by Rubenstein *et al.* (1977) in flocks of seed-eating finches. Here species converged in their diet, showing preferences for rare seeds, presumably as a result of interspecific copying.

Changes in foraging strategy. Foraging heterospecifics may provide the chance to employ a different and more profitable feeding strategy. This is particularly likely in species which are dietary opportunists. Kleptoparasitism is often an opportunistic feeding strategy used when other foragers procure food that can easily be stolen. House sparrows, which are normally granivorous, have been recorded stealing from digger wasps (*Sphex ichneumoneus*) as they returned to their egg chambers with paralysed katydids (Insecta: Orthoptera) (Brockmann and Barnard 1979). At least temporarily, the sparrows appeared to subsist on stolen katydids, procuring them with lengthy aerial chases and associating around the entrances to digger wasp tunnels. Other opportunistic feeders such as blackbirds (*Turdus merula*), cattle egrets (*Bubulcus ibis*), shrikes (*Lanius* spp.) and corvids may also adopt exploitative feeding strategies in the presence of other species (see e.g. Brockmann and Barnard 1979).

In their study of finch flocks, Rubenstein *et al.* (1977) found several changes in species-specific feeding strategies as a result of mixed association. In particular, the species composition of the flock affected the length of time for which birds of the three species (yellow-faced grassquit *Tiaris olivacea*, black seedeater *Sporophila aurita corvina*, and white-collared seedeater *S. torqueola*) fed on the ground in each feeding bout and perched in trees between bouts. Birds surrounded by heterospecifics tended to feed for longer and to spend less time perching than those surrounded by conspecifics or feeding solitarily. Another measure of feeding behaviour which was influenced by species composition was the horizontal distance covered during a feeding bout. In all three species, most solitary individuals showed virtually no horizontal hopping movement during feeding. Those birds (regardless of species) surrounded only by conspecifics showed a slight tendency to move about. Where neighbours included heterospecifics, however, both seedeater species appeared to feed without moving, whereas the grassquit showed considerable movement.

Changes in anti-predator benefits. There are several ways in which mixed association might influence individual predation risk. The simplest is that association between species is likely to mean more individuals in the group and therefore a greater dilution or 'safety in numbers' effect. Barnard (1982) has argued that this could be one reason for convergence in alarm-call characteristics between species. In some cases, however, heterospecifics may provide additional anti-predator benefits. For instance, they may provide more effective vigilance and therefore earlier warning of approaching predators. This is best documented among mixed groups of mammals, though these usually comprise members of a single family. Associations among cetaceans (Fiscus and Niggol 1965), bats (Bradbury 1977), ungulates (Elder and Elder 1970) and primates (Altmann and Altmann 1970, Washburn and DeVore 1961) are usually explained in terms of increased alertness. Washburn and DeVore (1961) found that baboon species tended to associate with herbivores such as impala (*Aepyceros melampus*) and bushbuck (*Tragelaphus scriptus*), and that both primate and herbivore warned of approaching predators and responded to each other's warning 'bark'. Of particular interest was an observation that impalas, when associating with baboons, did not flee when cheetahs (*Acinonyx jubatus*) (an important predator of impala) approached. On that occasion, Washburn and DeVore noticed baboons approach the cheetahs and drive them away. Similarly, zebras (*Equus* spp.) approached waterholes more quickly and confidently when giraffes (*Giraffa* spp.) were already drinking. Some of the best evidence for improved predator detection in mixed groups comes from Rasa's (1981, 1983) study of dwarf mongooses and hornbills.

We have already mentioned the feeding benefits accruing to hornbills from their association with mongooses, but what do mongooses get out of having hornbills around? The answer appears to be improved predator detection and warning. When mongooses feed, it is usual for a few individuals to act as 'guards'. This involves standing on some relatively high point such as a mound and keeping watch for predators, mainly raptors. When a predator is spotted, the guard gives a repetitive warning call and the group dashes for cover. 'Guarding' appears to be hazardous: the only animals Rasa observed being caught by raptors were guards. Intriguingly, Rasa (1983) found a negative correlation between the number of hornbills in the group and the number of mongoose guards. Earlier studies (Rasa 1981) had shown that hornbills responded to predators of mongooses even though they did not pose a threat to the birds. The birds appear to learn predator identification from the responses of mongooses. Hornbills spot and respond to raptors before mongooses, and the latter appear to capitalise on the birds' vigilance by reducing their own commitment to scanning.

Fieldfares and Redwings

So far, we have discussed some of the ways in which mixed- and single-species associations may differ. However, we have had to draw on data from a wide variety of species facing different ecological problems. One study which has compared the consequences of single- and mixed-species associations for the same species is that by Barnard and Stephens (1983).

Barnard and Stephens studied fieldfares (*Turdus pilaris*) and redwings (*T. iliacus*) feeding on agricultural pastureland near Nottingham, UK over winter. Fieldfares arrived in the study area from the beginning of November, redwings from December/January, and birds usually stayed until March. While fieldfares and redwings regularly formed mixed flocks, they also fed in single-species flocks, even when heterospecifics were present in the study area. They therefore provided a good opportunity to study the conditions under which the two types of flock built up.

Single-species flocks. Single-species flocks tended to build up on different types of pasture: fieldfares on 'new'/'young' pasture (<4 years standing) and redwings on 'old' pasture (>4 years standing). Fieldfares preferred new pasture where earthworms were apparently easier to detect. Although worm density was lower than in old pasture, grass density was also lower and there was no tendency for worm size to vary with depth. Both fieldfares and redwings did better (in terms of rate of energy intake) by taking large worms, but in old pasture these were deep down and required a lot of orientation (crouching) time to locate. Redwings appeared to be good at locating worms through dense grass and did much better than fieldfares on old pasture. It was only on new pasture that fieldfares achieved rates of energy intake similar to those of redwings.

The rate of energy intake in fieldfares was a function of the effects of flock size on individual time budgeting. Birds scanned less and crouched more in large flocks, thereby increasing their capture rate. Fieldfares fed in larger flocks away from the perimeter hedges. They did not, however, use the hedges as cover but avoided them because they provided a concealed approach for predators. While distance from cover emerged as an important predictor of energy intake, this was due to the flock-size effect rather than distance *per se*. Interestingly, fieldfares tended to feed nearer to cover when temperatures were low. Low temperatures increased feeding priority, reduced worm availability and forced birds to feed over a wider area. On warm days, when feeding priority was low, birds fed away from cover and tended to scan more.

While flock size had a direct effect on fieldfare behaviour, this was not the case in redwings. Redwings fed in much smaller flocks than fieldfares and tended to stay close to perimeter hedges. These were used as cover between feeding bouts. Individual rate of energy intake was influenced mainly by variation in the amount of time spent crouching. This was because, in contrast to fieldfares on new pasture, longer crouches resulted in larger, more profitable worms. There was no direct effect of flock size, or distance from cover, on time budgeting or feeding efficiency.

Mixed flocks. Mixed flocks were almost always (96%) formed by fieldfares joining redwings on 'old' pasture. This occurred in mid-late winter, when worm density in the more exposed soil of the 'new' pastures was reduced. Interestingly, *both* species did better in mixed flocks, although this turned out to be for quite different reasons.

Fieldfares appeared to benefit directly from associating with redwings. Their rate of energy intake increased significantly from that recorded in single-species flocks on old pasture without any change in worm size or density. Time-budget analysis suggested that fieldfares used redwings as indicators of locally profitable worm supplies. One piece of evidence for this was that *looking* behaviour became an important predictor of energy intake. 'Looking' refers to a side-to-side movement of the head while the body was held in the normal hopping posture, and appeared to reflect vigilance for the activities of other birds in the flock. Looks were often followed by a peck or by bursts of rapid hopping towards another bird. Looking was quite distinct from scanning, which was a sleek, erect posture associated with vigilance for predators. The amount of time spent looking increased with the number of redwings per fieldfare. The higher pecking rate of redwings meant that they tended to generate more potential information about the location of prey. Since worms were more clumped on old pastures and fieldfares appeared to have difficulty finding them in single-species flocks, the activities of redwings may have decreased their searching time and increased the time available for crouching. Supporting this is the increase in fieldfare crouching rate in mixed flocks and the positive correlation between crouch duration and worm size taken, which did not exist in single-species flocks. Furthermore, fieldfares usually fed in close proximity to redwings. When fewer redwings were present, they tended to be aggregated near to hedges. Under these conditions, fieldfares fed closer to cover.

In contrast to fieldfares, redwings did better in mixed flocks simply because worm densities happened to be higher than at times when birds were recorded in single-species flocks. Although redwing rate of energy intake correlated positively with the number of fieldfares, there was no indication that fieldfares influenced time budgeting. The relationship was apparently due to fieldfares accumulating in large numbers where feeding conditions were good. Again, the best predictor of energy intake in redwings was time spent crouching. There was, however, evidence that fieldfares interfered with redwing feeding efficiency. When fieldfares were present, redwing crouch duration no longer correlated with worm size and there tended to be very few long crouches.

Redwings may have reduced their tendency to crouch when fieldfares were present because fieldfares appeared to use redwings to locate food, and crouching indicated that a bird was likely to peck (Barnard and Stephens 1983, see also Chapters 5 and 6).

It appears, therefore, that the costs and benefits of both single- and mixed-species associations are very different in fieldfares and redwings. In fieldfares, both types of association influence time budgeting and feeding efficiency directly, although the influential factors are different in the two cases. Neither single- nor mixed-species associations influence individual time budgeting and feeding efficiency in redwings and their presence in mixed flocks appears to be due mainly to exploitation by fieldfares.

Individual Costs and Benefits and Flock Dynamics

The relationship between group size and composition and the costs and benefits of group living to individuals is obviously dynamic. Individuals which benefit by being in a particular group should join or remain in it; those which are disadvantaged should avoid or leave it. As individuals come and go, so the group, its attendant consequences and therefore its attractiveness change. This feedback between individual and group has recently been the subject of intense theoretical and empirical study because it provides a means of predicting how, when and where groups should develop. A particularly fruitful approach has been to look at the relationship between individual time budgeting (see earlier) and group dynamics.

Time Budgeting and Optimal Flock Sizes

The most comprehensive time-budget study is that by Caraco and co-workers (e.g. Pulliam 1973, Caraco 1979a,b, Caraco *et al.* 1980a,b, Caraco and Pulliam 1984) of winter flocking in yellow-eyed juncos. Here, birds are seen as trying to minimise two risks over the winter period, starvation and predation. Caraco (1979a,b) sees flocking as a means by which birds can modify their time budgets to achieve this. He assumes that birds allocate their time between three mutually exclusive behaviours: *scanning* (a head-up posture looking for predators), *feeding* (a head-down posture involving pecking) and *aggression* (aggressive disputes over food). Aggression involves both conflicts over access to food and the eviction of subordinates by dominants to ensure their long-term supply of food (in this case there is a finite amount of seed which has to last the winter). He further assumes that scanning takes precedence over feeding, because death from predation is more deleterious than missing a few items of food, and that dominants give higher priority to satisfying their immediate energy needs than to safeguarding long-term food supplies. Subordinates, however, are assumed to give aggression the priority over feeding out of necessity (they are often attacked and cannot feed until attacks are over). Following Krebs and Davies (1981), the relationship between time budgeting and flock size in juncos is modelled in Figure 1.1.

The model assumes the following: (a) in common with other flock-feeding species, birds spend less time scanning in larger flocks (see earlier); (b) birds spend more time in aggression as flock size increases, because encounters over food are more frequent; (c) as a consequence of (a) and (b), birds spend most time feeding in medium-sized flocks.

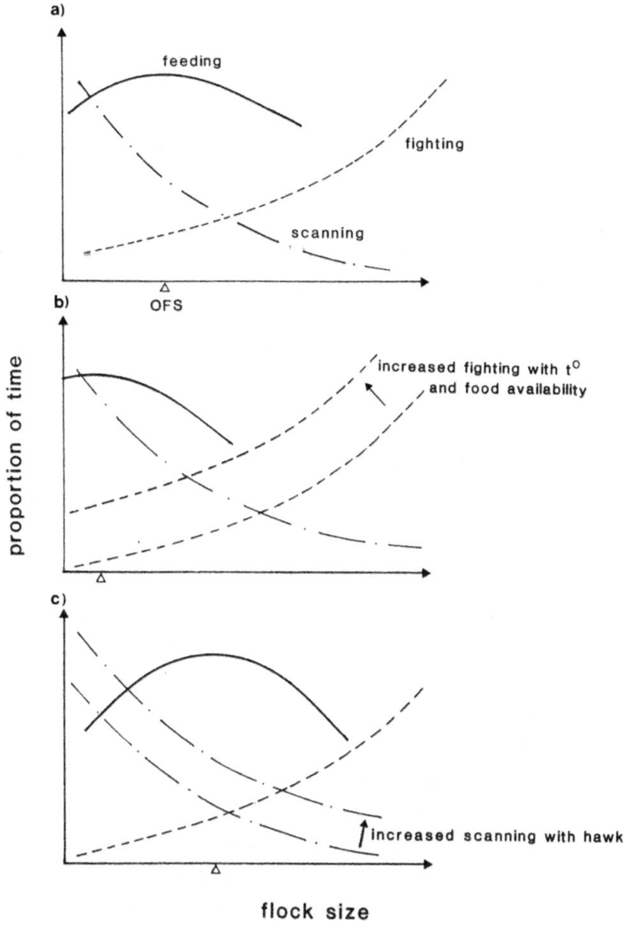

Figure 1.1: A model of the relationship between individual time budgeting and flock size. (a) The effect of flock size on time spent feeding, fighting and scanning. As birds scan less but fight more in large flocks, intermediate-sized flocks allow most time for feeding. (b) Dominant birds fight more at high temperatures (t°) and when more food is available, so birds do better in smaller flocks. (c) Birds scan more when a hawk is present so do better in larger flocks. Open triangle indicates optimal flock size (OFS)

Source: modified after Krebs and Davies (1981), based on data in Caraco et al. *(1980a, b)*

Modelling flock-size effects on time budgeting in this way allows us to make some predictions about how big real flocks will be. If we are right in thinking that birds are trying to minimise the risks of starvation and predation, then OFS in Figure 1.1a represents the flock size in which birds do best and the size to which flocks should therefore converge. Altering the effects of flock-size on scanning and aggression will alter the OFS, and flock size should change accordingly. Caraco and co-workers tested this prediction by manipulating the relationship between flock size and time budgeting.

Changes in temperature. As ambient temperature rises, feeding priority decreases. Dominant birds should therefore have more time to chase off subordinates and the OFS will decrease (Figure 1.1b). We should therefore expect smaller flocks on mild days. Figure 1.2a shows that this is the case.

Changes in food availability. On certain days, Caraco (1979b) made more seed available to birds. He predicted that higher seed densities would, like higher temperatures, decrease feeding priority and allow dominants to spend more time fighting. Once again the OFS will decrease (Figure 1.1b) and we should expect to see smaller flocks. Figure 1.2b shows that flocks were indeed smaller when more seeds were available.

Changes in the risk of predation. An increase in predation risk should result in birds scanning more and therefore the OFS increasing (Figure 1.1c). To test this, Caraco *et al.* (1980a) flew a trained Harris' hawk (*Parabuteo unicinctus*) over flocks of juncos. As predicted, individual scanning rates and flock size were greater at times when the hawk was present (Figure 1.2c). As we have seen earlier, distance to cover is an important factor influencing predation risk in small birds. In Caraco's model, decreased distance should result in decreased scanning rates but *increased* flock size because more time is now available for feeding. Caraco *et al.* (1980b) placed artificial bushes near feeding flocks and found that, as expected, flocks became larger (Figure 1.2d).

While Caraco and co-workers' model is extremely simple, it shows that, at least in some cases, flock dynamics can be predicted from changes in individual costs and benefits. Showing that flock size changes in the predicted direction, however, is not the same as showing that it is optimal. Indeed, the very concept of an OFS is fraught with difficulties (Sibly 1983, Pulliam and Caraco 1984, Kirkwood 1986).

Optimal Versus Actual Flock Size: single-species flocks

While an optimal group size is a useful heuristic concept, there are good reasons why it might not prevail in the real world. In a simple model, Sibly (1983) showed that, more often than not, we should expect groups to be larger than the optimal size. He argued that, while a certain group size might be optimal in terms of individual fitness, individuals might still do better by joining the group than by feeding alone. Group size will therefore grow beyond the optimal size. To use Sibly's example, suppose a number of birds leave their overnight roost singly and, seeing all the existing flocks which have

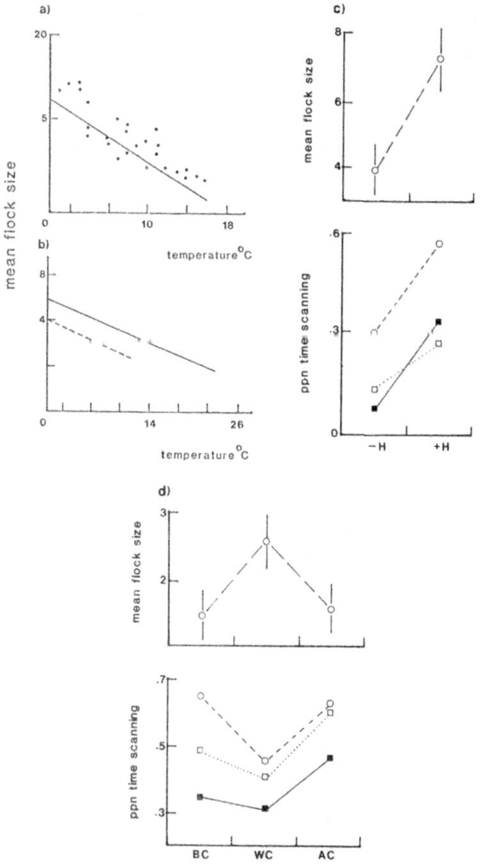

Figure 1.2: (a) Mean flock size ($\log_{10}$) in yellow-eyed juncos decreases with increasing ambient temperature ($r = -.89$). (b) At any given temperature, mean flock size is lower after the addition of food (dashed regression line, $r = -.53$) than before (solid regression line, $r = -.67$). (c) Flock size (above) and time spent scanning (below) increase when a hawk is present (−H, hawk absent; +H, hawk present). Open circles in lower figure, solitary birds; open squares, flocks of 3-4 birds; closed squares, flocks of 6-7 birds. (d) Changes in flock size (above) and time spent scanning (below) when cover is present (BC, before cover available; WC, with cover; AC, after cover removed. Symbols as in (c)). Data for temperatures between 3° and 6° C. Bars represent standard errors

Source: (a) and (b) modified after Caraco (1979a); (c) plotted from data in Caraco et al. *(1980a); (d) plotted from data in Caraco* et al. *(1980b)*

formed, join the flock where their fitness advantage is greatest. Thus the first 20 birds to leave will form an optimal flock of 20. The 21st bird will still do better by joining the flock than by feeding alone, so flock size increases away from the optimum to 21. This process continues until the flock reaches, say, 55 birds, when the benefit of joining becomes less than that of feeding alone. The 56th bird feeds alone, the 57th joins it and so on. In a series of such simple examples, Sibly shows that optimal group size will tend to be unstable and groups will grow beyond their optimal size.

Kirkwood (1986) proposed a more comprehensive model which suggests that the discrepancy between optimal and actual group size might be more complex. His model is worth describing in some detail. Like Sibly (1983), Kirkwood tailored his model to bird flocks. He considered that flock size would influence individual survivorship positively by increasing food intake and protection from predators and negatively by increasing competition and aggression. As in Caraco's model, food intake was greatest in medium-sized flocks. The three factors (anti-predation, food intake and competition/aggression) were modelled as the probability that a bird would perceive a benefit (or cost) in terms of each of them as a result of being in a given flock. The probabilities were then synthesised to give an overall probability that a bird would gain a net benefit in terms of survivorship from any given flock size. Kirkwood assumed that a bird would not *join* a flock unless there was a net advantage in doing so, and would not *leave* a flock unless there was a net cost in remaining. He also assumed that all feeding sites had some attraction value in the absence of already-feeding birds and that all sets of birds encountering feeding sites (which may or may not contain a flock) were similarly heterogeneous in terms of dominance status, foraging strategies and other phenotype-limited (Parker 1982) characteristics. The probability that any given individual would join/stay in/leave a flock could therefore be translated into the proportion of any given set of birds doing any of these. The OFS was defined as that flock size where the probability of a net benefit being perceived by a bird was greatest.

In a simulation, the model was run over 80 time periods (t) during each of which a number of birds became available to join the flock and birds already in the flock decided whether to stay or leave. If the flock reached an equilibrium size (EF size — where arrival and departure rates were equal, see later) before $t = 80$, the simulation was stopped. Simulations were run varying the flock size at which maximum anti-predator (*pmax*) and feeding benefits (*fmax*) accrued, the total number of birds in the population (from which flocks formed) and the form of arrival rate (constant or random). Figure 1.3 shows some sample results from the simulation presented as ratios of OFS:AFS, where AFS is the actual flock size pertaining at the end of a simulation.

When arrival rate is constant (Figure 1.3a-c), AFS>OFS in 90% of cases. In the remaining 10%, anti-predator and feeding benefits were in conflict and, depending on their relative strengths, maintained AFS below the OFS. If the

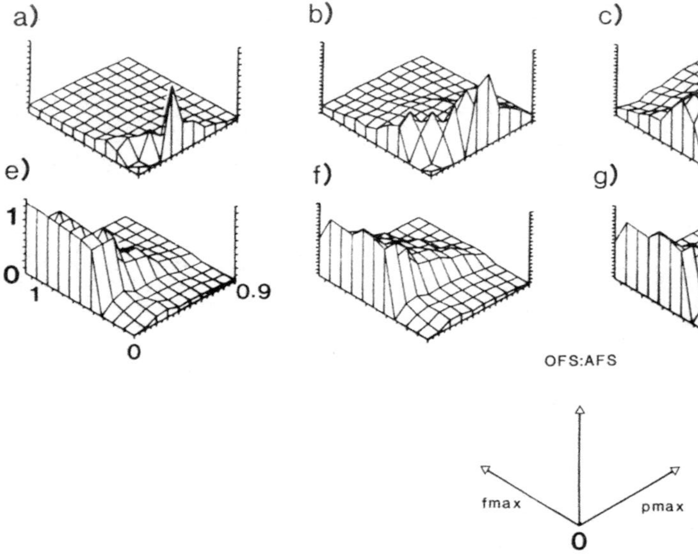

Figure 1.3: Plots showing the ratio of optimal to actual flock size (OFS:AFS) in relation to pmax *and* fmax *(see text).*
(a) Constant arrival rate, flock size associated with maximum feeding benefit (N_f) *= 10,* gmax *= 0.1. (b) As (a) but* gmax *= 0.5. (c) As (a) but* gmax *= 0.9. (d) As (a) but with pool of birds limited to 100. (e) As (b) except* N_f *= 53 and feeding benefit decreases with increasing flock size. (f) As (e) except pool of birds limited to 100 and arrival rate declines exponentially with time. (g) As (f) but with limited pool of birds*

Source: from data in Kirkwood (1986)

flock size at which maximum competition cost (*gmax*) accrued was varied, the maximum OFS:AFS ratio decreased, but this was a consequence of smaller flocks. Figure 1.3d shows results for simulations identical to those in Figure 1.3a except that the number of birds in the population was limited to 100. At low *pmax*, the results were similar, except where *fmax* was low and OFS:AFS consequently greater. As *pmax* increased, however, the effects of the limited pool of birds became apparent (e.g. for *pmax* = *0.9*, *gmax* = *0*, OFS:AFS = 0.4 instead of 1.0 with limited pool size, Figure 1.3a). Figure 1.3e shows a case where there is an unlimited pool of birds but feeding benefit is a decreasing function of flock size. Under many conditions OFS is now >AFS. At low *pmax*, OFS is low (sometimes 1) and most flocks build up past it. When it is high, OFS<AFS in all cases. In between these extremes, however, is an interesting area. Here, as *fmax* increases it quickly becomes the dominant factor determining the OFS, but *pmax* and values of feeding benefit at low flock sizes prevent flocks building up to the OFS.

Figure 1.3f,g shows results of simulations in which the pattern of arrival rate was varied. In Figure 1.3f, an exponentially declining arrival rate was used. This assumes that the first birds to occupy a feeding site will arrive in large numbers, but that birds will subsequently arrive in ones and twos. This is a reasonable approximation for species which tend to leave roosting sites and search for feeding areas in large groups, but which then move between feeding sites as individuals or small groups. A similar pattern to Figure 1.3e emerged, except that, in intermediate cases, flocks did not build up to the OFS (45-53 birds) despite the initial influx of birds (here, 30). Many of the initial birds left to search elsewhere and flock size became increasingly suboptimal until it reached an EF size. Note however that, as *pmax* increased, the AFS approached the OFS more closely. That these results are not simply a consequence of small numbers of birds arriving in each time period (because of the limited pool size of 100) is shown by Figure 1.3g. Here, the trends are similar to those in Figure 1.3f except that, at medium and high *pmax* values, OFS:AFS is lower because flocks can build up for longer. Nevertheless, it must be pointed out that these are EF sizes. Flock build-up has not stopped below the EF size simply because of time limitations.

Kirkwood's (1986) model, therefore, points to the complex way in which observed and optimal group sizes may differ. Individual costs and benefits, population size and pattern of arrival at feeding sites interact in such a way as to make simple predictions about flock size very difficult. Clearly variation in other factors, such as the initial attractiveness of the feeding sites, will complicate the picture still further. Under given conditions, one factor may predominate in determining the OFS, but another may predominate in determining how or whether flocks build up. Later, we shall see how this approach can be used to examine observed and optimal flock composition in mixed groups.

Group stability. During the discussion of actual versus optimal flock sizes, we referred to the formation of equilibrium flocks (*EFs*). We defined an EF as a flock in which arrival and departure rates are equal. This is the definition used by Krebs (1974) and Barnard (1980c). While many of the flocks in Kirkwood's (1986) simulations and some of those in the field (Barnard 1980c and see Chapter 7) reach an equilibrium size, many do not, and it is worth looking briefly at the different end-points of group formation and the factors which may determine them. Kirkwood's simulations and field observations by various workers suggest that, in birds at least, group build-up may end in one of three ways: (a) the formation of an equilibrium group size, (b) the formation of a statically stable group and (c) the formation of a continuously growing group which is limited in size by the time available for growth.

(a) Equilibrium flock size. In birds, EFs occur where more individuals are attempting to feed at a site than can be accommodated at any one time (Barnard 1980c). Thus, flocks build up as new birds arrive, but reach a point where no more can be accommodated. As new birds continue to arrive, therefore, others depart. The result is an EF which oscillates about a mean size (Figure 1.4a). In a study of EFs in house sparrows, Barnard (1980c) examined the environmental and behavioural factors which influenced their formation.

Barnard studied flocks of sparrows feeding on localised seed spillages on a farm. He found that, under these conditions, several flocks reached an EF size. Analyses of the relationship between EF size and environmental variables showed that EF size was most affected by the area of the densest seed patch at the site and the density of seeds outside the patch. EFs were bigger on bigger patches and when there was a higher density of seeds outside them. Birds aggregated at the densest patch because they pecked more, hopped less and turned more as seed density increased. They therefore moved more quickly over low density areas and gravitated to the dense patch. The tendency for newly arriving birds to land near others already feeding (Barnard 1978) also contributed to the build-up of birds in the densest patch. Seed density outside the patch was important because birds appeared to base their decision to leave on their pecking rate over the three pecks prior to departure. As flocks built up, fighting increased and more birds were ousted from the patch. If, on being ousted, they encountered a low seed density, they would tend to leave quickly and the EF would therefore be small. If seed density was high, birds would leave less quickly and the EF would be bigger.

(b) Statically stable flocks. Sometimes groups stop growing simply because no more potential joiners encounter the feeding site. Group size is no longer limited by equal arrival and departure rates, but by the fact that fewer individuals are attempting to feed than could be accommodated. Several factors may contribute to the formation of statically stable groups, including a small population size, a large number of available feeding sites and differences in

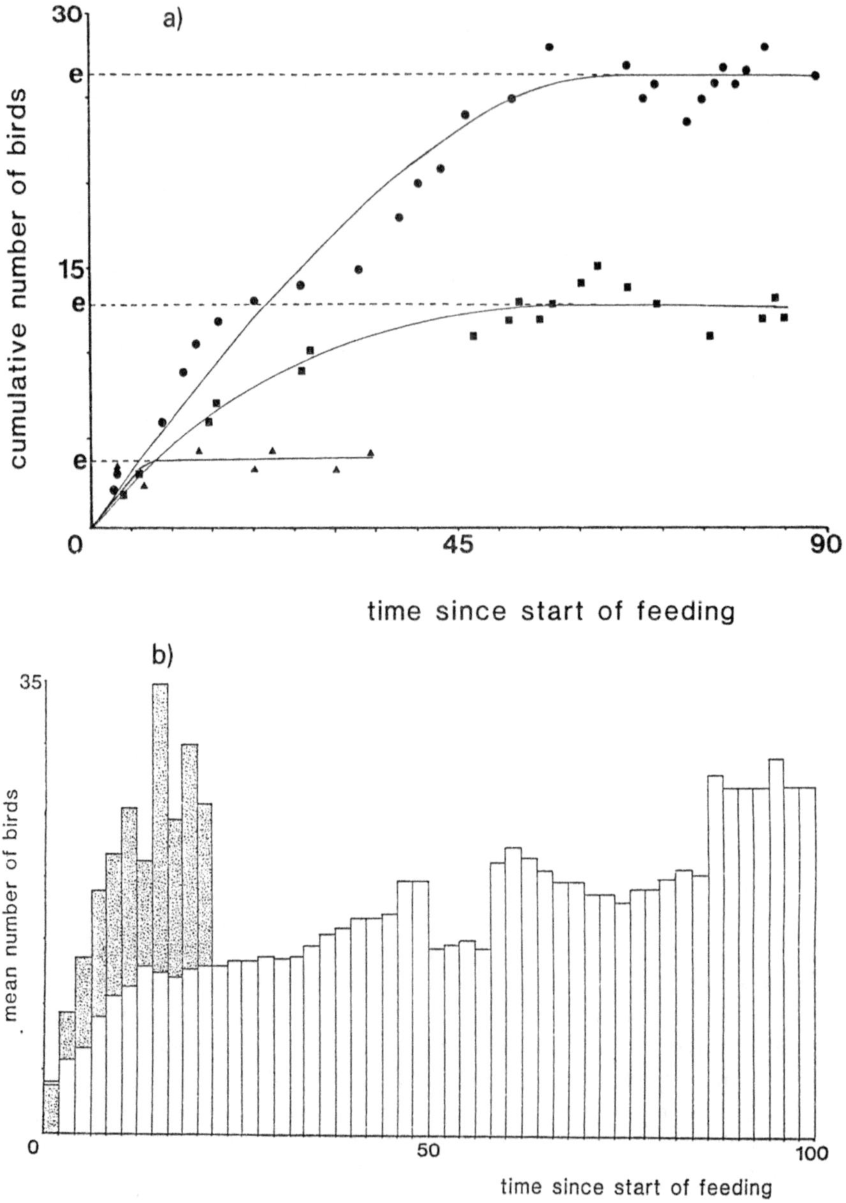

Figure 1.4: (a) Examples of equilibrium flock sizes (e) among house sparrows feeding on a farm (see text). (b) Time-limited flock size in house sparrows; flocks feeding in a protected cattleshed habitat (open histogram) build up for longer than those feeding in an exposed open-field habitat (shaded histogram), but note that flocks are generally larger in the open fields (see text)

Source: modified after Barnard (1978, 1980c)

feeding priority between individuals. Clearly, many, if not most, groups encountered are likely to be of this type, even if only temporarily.

(c) Time-limited flock size. In other cases, flock size may be restricted not by limited input or reaching an equilibrium, but by time limitations on growth. Food availability at a site and population size may be such that a group could go on growing more or less continuously, but its size is ultimately limited by, for instance, the length of time individuals are prepared to expose themselves to predation or repeated disturbance. Such groups differ from the last category because growth is usually continuous to the point of group departure. Figure 1.4b shows an example from feeding flocks of house sparrows (Barnard 1978). The plots show flocks building up in two habitats on a farm. In cattlesheds, where birds are protected from predators, flocks persisted for much longer than those in open fields where predation risk was apparently higher (Barnard 1980a).

When do different types of flock form? In the last section, we saw that changes in the relationship between individual costs and benefits and group size could influence group dynamics. Some groups tend to reach an equilibrium size, while others become statically stable or have the potential to grow continuously. What are the conditions under which different types of group form? Kirkwood (1986) looked at this in a further simulation model.

Kirkwood's (1986) model was again designed with bird flocks in mind, and it considered the formation of equilibrium and potentially continuously growing flocks. Conditions under which birds joined, left or remained in flocks and the nature of the source population and arrival rate were similar to those in Kirkwood's earlier model discussed above. An example of the model's predictions are presented in Figure 1.5, which shows the proportion of flocks becoming EFs or 'continuously growing' in relation to the probability that a given bird would perceive an anti-predator or feeding benefit from joining them. Figure 1.5a-c considers the effect of increasing *gmax* with constant arrival rates. As *gmax* increases, the proportion of flocks which attain EF sizes also increases. As might be expected, therefore, competition costs ultimately limit flock build-up. Except at very low values of *gmax* (Figure 1.5a), there was no effect of *pmax* and *fmax.* Figure 1.5d is comparable with Figure 1.5b except that arrival rates are now random instead of constant. The effect of random arrival is to increase the proportion of flocks becoming EFs, especially at higher values of *pmax* and *fmax.*

The type of flock which develops is thus likely to depend on a combination of individual costs and benefits, patterns of arrival at feeding sites and the size of populations from which feeding flocks can be drawn. Relationships between these variables, however, may not be simple. For any given relationship between flock size and competition, for instance, a change in the pattern of arrival can alter the importance of anti-predator and feeding benefits in determining flock stability.

Optimal Versus Actual Flock Size: mixed-species flocks

Kirkwood (1986) also considered the dynamics of mixed flocks. In a direct extension of his single-species model, Kirkwood simulated patterns of arrival

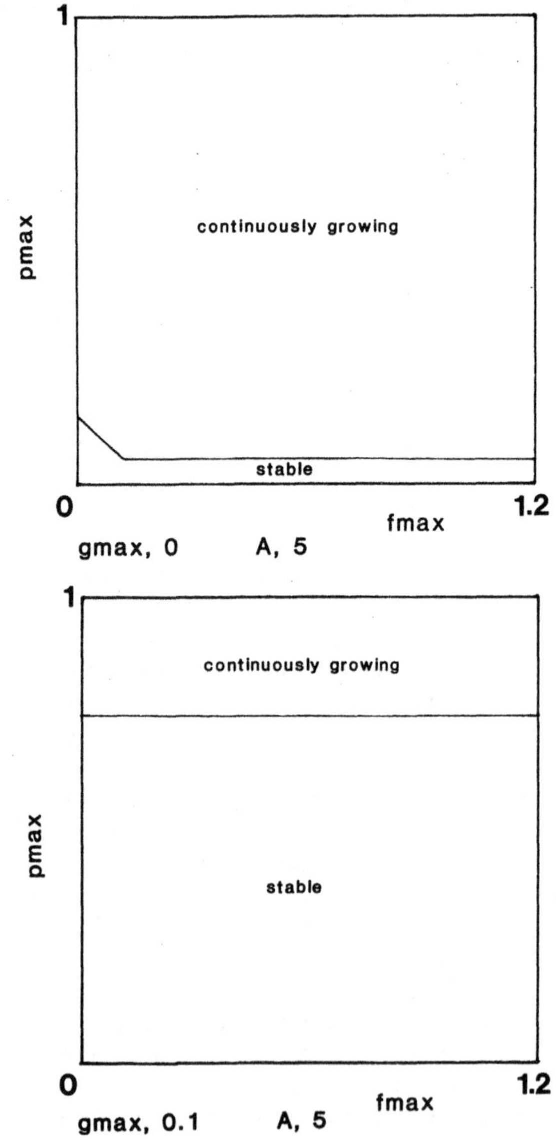

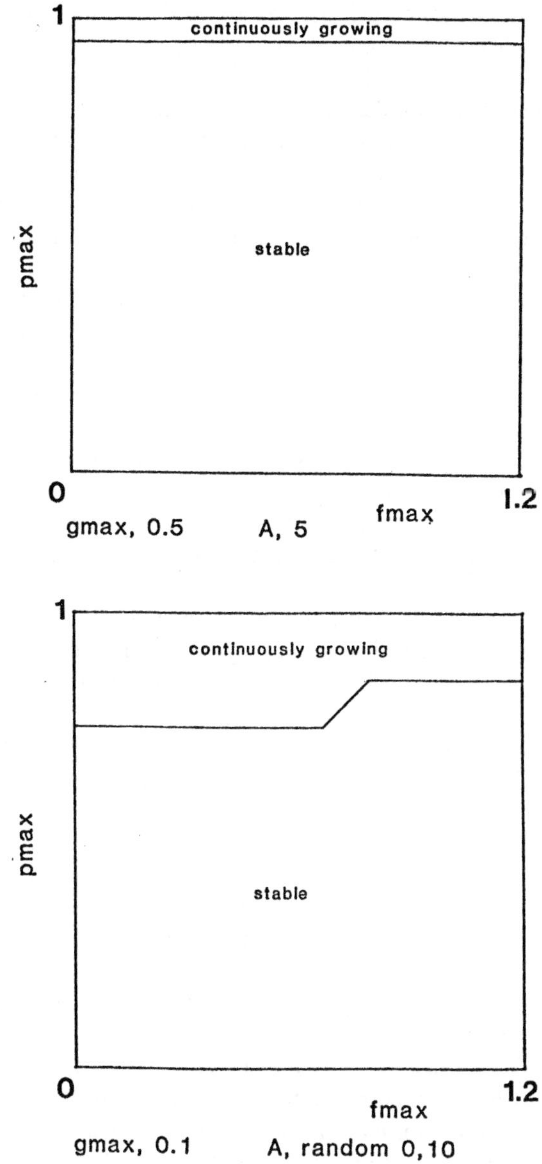

Figure 1.5: Effects of pmax *and* fmax *on flock stability under different conditions of* gmax *and arrival rate (see text)*

Source: from data in Kirkwood (1986)

and departure in flocks comprising two species. The shape of the curves relating the probability of a given bird perceiving a net benefit from joining a mixed flock (on the basis of the number of birds of each species) was the same as in the single-species model, and it was assumed, for simplicity, that anti-predator and aggression costs were simple functions of total flock size, although, of course, this will not always be the case. The crucial extension in the two-species model was synthesising the different benefits arising from the presence of two species. Kirkwood considered four basic methods of synthesis:

1. Individuals perceive a benefit from associating with one species or the other, depending on the species of their nearest neighbour.
2. There is a threshold of benefit on a common scale for both species, so that, if the higher threshold is passed, the lower one must also have been passed.
3. The perception of benefit from mixed association to species *i* is independent of that to species *j*.
4. Both species are regarded by each other as equivalents, but species *i* and *j* may be the same or different in their perception of this equivalent benefit.

The model also catered for variation in perceived benefit as a function of the proportion of each species in the flock by weighting the contributions of the two species appropriately. This is especially important for the first method of synthesis.

As with the single-species model, Kirkwood considered the consequences of different patterns of arrival and departure for flock dynamics. In this case, he recognised seven outcomes: (a) flocks achieve a constant size, regardless of species composition; (b) they achieve a constant species composition (i.e. a constant number of birds of each species); (c) they achieve a constant species ratio (but numbers can vary); (d) flocks become ever-increasing (see single-species model); (e) they become persistent (but not at an equilibrium size); (f) they decline to extinction; and (g) they fail to build up at all.

As an example of the model's output we can consider one particular set of circumstances. Here, the arrival rates of each species and the parameters of the model are kept constant and only the type of perception (see 1-4 above) varied. Figure 1.6 shows plots of actual and optimal species flock sizes, arising from the simulation, plotted against the flock size at which birds of each species (*i* and *j*) perceive maximum benefit (*bmax*). Clearly, there are differences between the two species' perception of an optimal flock size and what actually results. As in the case of single-species flocks, observed mixed flocks are likely to deviate from the optimal size for any of their associating species. Variation in perceived benefit for the two species, however, produced very different types of flock. The conditions of simulation in Figure 1.6a resulted in flocks of constant species composition (type (b) above), while those in Figure 1.6b and c produced both flocks of constant species ratio (type (c)) and flocks

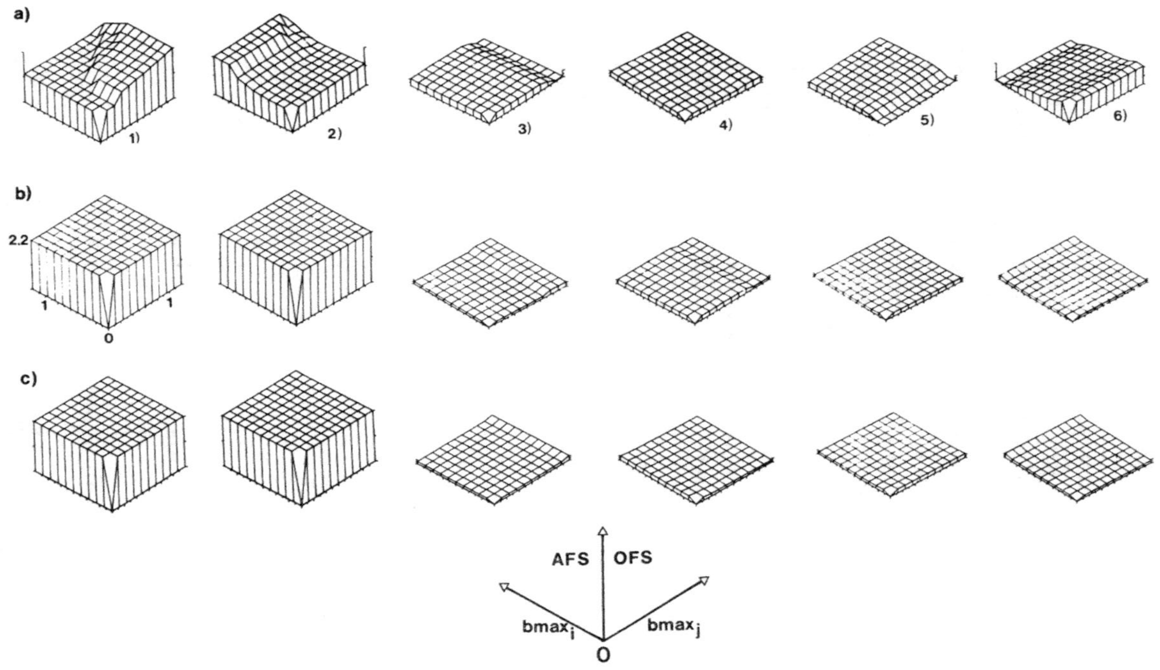

Figure 1.6: Plots of actual (AFS) and optimal (OFS) flock sizes plotted against bmax *for each species in simulated mixed-species flocks. (a)* fmax *fixed at 0.9 for species* i *and 0.2 for species* j, pmax *fixed at 0.5 and* gmax *at 0.1 for both species, perception of benefit was type 2 in the list above. (b)* fmax *fixed at 0.9,* pmax *at 0.9 and* gmax *at 0.1 for both species, perception was type 3. (c) As (b) except that perception was type 2 in species* i *and type 3 in species* j. *See text for details of parameters. For (a)-(c): 1) AFS of species* i; *2) AFS of species* j; *3) OFS of* i *as perceived by* i; *4) OFS of* j *as perceived by* i; *5) OFS of* i *as perceived by* j; *6) OFS of* j *as perceived by* j

Source: from data in Kirkwood (1986)

which increased in size indefinitely (type (d)). The proportion of the two types of flock differed between conditions (8% and 92% respectively in (b), 73% and 27% in (c)). Interactions of the perceived costs and benefits of association in different species thus affect not only the relationship between OFS and AFS, but also the nature of flock build-up and stability.

Herrera's threshold model. Herrera (1979) also proposed a graphical model to predict the formation of single- and mixed-species flocks. He was attempting to explain the existence of so-called 'flock-positive' and 'flock-negative' species in a Mediterranean woodland community (see earlier). The essence of his model is illustrated in Figure 1.7. The curves I_1 to I_3 represent respectively the increase in survival as a result of flocking due to improved protection from predators, increased amount of food intake and increased rate of food intake, plotted along a gradient of environmental conditions (*y*-axis). I_4 represents the decrease in survival from flocking due to aggression and competition. Where the sum of the benefits $I_1 + I_2 + I_3$ exceeds I_4 there is a net advantage to flocking. Clearly, species will vary in the environmental conditions under which $I_1 + I_2 + I_3 > I_4$, and Herrera accounts for the different tendencies towards association in his species by envisaging their environments as different points along the gradient. Species in environment A will not form

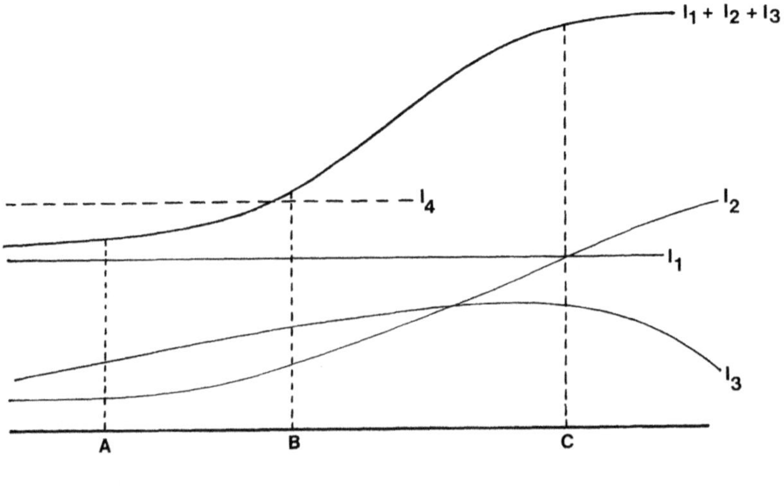

Figure 1.7: A threshold model for the formation of mixed flocks of birds. I_1 *to* I_4 *as in the text. A, conditions under which flocks are unlikely to form; B, conditions under which species are likely to be 'flock-negative'; C, conditions under which species are likely to be 'flock-positive'*

Source: modified after Herrera (1979)

flocks at all, because $I_1 + I_2 + I_3$ is never greater than I_4. Those in environment C will often form flocks because the payoff in doing so is high. Those in environment B, however, will sometimes form flocks but, because the net benefit is marginal, they will do so less often than expected on a random basis. In Herrera's scheme, environments B and C select for flock-negative and flock-positive species respectively.

Summary

1. A number of advantages and disadvantages accrue from forming single-species flocks. The most important advantages seem to be enhanced feeding efficiency and greater safety from predators.
2. Association provides the opportunity for some individuals to exploit, or otherwise capitalise upon, the time and energy investment of others.
3. Advantages and disadvantages accruing from mixed association may be qualitatively different from those experienced in single-species flocks. In particular, individuals may be able to expand their foraging niche and capitalise on the greater vigilance of other species.
4. The advantages and disadvantages of single- and mixed-species flocking interact to determine flock dynamics. While optimal flock sizes can be envisaged, they are unlikely, for several reasons, to persist in nature.

Chapter 2

Gulls and Plovers

So far, we have looked at the ecology and evolution of single- and mixed-species feeding flocks and the relationship between individual costs and benefits and flock dynamics. We now turn to a detailed analysis of one particular feeding association to see how the points we have raised can be integrated within a single study. While the remainder of the book deals almost exclusively with three bird species, its approach and methods of analysis are applicable to feeding associations in general. The association we shall be discussing is a familiar one in temperate European regions between three charadriiform species: lapwing, golden plover and black-headed gull (see below). These commonly associate on agricultural pastureland during the non-breeding season to feed on soil and surface invertebrates. Lapwings form single-species flocks, but are joined regularly by golden plovers and gull species. In this chapter, we introduce the three species, discuss the importance of pastureland in their ecology and survival and describe the area in which our study took place.

The Species

Here, we outline some of the basic characteristics of the species with which we shall be concerned. We restrict the profiles to those aspects that are relevant to the context and time of year in which the birds were studied. For fuller details of each species, the reader is referred to the recent, definitive reviews of Glutz *et al.* (1975) and Cramp and Simmons (1983), on which the following is based.

Lapwing (Subfamily Vanellinae, *Vanellus vanellus* L.)

The lapwing (Plate 1a) is a familiar, medium-sized plover (average length and wingspan 30cm and 85cm respectively) with characteristic broad and rounded wings. The species' range extends from Britain through Continental Europe and the USSR, occurring as far north as the northern coast of Norway and as far south as Portugal, Morocco and Turkey. It is strikingly patterned, predominantly black and white but, in winter, a dark, iridescent, olive-green dorsally. There is a distinctive thin crest on the head. Juveniles and adults of the two sexes can both be distinguished on the basis of plumage characteristics, and these will be discussed in detail in Chapter 6. In non-breeding adults, the face is barred and buff above the cheek-line, and the feathers of the chest and scapulars have pale, virtually white, margins. The flight is equally striking, with a rapid take-off and great agility in the air. In spring, the display flights of males are a dramatic feature of lapwing flocks. On migration, or travelling between winter feeding grounds, lapwings fly in characteristically loose flocks and with slow, erratic wingbeats.

As Cramp and Simmons (1983) point out, the habitat range is wide, from continental to oceanic and from boreal and temperate to steppe and Mediterranean regions. Lapwings require access to soil with an appreciable biomass of surface or immediately sub-surface invertebrates. Almost all prey are taken from the ground, although birds sometimes 'hawk' for insects in the air. They are catholic feeders, with prey ranging from beetles, flies, crickets, moths and earthworms to frogs and small fish (see also Table 2.1). Small amounts of plant material are also occasionally taken. Not surprisingly, there is seasonal variability in the diet. In the Netherlands and Scandinavia, earthworms appear to be the main summer diet, though insects assume greater importance during dry spells. Earthworms and a small number of insects appear to be the most important items for birds wintering inland in Europe. In exceptionally cold and frosty conditions, lapwings have been recorded eating cattle dung, perhaps because this is usually a rich source of earthworms and other invertebrates.

Lapwings are highly gregarious outside the breeding season, with flocks beginning to form in May. As in many wader species, adult females are usually the first to join, but early flocks often include immatures and adults of both sexes which have failed to pair or breed successfully (e.g. Spencer 1953). Depending on the region, later flocks (early summer) may be divided into those comprising mainly juveniles and those comprising adults. Migrating flocks may be seen at any time between late May and late August. These are mostly small (around ten birds) but may be enormous (a 5-km flock was recorded over the Isles of Scilly in 1953). Overwintering feeding flocks inland in Britain frequently contain 100 or more birds and have been recorded with over 5,000 (Oxfordshire).

Lapwing populations have undergone complex changes over the past 150 years and have declined in many areas of Europe, apparently as a result

Table 2.1: Stomach contents of lapwings shot on pastureland in winter

	Areas (and sources)					
	(1)	(2)	(3)	(4)	(5)	(6)
	Britain	USSR	USSR	N. France	Netherlands	Finland
Food items	**n=69**	**n=20**	**n=111**	**n=54**	**n=56**	**n=12**
ANNELIDA		—		—	'P'	83 (?)
Lumbricus	10 (?)		+			
Allolobophora	—		+			
INSECTA	64 (?)					
Coleoptera		? (20)	60 (60)	? (72)	? (25)	—
Diptera		? (60)	20 (>11)	—	? (29)	17 (?)
Others		?	20 (>40)	? (20)	? (30-60)	—
MOLLUSCA	10 (?)	—	+	? (20)	+	—
OTHER INVERTEBRATES	5 (?)	—	+	? (20)	+	—
PLANT	11 (?)	?	+	? (20)	+	—
OTHER	? (?)	—	+	—	—	—

KEY:
Numbers without parentheses = % of contents; numbers in parentheses = % of birds with prey type.
OTHER = stones, grit, shell fragments (2-9mm).
? = unknown, 'P' = principal item, + = present but <1% of contents, — = not recorded, n = no. birds sampled.

Sources: (1) Collinge (1924-27); (2) Ryabov and Masalova (1966); (3) Kistyakiviski (1957); (4) Madon (1935); (5) Klomp (1954); (6) Vepsäläinen (1968). The date from the Russian studies were taken from Cramp and Simmons (1983).

of habitat changes, severe winters and egg-collecting. The conversion of established pasture to arable with the spread of intensive farming appears to have been particularly deleterious. The British Trust for Ornithology Bird Census Index for the species was 196 in 1962, dropping to 87 after the hard winter of 1962/1963, 167 in 1970, 129 in 1971 and 153 by 1977. Declines have been widespread, especially in southern England, though western England and Wales appear to have suffered least.

Golden Plover (Subfamily Charadriinae, *Pluvialis apricaria* (formerly *Charadrius apricarius* L.))

The golden plover (Plate 1b) is a small to medium-sized plover (average length 30cm with a wingspan of around 70cm). Its range is more northerly than that of the lapwing, especially in Scandinavia and the USSR. Its straight bill, rounded head, oval-shaped body and running gait are similar to those of lapwings, but it is very distinct in overall appearance. At all ages and seasons, the upperparts are black or brown with a gold flecking which gives a spangled appearance at close quarters. Further away, the bird takes on a tawny-green hue. The underparts change from virtually all black in summer to mainly white in winter. The sexes are similar, but females tend to be duller in summer. At one time, the species was regarded as having northern and southern races, but is now assumed to be monotypic with marked plumage variation across latitudes.

Golden plovers occur mainly in higher latitudes, in continental-arctic or arctic-subalpine and boreal tundra regions. However, they also occur on temperate oceanic, unenclosed upland moors and peatlands. They live independently of marine or fresh water bodies. Golden plovers prefer flat or gently sloping topography and often use raised points as look-outs. In general, they prefer well-drained surfaces. After breeding, birds tend to move with their young to mown grass, close-cropped pasture, stubble, fallow and harvest fields or other open farmland (see the excellent account by Ratcliffe 1976). Golden plovers take a wide variety of invertebrates, but mainly earthworms and beetles. The diet ranges from these to flies, moths, millipedes, spiders and molluscs (see also Table 2.2). Like lapwings, they also take some plant material, including seeds, grasses and berries. Most food is taken from the ground surface (0-3cm depth).

The species is gregarious throughout the year, particularly outside the breeding season. Flocks are generally smaller inland in mid-winter than at other times on the coast and during migration. Inland flocks tend to occupy traditional ranges within which they move often and over large distances. Some ranges hold their largest flocks in winter, others in spring (see Fuller and Youngman 1979, Fuller and Lloyd 1981). Flock density tends to be greatest when birds are sleeping, intermediate when they are 'loafing' (resting, often on feeding grounds), and lowest when they are feeding (see later). Birds land in a 'wings-high' posture which exposes white axillaries, and there is usually much interaction between individuals with the wings being lowered only when the flock spreads out. Birds often make several approaches before they land to feed in a field, flying fast and low or wheeling over the area before climbing steeply again (see Chapter 4).

There has been a marked contraction of the southern limits of the golden plover's geographical range in northwest Europe since the mid-nineteenth century, caused mainly by habitat changes (heathland 'improvement' and afforestation). Coupled with this, there have been declines in southern populations (from Ireland to Finland). In some cases, the decline has been dramatic, especially in the plains extending from the Netherlands to West and East Germany and Poland where, a once extensive population is now all but extinct. The species has decreased in several places in Britain, especially in southern Scotland and northern England, apparently as a result of afforestation and reduced productivity of moorlands.

Black-headed Gull (Family Laridae, *Larus ridibundus* L.)

A small, agile gull, the black-headed gull (Plate 1c) measures some 40-45cm in

(a)

(b)

(c)

Plate 1. Lapwing male (a), golden plover (b) and black-headed gull (c) in winter plumage. (a) Courtesy of the copyright holder, Dr Tony Holley; (b) courtesy of Professor Richard Vaughan; (c) courtesy of Sam Grainger

length with a wingspan of 100-110cm. It is broadly similar in distribution to the lapwing, though less extensive in the north. In post-juvenile plumage, it has a brilliant white leading edge to the outer wing and a dusky lining to all but the outer primaries. In the breeding season, adults have a dark, chocolate-brown head sharply delineated at the throat and near the base of the skull. In the non-breeding season, the dark head plumage is completely lost except for a small, round, dark brown or grey ear-spot. One or two dusky bands extend over the crown from the eye and ear-spot. Irregular moulting and feather wear results in wide variation in the pattern of head plumage during winter; the ear-spot and bands may be indistinct and the back of the crown grey. Juveniles are initially much more richly coloured with variegated browns and greys.

The species breeds across the middle western Palaearctic from steppe and Mediterranean to boreal and peripheral subarctic regions. It extends from the continental interior to coasts and islands. It is always found near calm, shallow water bodies (fresh, brackish or saline) ranging from ponds, lakes and slow-flowing rivers to canals, gravel pits and sewage farms. Birds forage on nearby grasslands and plough, but also over water, including the sea. Outside the breeding season, birds move partly to inshore tidal waters and lowland sites, especially plough, moist grasslands, playing fields, parks and rubbish

Table 2.2: The stomach contents of golden plovers shot on pastureland in winter

	Areas and (sources)					
	(1)	**(2)**	**(3)**		**(4)**	**(5)**
	England (Essex)	**England (North)**	**Scotland Benbecula, N. Uist**		**France**	**Yugoslavia**
Food items	**n=28**	**n=13**	**n=40**		**n=9**	**n=3**
ANNELIDA		+ (93)	4 (?)	3 (?)	? (24)	—
Lumbricus	1 (25)					
Allolobophora	7 (46)					
INSECTA						
Coleoptera	10 (36)	68 (28)	+	+ (?)	? (100)	67 (67)
Diptera	37 (50)	8 (14)	49 (?)	37 (?)	? (100)	—
Others	+ (01)	4 (14)	+	+	—	—
MOLLUSCA	20 (11)	21 (?)	25 (?)	36 (?)	—	33 (33)
OTHER INVERTEBRATE	+ (02)	+	+	+	? (22)	—
PLANT	2 (57)	+	23 (?)	24 (?)	? (77)	—
OTHER	25 (92)	+	+	+	—	—

Sources: (1) Burton (1974); (2) Ratcliffe 1976; (3) Campbell (1935, 1946); (4) Madon (1935); (5) Cvitonic and Novak (1966, given in Cramp and Simmons 1983). For other details see Table 2.1.

tips. They may even move into cities. The species is mainly migratory to the east and north of the zone of winter freezing, and dispersive or partially migratory elsewhere.

The diet consists mainly of animal material, especially earthworms and insects, supplemented by plant material and domestic and other waste. Diet and feeding methods vary considerably with locality, season, fluctuations in food availability and individual birds, and feeding plasticity is no doubt another factor facilitating the spread of the species in recent decades (Vernon 1970, Curtis *et al.* 1985). Following Cramp and Simmons (1983), the main foraging methods include the following. (a) Walking and pecking: used mainly on arable and grassland where vegetation is low, also on mud and in shallow water; 'foot-paddling' is often used on moist surfaces, including pasture (see discussion in relation to plovers in Chapter 6). (b) Kleptoparasitism: stealing food, especially earthworms, from other species (see later). (c) Low searching flight: flying 1-2m above the ground, mainly over ploughed, sown or harvested fields and water bodies; the bird dips down to pick up food with deft hovering, braking and turning movements. (d) Medium-height searching flight: hovering over hedges and treetops to feed on fruit and insects. (e) High searching flight: using thermals to locate airborne insects such as ants and flies. (f) Dips and dives: used to take food from the water surface. (g) Scoop-feeding: used in saltmarsh pans and involving swimming or running with the head partly submerged to scoop in prey.

Unlike the two plover species, black-headed gulls have undergone a marked extension in their northern European range since the early nineteenth century, with the additional recent colonisation of Spain, Italy, Greenland and Newfoundland. The population is still increasing, especially in western Europe, though now at a reduced rate. Factors involved in the increase probably include a moderating climate (facilitating the spread north), reduced persecution, and increased food supplies (e.g. from rubbish tips and extending agricultural activity).

Single- and Mixed-species Associations in Plovers

Plover species vary considerably in their social behaviour. Very few detailed studies have been made, however, and there is no thorough comparative survey (though see Vaughan 1980). The best studies of foraging associations are those of Baker (1973), Baker and Baker (1973) and Pienkowski (1981, 1982, 1983a,b). Table 2.3 summarises data from the literature and the authors' observations. The majority of the 16 species occurring in the western Palaearctic are intraspecifically gregarious (flocks regularly exceed 100 individuals), although many tend to form loosely-packed flocks. Of these, however, 12 sometimes form mixed associations and five regularly do so. In the closely-related Scolopacidae, with which five species of plover associate, the tendency to form single- and mixed-species flocks is much greater (e.g. Goss-Custard 1980, 1983). The most likely reason for the difference is that plovers appear to detect their prey visually rather than by tactile means (as in many scolopacids) (see especially Pienkowski 1983a, Metcalfe 1984a).

Feeding rate is therefore sensitively dependent on factors which disturb surface-dwelling prey. Because the activities of close neighbours are likely to be a serious source of disturbance, large inter-neighbour distances are generally maintained.

The Evolution of Pasture Use

Throughout the year, lapwings and golden plovers depend in different ways on farmland for foraging and breeding. How has this association between shorebirds and farmland evolved? An excellent account of the shorebird fossil record is provided by Hale (1980).

Fossil records of plovers date back to the Oligocene (40my b.p.), when the climate was milder than at present. From the mid-Miocene, the climate gradually cooled until, by the late Pliocene (2my b.p.), conditions were similar to those of today. At this time, hardier wader species such as knot (*Calidris canutus*) and turnstone (*Arenaria interpres*) probably bred in the Holarctic tundra. Cooling continued, culminating in the series of 'Ice Ages' (0.8my b.p.). The four major periods of glaciation during this time had a profound

Table 2.3: Group foraging and social behaviour among western Palaearctic plovers (Charadriidae) in the non-breeding season

(1) Species body length in cm	(2) Gregariousness	Heterospecific associates	Territoriality	Aggression	Notes
Little ringed plover [15]	Intermediate	Sometimes: small plovers, stints and sandpipers	None	None	
Ringed plover [19]	High	Sometimes: small plovers and stints	None	Sometimes for space	(3)
Kittlitz's plover [13]	Intermediate	Rarely	None	None	(4)
Kentish plover [16]	High	Frequently: ringed plovers and greater sand plovers or small Charadrii, especially dunlin	None	Frequent for space	
Lesser sand plover [20]	High	Sometimes	None	None	
Greater sand plover [25]	High	Frequently: lesser sand plovers and some Kentish plovers; or ringed plovers and Kentish plovers on grassland	Frequent	Frequent, usually for space, including kleptoparasitism	(5)
Caspian plover [19]	High	Sometimes: e.g. ringed plover	None	None	
Dotterel [22]	Occasional	None	None	Some	
Turnstone [23]	High	Sometimes: e.g. sandpipers	None?	Sometimes?	

Golden plover [28]	High	Frequently: lapwings. Sometimes: oystercatchers, redshank, curlew, more rarely dunlin	Sometimes: 5m diameter	Sometimes	
Grey plover [28]	Intermediate	Sometimes: e.g. Ringed plover	Frequent: approx. 40m diameter	Sometimes	(6)
Spur-winged plover [27]	Solitary Intermediate	Sometimes: e.g. lapwings	Sometimes (all)	Sometimes	
Red-wattled plover [33]	Intermediate Solitary	None	Sometimes	Sometimes	
Sociable plover [29]	High	?	?	?	
White-tailed plover [28]	High	Frequently: e.g. redshank	None ?	None ?	
Lapwing [30]	High	Frequently: e.g. golden plovers. Sometimes: oystercatchers, redshank, curlew, more rarely dunlin	Rare: diameter of 2-3m recorded	Sometimes	(7)

See over for notes

Notes to Table 2.3:

(1) Species observed breeding in the western Palaearctic region (values in square brackets give body length in cm).

(2) Gregariousness: High = flocks exceeding 100 individuals throughout most of non-breeding season. Intermediate = flocks containing 25-100 individuals throughout non-breeding season. Occasional = small flocks 10-25 sometimes during the non-breeding season. Solitary = rarely feeds in groups as large as 5 individuals.

(3) Possibly excluded from good feeding sites by dunlin and larger wader species, e.g. grey plover.

(4) Forages with other species only when tidal movements bring birds together on localised feeding sites. The associations are temporary.

(5) Holds territories of approximately 40m diameter in tidal flats; these may be contiguous or loose and scattered. Aggression occurs frequently among foraging birds, particularly when inter-individual distance drops below 10m, in which case the encroacher is attacked. When large prey (e.g. molluscs) taken to water for washing, conspecifics may approach carrier and attempt to steal prey item.

(6) Both long-term (several months) and short-term territories defended during autumn/winter period. Long-term territory owners include smaller proportion of juveniles than short-term owners (Townshend *et al.* 1984).

(7) Territories, sited contiguously along muddy shore, recorded for birds in autumn. Laven (1941) observed 'ground fights' more frequently in flocks than in nesting territories. Kleptoparasitism recorded infrequently (McLennan 1979).

Sources: Cramp and Simmons (1983), Dittberner and Dittberner (1981), Dugan (1982), Lind (1957), Pienkowski (1981, 1982, 1983a, b), Stinson (1980), McLennan (1979), Baker (1974), Baker and Baker (1973), Witherby *et al.* (1940), Townshend *et al.* (1984), Thompson (1984).

Cramp and Simmons (1983, pp. 4-6) do not quantify some of their descriptive terms (e.g. 'frequently', 'sometimes', 'occasional'), so we have used them as they occur in their text.

effect on the distribution and speciation of present-day waders (see Hale 1980). As the ice sheet advanced and contracted, however, the distribution of tundra changed accordingly. During the last period of glaciation, the tundra refuge for waders was restricted to northern Greenland, Alaska and parts of Siberia. Such species as knot, turnstone, dunlin (*Calidris alpina*) and golden plover were probably restricted to northern Greenland at this time. As the ice sheet spread southwards, wader populations moved into North America, the northeastern and western corners of the USSR and down as far as northern France. Whereas some species now moved into new habitats, others remained in the former inter-glacial zones. The wader community in the northern hemisphere was probably at its peak, in terms of distribution, numbers and species diversity, during the last great period of glaciation (about 20,000 years ago).

In continental-arctic and arctic-subalpine boreal tundra, the wader community shows a marked diversity in foraging and breeding ecology, despite species frequently occupying similar biotopes. Lapwings forage and breed in a wide range of farmland habitats, and breed alongside golden plovers only very locally on some hill pastures and blanket bog. Ratcliffe (1976) suggests that heavier egg predation suffered by golden plovers (whose eggs are less cryptically marked than those of lapwings) has prevented them from expanding their breeding range onto pastureland. While golden plovers breed in non-agricultural upland areas, they may travel 10km or more from their nesting site to feed on pasture (Ratcliffe 1976, authors' pers. obs.). Invertebrate prey (e.g. lumbricid worms, dipteran larvae, coleopterans) occur in higher densities on pasture than on neighbouring high ground (Coulson 1959, 1962, Coulson and Whitaker 1977). We compared the surface availability of invertebrates in moorland and hill pasture used for nesting and feeding by a breeding population of golden plovers in northwest Sutherland. We found that hill pasture contained between four and ten times more invertebrates than surrounding moorland.

Although the first formal records of golden plovers foraging on pasture are given by Gilbert and Brook (1924), anecdotal evidence goes back to the sixteenth century when land was first enclosed for agricultural purposes (Fitzherbert 1534). In Britain, it is possible that the use of pasture by golden plovers goes back to around 2,000 BC when man switched from a nomadic to a pastoral lifestyle (Ratcliffe 1976, Thompson 1983a). Before that, golden plovers were probably widespread on upland moors and fells of northern Europe. In Britain, much of present-day moorland was covered by forest, which first began to be reduced by man about 10,000 years ago. Forests were cleared to produce heaths and many of these were subsequently converted into agricultural land. The Anglo-Saxons and Romans continued the forest clearance and there is evidence that fire cleared extensive areas (McVean and Lockie 1969). Later human influence, such as the clearing of glens and hillsides in the Scottish highlands and islands for sheep and the increasing popularity of grouse shooting during the nineteenth century, extended the moorland habitat. Ratcliffe (1976) suggests that the virtual culmination of forest destruction by the 1860s correlated with the maximum breeding-population sizes of golden plovers.

The Use of Pasture during Winter

During the summer, after the breeding season, lapwings and golden plovers move south from their breeding grounds to winter on lowland pasture. By as early as July, many golden plovers in northern England and eastern Scotland have moved onto farmland (e.g. Parr 1980). Recoveries of individuals breeding in Britain showed that birds move an average of 115km to their winter grounds, with very few emigrating (Cramp and Simmons 1983). Further north, birds from Iceland and Fennoscandinavia may travel over 1,000km to central and southern England (Fuller and Youngman 1979, Fuller and Lloyd 1981) and the Netherlands and Germany (van Eerden and Keij 1979, Cramp and Simmons 1983) respectively. Lapwings show similar migrations, with those breeding in Britain travelling shorter distances

(<100km) than those breeding elsewhere in the western Palaearctic (Cramp and Simmons 1983).

Since lapwings often breed on farmland (Spencer 1953), it is not surprising that they move to farmland over winter (Crooks and Moxey 1966). Wader species which associate with golden plovers on moorland, however, all resort to coastal areas in winter (see Hale 1980, Pienkowski 1981, Prater 1982, Cramp and Simmons 1983). Golden plovers may have used lowland pasture initially because they resemble the hill pastures where the birds prefer to feed during the breeding season. Golden plovers appear to prefer hill pasture to other, apparently equally profitable feeding habitats (D.A. Ratcliffe pers. comm.). As we mentioned earlier, one reason for this may be that, like lapwings, golden plovers are principally visual foragers and the short, grazed grass of hill pasture may make prey easier to detect. There are approximately 400 'winter ranges' used by a total of about 200,000 golden plovers in Britain (Fuller and Youngman 1979, Fuller and Lloyd 1981). Lapwings are more widely distributed, and Fuller and Youngman suggest that they feed on arable land more frequently than do golden plovers.

Winter associations between lapwings, golden plovers and gulls. Throughout the winter, golden plovers are usually associated with lapwings, mainly on pastureland but occasionally also on arable land (e.g. Källander 1977, Fuller and Youngman 1979, McLennan 1979). In some areas, particularly along the east coast of Scotland and northern England, lapwings and golden plovers also associate with oystercatchers (*Haematopus ostralegus*), redshanks (*Tringa totanus*) and curlews (*Numenius arquata*) (McLennan 1979). Although single-species flocks of plovers in general are common, they are rarely observed in golden plovers. Fuller and Youngman (1979), for instance, recorded golden plovers feeding in conspecific flocks in only one out of six winter ranges, and this was because lapwings had failed to return after a prolonged cold spell (see Chapter 3). McLennan (1979) also rarely observed golden plovers feeding in single-species flocks (see Chapter 3). He also observed a small number of attacks by lapwings against golden plovers, but none the other way round (see also Wallace 1983). Indeed, rates of interaction were generally lower among golden plovers. Analysis of interspecific dispersion showed non-random distributions, suggesting active avoidance or deterrence by the two species.

Black-headed and common (*Larus canus*) gulls often join flocks of lapwings and golden plovers (see later, and also Källander 1977, 1979, Fuller and Youngman 1979, McLennan 1979) and feed kleptoparasitically (see Chapter 1). Källander (1977) has calculated that gulls can obtain in excess of their daily energy requirement solely by stealing earthworms from lapwings. Both Källander (1977) and McLennan (1979) found that lapwings were about twice as vulnerable to attack as golden plovers. McLennan also found that lapwings in small to medium-sized flocks (2-10 birds) were much more likely to be attacked than those in large flocks ($>$120 birds). Gulls joined 60% of lapwing flocks but only 2% of redshank flocks. They never joined oystercatchers. Characteristically, the gulls observed by McLennan were dispersed evenly within plover flocks and maintained ratios of approximately 1 gull:20 plovers (common gull) and 1 gull:10 plovers (black-headed gull) through the winter. Källander (1977) recorded distances of between 5m and 20m between gulls. Dispersion among lapwings and golden plovers was variable. Golden plovers were more densely aggregated (usually $<$5m between neighbours), but both species tended to be regularly distributed in moderate- and low-density flocks and randomly distributed in high-density flocks.

Kleptoparasitism in gulls. A detailed description of kleptoparasitism by common and black-headed gulls in lapwing/golden plover flocks is given by Källander (1977) (see e.g. Brockmann and Barnard 1979, Mudge and Ferns 1982, Burger and Gochfeld 1981, Greig 1985 for broader descriptions).

Gulls stand in a characteristically alert posture observing the foraging activity of plovers. They often take up position on a raised look-out point, such as a mole hill, fence post or tussock. When a lapwing or golden plover finds a prey item, usually an earthworm, one or more gulls orientate or fly towards it. If the plover has already extracted the item, it will usually take off and the gull(s) may follow. Sometimes other gulls join the chase at this stage. Both lapwings and golden plovers tend to make erratic turning movements in the air while being chased, but chases usually end with the prey being dropped and picked up by a gull. Sometimes gulls attack before a plover has fully extracted its prey from the ground. In these cases, the plover usually abandons the item and runs a few feet to begin foraging elsewhere. Both Källander (1977) and McLennan (1979) found that gulls took predominantly the larger prey items procured by plovers. Distances between the well-spaced gulls are maintained by threat and aggression (using the 'oblique', 'forward' and 'aggressive upright' (Moynihan 1955) postures), so that gulls appear to partition plover flocks into mobile feeding territories. As the distribution of plovers within the flock changes, gull territories also change. In large, dense flocks, gulls may be only 3-10m apart (see Plate 2).

While lapwings and golden plovers are the commonest hosts of kleptoparasitic common and black-headed gulls, they are not the only ones. Table 2.4 summarises the range of charadriiform host species which have been documented. Although common gulls do steal from plovers on farmland, they are far less widespread than black-headed gulls (Vernon 1972, Mudge and Ferns 1982). Vernon (1972) suggests that, of the two, common gulls frequent drier, well-drained soils at higher elevations. Nevertheless, it is common for both species together to kleptoparasitise plovers in the same flock. Where they do, common gulls are much the more successful by virtue of their size (Källander 1977). Although both gull species often steal food, it is more usual

(a)

(b)

Plate 2. Changes in inter-neighbour distance between birds as flock size increases from (a) to (b). (a) Note the direction in which lapwings are facing in relation to their distance from the gull (see Chapter 8). Photographs courtesy of Dr R.J. Fuller

Table 2.4: Known wader hosts of kleptoparasitic black-headed and common gulls

Black-headed gull victims	Food stolen	Habitat	Sources
Lapwing	Earthworm	Pasture	1, 2, 5, 6, 7, 10, 11, 12, 14, 15, 17, 18
Lapwing	*Nereis* sp.	Mudflats	5, 6
Golden plover	Earthworms	Pasture	1, 2, 6, 9, 10, 17, 18
Oystercatcher	Mussels	Mudflats	6, 8, 16, 17
Redshank	Earthworm	Pasture	6, 9
Redshank	*Nereis* sp.	Mudflats	5, 6
Ruff and reeve	Earthworms (?)	Pasture	4
Curlew	Earthworms	Pasture	6

Species usually avoided: grey plover, dunlin, knot, snipe, turnstone, purple sandpiper, bar-tailed godwit, (curlew)

Common Gull victims	Food stolen	Habitat	Sources
Lapwing	Earthworms	Pasture	8, 10, 12, 17
Golden plover	Earthworms	Pasture	17
Oystercatcher	Mussels	Mudflats	6, 8
Redshank	Earthworms	Pasture	9
Ruff and reeve, Curlew, Bar-tailed godwit	?	?	4

Species usually avoided: grey plover, ringed plover, dunlin, snipe, turnstone, purple sandpiper

Sources: 1 Barnard and Stephens 1981; 2 Burgess 1975; 3 Clegg 1944; 4 Cramp and Simmons 1983; 5 Curtis *et al.* 1985; 6 D.B.A.T. pers. obs.; 7 Evans 1908; 8 Gillham 1952; 9 Hamilton and Nash; 10 Källander 1977, 1979; 11 Laidlaw 1908; 12 Mudge and Ferns 1982; 13 Sage 1963; 14 Selous 1927; 15 Tinbergen 1953b; 16 Tinbergen and Norton-Griffiths 1964; 17 Vernon 1972; 18 Wallace 1983.

for them to search for their own (Mudge and Ferns 1982, Curtis and Thompson 1985, Curtis *et al.* 1985).

Few other gull species are known to kleptoparasitise waders. Payne and Howe (1976) observed ring-billed (*Larus delawarensis*) and Bonaparte's (*L. philadelphia*) gulls stealing earthworms from dunlins and black-bellied plovers (*Pluvialis squatarola*) in a ploughed field. Johnston (1945) recorded herring gulls (*L. argentatus*) stealing earthworms from lapwings, and Dummigan (1977) observed Iceland gulls (*L. glaucoides*) stealing food (presumed to be mussels) from oystercatchers.

Within the Laridae as a whole, kleptoparasitism is only one of a variety of opportunistic feeding strategies (Brockmann and Barnard 1979, Mudge and Ferns 1982). Which strategy gulls prefer to adopt depends at least partly on body size. Table 2.5 shows that smaller species (e.g. Ross's gull, *Rhodostethia rosea*) tend to search for their own food. The slightly larger sooty gull (*L. hemprichii*) is exclusively a scavenger, feeding on refuse tips and other sources of food. Kleptoparasitism is common only in intermediate-sized species such as those mentioned above. The largest species, such as the great black-backed (*L. marinus*) and glaucous gulls (*L. hyperboreus*), are mostly predatory on the eggs, chicks and adults of smaller seabirds.

The Study

The Study Area

The area chosen for our study of lapwing/golden plover/black-headed gull associations is a stretch of mixed farmland 20km southeast of Nottingham, lying between the villages of Wysall, Rempstone, Bunny and Keyworth (Figure 2.1). It is part of the catchment area for the River Soar, and the fields which are used by plovers are between 50m and 100m above sea level. The area is situated on Triassic and Liassic (lower Jurassic) strata with a discontinuous cover of boulder clay. Many of the fields contain 'brown earth' (see

Table 2.5: Body size, weight and wing length of gull species in the western Palaearctic in relation to foraging strategy, flocking tendency and preferred prey

Main feeding strategy	(1) Mean body size (cm)	(1) Mean body weight (g)	(1) Mean wing length (cm)	(2) Flocking tendency	(3) Prey size
Predatory (large prey) (n=4)	64 ± 2.7	1254 ± 159	446 ± 12.1	2-3	3 mainly
Kleptoparasitic (n=8)	52 ± 4.8	651 ± 138	390 ± 16.5	2-3	2 mainly
Scavenger (n=1)	43	455 ± 55	118 ± 73.0	1-2	2
Predatory (small prey) (n=15)	39.8 ± 2.49	410 ± 85	329 ± 11.7	2 mainly	1-2

Data from Cramp and Simmons (1983), n = no. species.

(1) mean ± standard error for gulls collected from all regions of the western Palaearctic.

(2) 1 = solitary; 2 = intermediate; 3 = gregarious.

(3) 1 = invertebrates; 2 = fish and large invertebrates, including mussels and earthworms; 3 = large prey items, including birds and small mammals.

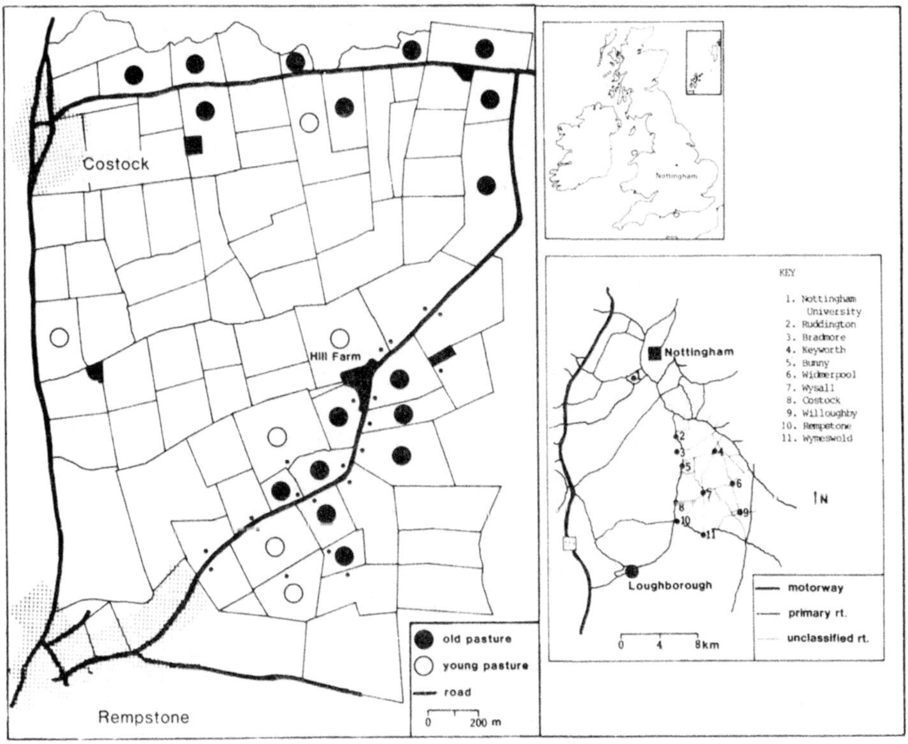

Figure 2.1: The study area. Left: the Hill Farm area; fields not shown as pasture were plough, stubble or contained winter crops. Bottom right: the locality of the study area in south Nottinghamshire

Edwards 1966) soils which are characteristically leached with movements of clay into the B horizon, but, on Keuper marl, the soil is mainly clay-loam. The majority of the soil cover in the area is sandy clay-loam with a subangular blocking structure, resulting in many fields being poorly drained and two prone to flooding. These details are important because it has been suggested recently that, in Britain, it is only the dry agricultural grasslands that are used by wintering lapwings and golden plovers (Fuller and Lloyd 1981).

Land use within the study area is shown in Figure 2.1. By far the biggest single land-use category is temporary pasture (those areas which were pasture at the time of sampling but which had been ploughed within the previous four years). The only comparable data for a range occupied by golden plovers are those given by Fuller and Youngman (1979). They found that 70% of a 2,084ha winter range in Oxfordshire was grassland, and that 70% of this was temporary. Pasture size in the study area averages 6.72 ± 3.46ha, with a range of 1.04-17.53ha (n = 47 fields). Most fields are bordered by 1.5-3m-high mixed hedgerows, but some which regularly contain livestock (dairy cattle, horses and sheep) are bordered by fences. Pasture age varies discontinuously. Some fields have been pasture for only one or two years, others have been so for over 400 years. Most, however, are of either one to four years' or greater than 25 years' standing.

The area around Hill Farm (Figure 2.1) was the most intensively studied because it comprises mainly permanent pasture and birds feed there regularly in a wide range of flock sizes. The pasture area is surrounded by arable land consisting mainly of cereal (barley and wheat) crops, plough and stubble or fallow. It was thus possible to make complete circuits of the pastures and observe all feeding birds. During the six years of the study, this area contained winter populations of about 1,000 lapwings, 800 golden plovers and an average of 50 (maximum 215) black-headed gulls; it has been used by these species since at least the turn of the century (Dobbs 1979, D. and J. Allsopp pers. comm.). Depending on weather conditions, plovers roosted in a number of different stubble fields within the Hill Farm area. However, they were also joined here by other birds from the Keyworth, Bunny and Wymeswold areas, so the number at the roost was usually greater than the Hill Farm feeding populations (see also Fuller and Youngman 1979).

As elsewhere, the number of pasture fields within the study area has begun to decline within the last two or three years as more land is ploughed for cereal production. This reflects a general reduction in the amount of permanent pasture within the county since the Second World War (Figure 2.2, and see Dobbs 1979, HMSO 1980). While, as Parslow (1974) points out, modern arable farming has provided lapwings with more nesting habitat, it has simultaneously reduced the habitat on which they and golden plovers depend for food during winter. Loss of permanent pasture appears to be partly responsible for the marked decline in the number of lapwings and golden plovers wintering in Britain (Ratcliffe 1976, 1977, Fuller and Youngman 1979, Fuller 1982, Thompson 1983b, Thompson and Fuller in prep.).

The Study Period

The study was carried out from October to March of 1978/1979 to 1983/1984 inclusive. This period in the year was chosen to simplify the problem of identifying prey. While lapwings and golden plovers are known to take a wide range of invertebrates (e.g. Burton 1974, Glutz *et al.* 1975, Tables 2.1, 2.2), the availability of species other than oligochaete worms declined dramatically in mid-late October and observations suggested that birds were taking exclusively earthworms. All identifiable items (over 1cm in length) taken by birds during the study period were earthworms, and samples of surface (0-3cm deep) soil assayed by hand-sorting and Tulgren funnel (see Edwards and Lofty 1972, Southwood 1978) showed that between 97.5% and 99% of surface-layer invertebrates were earthworms. The remainder were staphylinids or the larvae of other coleoptera and diptera.

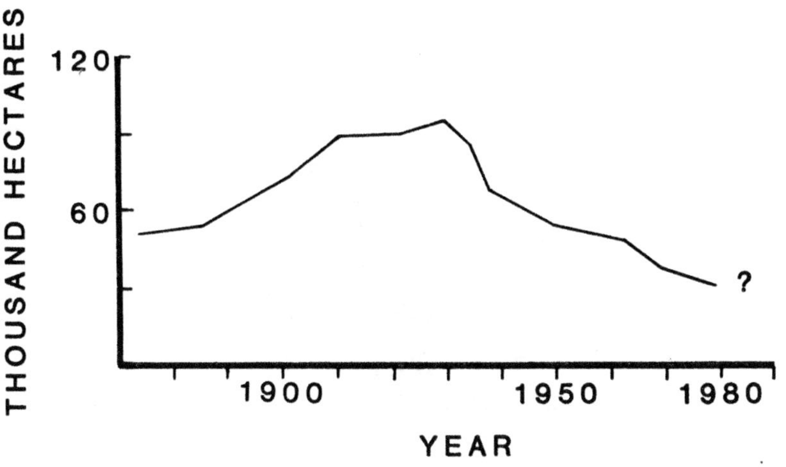

Figure 2.2: The change in the area of permanent pasture in Nottinghamshire since 1880. Data from Dobbs (1979) and HMSO (1980)

The period of data collection ended in early March each year, because lapwings then began courtship behaviour and flocks became much larger than feeding flocks (upwards of 500 birds in some cases) and because the number of surface invertebrate species other than earthworms began to increase. Since gulls stole only earthworms, the October-March observation periods coincided with the peak of gull attendance in plover flocks.

Summary

1. Lapwings, golden plovers and black-headed gulls commonly associate on agricultural land during winter.
2. Variation in group foraging among waders may be related to their mode of foraging. Plovers may be less likely to form foraging aggregations, or more likely to forage at low densities, than other waders because they tend to hunt visually.
3. The use of pasture by lapwings and golden plovers during winter may have evolved initially as a consequence of food requirements during the breeding season.
4. The continuing decline in plover populations in western Europe may be due in part to the conversion of established pasture to arable.
5. Kleptoparasitism is one of several feeding strategies adopted by black-headed gulls. Among gulls in general, primary feeding strategies appear to be related to body size.

Chapter 3

Choosing Where to Feed: Choice of Field

As we have seen in Chapter 1, several factors influence the spatial and temporal distribution of predators around feeding sites. Among birds, variation in the availability of prey, climatic factors, predation risk and social factors have all turned out to be important. The interaction between different factors in socially feeding species may be complex, even in those feeding in single-species flocks. In mixed flocks, the situation may be further complicated by frequency and density-dependent effects of the distribution and behaviour of one species on those of others. As we shall see later, the influence of conspecific and other companions on the behaviour of birds can vary with both flock size and flock-species composition. In this chapter, we examine the factors which determine the distribution and composition of mixed-species charadriiform flocks around our study area and ask why certain feeding sites are occupied more regularly than others.

Choice of Field and the Distribution of Foraging Flocks

The distribution of lapwings, golden plovers and black-headed gulls around feeding sites was recorded over three winters (1980/1981, 1981/1982 and 1982/1983) by making regular circuits of the study area. Circuits of 62 fields along a route from Keyworth, through Wysall to Rempstone and Costock (Figure 2.1) were made weekly between November and March of 1980/1981 and 1981/1982, but more sporadically in 1982/1983 owing to increased farming activity in the area. In addition, more intensive circuits (on 97 days) were made of the Hill Farm area (Figure 2.1). These circuits covered 21 fields and took approximately 30 minutes. Two sets of counts were taken per day between 10.00 and 11.15 and between 14.30 and 15.15 GMT. Counts were made from gateways or other vantage points from which entire flocks were visible. For very large flocks (>100 birds) or flocks in large fields (>6ha), replicate counts were made. Errors in replicate counts were within the 95% confidence limits of ± 5% (cf. margins for wader counts given by Prater 1979). In the few cases where entire flocks were not visible from any vantage point, birds were alarmed and counted in the air. Here, the error margin was greater (± 10%). Since counting in the air was necessary only on large fields, birds usually returned to the same field after alarm (see Chapter 9) and did not confound counts in other fields. During circuits, the number of birds of each species (= species *subflock* size in mixed flocks) in each field was counted and their activity (see below) recorded. Throughout the book, the term 'flock' will be used to refer to *all* the birds within a flock (single- or mixed-species), while 'subflock' will refer to the birds of one species within a mixed flock. In addition to head counts, we also recorded ambient temperature and the maximum and minimum temperatures for the preceding 24 hours, windspeed, rainfall and frost and snow cover on each circuit. All of these climatic factors have been shown to affect survivorship, metabolic rate, feeding rate and/or breeding success in charadriiforms (e.g. Dugan *et al.* 1981, Pienkowski 1981, 1984). Daylength and moonphase were also recorded because: (a) several studies have shown that, during short mid-winter days, some charadriiform species spend a greater proportion of the day foraging and even feed into the night (e.g. Goss-Custard 1969, Baker 1981, Sutherland 1982, and see Chapter 6); and (b) it has been suggested that moonphase influences the temporal distribution of roosting and feeding in some charadriiforms which feed inland (Spencer 1953, Hale 1980, Milsom 1984) and the diurnal activity rhythm of certain earthworm species (Ralph 1957).

From the circuits, it was clear that flocks could be divided into four main categories: foraging, pre-foraging, post-foraging and roosting, although it was usual for a small number of birds in each type of flock to perform activities more characteristic of other types.

Foraging Flocks

Foraging flocks are defined as flocks in which 90% or more of birds are searching for or handling prey. This is easy to determine because foraging lapwings and golden plovers are characterisically widely spaced and individuals variably orientated (long axis of the body) with respect to one another. Foraging birds also move about in clearly recognisable bursts of stepping interspersed with scanning or orientation postures (see Chapters 5 and 6). Table 3.1a,b and Table 3.2 summarise the species composition and the size of foraging flocks across the winter period. The most immediately striking point is that golden plovers hardly ever forage in single-species flocks, but are usually associated with lapwings. Lapwings, on the other hand, are often recorded in single-species flocks, though most occur with golden plovers and/or gulls. Gulls are almost always associated with mixed flocks of both lapwings and golden plovers. Within the study area, therefore, most birds of each species tend to forage in mixed flocks. Furthermore, the number of each plover species increases with the number of species in the flock (Table 3.2). The same is true for gulls in flocks of two and three species ($t = 2.61$, $p < .05$ in 1980/1981, and $t = 5.89$, $p < .001$ in 1981/1982).

Pre- and Post-foraging Flocks

Shortly after dawn birds move from their roosting sites on recently-ploughed fields to fields of newly-sown barley crop. Lapwing and golden plover flocks which build up on sown fields usually contain a small proportion of foraging individuals (6.5 ± 3.72% of lapwings, n = 54 flocks; 4.3 ± 0.62% of golden plovers, n = 37 flocks). Non-foraging birds tend to orientate in one direction (facing into the wind), as in roosting flocks. Birds remain in these *pre-foraging flocks* for up to two hours before moving off to forage on pasture. Table 3.3

Table 3.1: The percentage of foraging lapwings, golden plovers and black-headed gulls in single- and mixed-species flocks in winter. F = the percentage of flocks, I = the percentage of individuals. Data for two winters (see text)

(a) *1980/81*		Species composition of flock						
		L	GP	L + GP	L + B–H G	GP + B–H G	L + GP + B–H G	n
L	F	58.9	—	25.3	04.0	—	11.8	569
	I	32.7	—	38.2	06.6	—	22.5	5488
GP	F	—	00.9	67.8	—	00.0	31.7	211
	I	—	00.2	55.4	—	00.0	44.4	10012
B–H G	F	—	—	—	25.6	00.0	74.4	90
	I	—	—	—	17.6	00.0	82.4	179

(b) *1981/82*		Species composition of flock						
		L	GP	L + GP	L + B–H G	GP + B–H G	L + GP + B–H G	n
L	F	50.2	—	23.8	04.8	—	21.2	189
	I	18.9	—	31.1	06.4	—	43.5	6787
GP	F	—	01.2	51.1	—	00.0	45.5	88
	I	—	00.1	45.6	—	00.0	54.3	4720
B–H G	F	—	—	—	18.4	00.0	81.6	49
	I	—	—	—	5.6	00.0	93.2	161

L = lapwings, GP = golden plovers, B–H G = black-headed gulls.
n = total number of flocks or birds.

Table 3.2: The mean (± standard error) flock/subflock size for lapwings, golden plovers and black-headed gulls in single- and mixed-species flocks

		Species composition of foraging flock							
					L+	L+GP			
		L	GP	L+GP	B–H G	+B–H G	(i)	(ii)	n
L	80/81	14.3	—	34.3	40.2	48.4	***	***	
		±00.93		±03.20	±05.31	±03.04	3.19	5.99	569
	81/82	17.4	—	43.1	50.1	81.9		*	
		±01.86		±09.32	±11.30	±14.46	1.89	2.01	189
GP	80/81	—	25.0	36.2	—	63.7	*		
			±00.00	±11.30		±08.31	1.97	—	211
	81/82	—	03.0	24.9	—	62.0	*		
			±00.00	±09.41		±12.71	2.33	—	88
B–H G	80/81	—	—	—	01.30	02.21			
					±00.11	±00.3	—	—	90
	81/82	—	—	—	01.12	03.90			
					±00.17	±00.44	—	—	49

(i) t-test comparisons between L + GP and L + GP + B–H G.
(ii) t-test comparisons beween L and L + GP.
* $p < .05$, ** $p < .01$, *** $p < .001$; t-test. Data from circuits (see text and see Table 3.1).

Table 3.3: The species composition and mean (± standard error) flock/subflock sizes (no. birds) for pre-foraging and post-foraging flocks

	Pre-foraging flocks		Post-foraging flocks	
	Single spp.	Mixed spp.	Single spp.	Mixed spp.
LAPWINGS				
%	43.8	56.3	38.2	61.7
x̄ mean ± s.e.	43.9 ± 10.3	64.9 ± 15.5	48.2 ± 21.6	31.8 ± 4.3
no. flocks	21	27	13	21
% all flocks	58.5		41.4	
GOLDEN PLOVERS				
%	11.1	88.9	0	100.0
x̄ mean ± s.e.	75.0 ± 67.0	44.8 ± 8.64	0	60.9 ± 14.7
no. flocks	3	24	0	20
% all flocks	57.4		42.6	
B–H GULLS				
%	0	100.0	0	100.0
x̄ mean ± s.e.	0	2.63 ± 1.08	0	1.33 ± 0.1
no. flocks	0	11	0	3
% all flocks	78.5*		21.5	

* $p < .05$; X^2 test comparing pre- and post-foraging flocks.
Data for the December-January (1980/81) period only.

shows that pre-foraging flocks are usually mixed associations, but that golden plovers are now more likely to occur in single-species flocks. There are also likely to be two or three gulls associated with the flock.

Towards the end of the day (up to two hours before dusk), birds form *post-foraging flocks* (also called 'sub-roosts' by Spencer (1953) and Hale (1980)), once again on sown fields. Fewer individuals forage in post- than in pre-foraging flocks (4.1 ± 1.21% of lapwings, n = 18 flocks; 3.2 ± 0.33% of golden plovers, n = 15 flocks). Foraging by all individuals, as reported by Spencer (1953) at his sub-roosts, was not observed in our post-foraging flocks. Table 3.3 shows that pre- and post-foraging flocks do not differ consistently in terms of the number of plovers, but that post-foraging flocks contain significantly ($p<.05$) fewer gulls. Golden plovers occur only in mixed post-foraging flocks.

Roosting Flocks

Roosting flocks usually occur only in newly-ploughed fields. Birds are almost never observed foraging and are always orientated in the same direction into the prevailing wind (see also Fuller and Youngman 1979). Many birds sleep with the head tucked into the feathers on the back. An interesting feature of roosting flocks is the tendency for lapwings and golden plovers to separate and occupy different areas of the field. Gulls also tend to roost away from plover species. While roosting flocks develop mainly as dusk approaches, they sometimes build up earlier in the day.

Pre- and post-foraging flocks will be discussed again in more detail later (in Chapter 7). Here, we shall consider the distribution of foraging flocks and their preference for certain types of field.

The Distribution of Foraging Flocks

To begin with we simply scored pasture fields, which contained only foraging flocks, for the presence or absence of birds on each circuit day. We then used a preference index similar to that of Heppleston (1971a) except that our data were obtained from regular circuits rather than random counts. For each species we calculated field preference as:

$$I_c = \frac{D_i}{D_0} \quad (3.1)$$

where D_i is the number of circuit days on which at least two individuals of species i were recorded in field n and D_0 is the number of circuit days on which field n was observed. The value of I_c therefore varies from 0 (birds never present) to 1 (birds present every day). I_c values for fields observed on the circuits are shown for two winters in Table 3.4. It is clear from the table that there is enormous between-field variation in the probability of birds being present. Some fields regularly contain birds, others seldom do. Are there any physical qualities of the field which correlate with their probability of containing birds?

Foraging flocks and pasture age. As we mentioned in Chapter 2, pasture fields in the study area vary considerably and discontinuously in age. Pastures fall into two categories, which we shall refer to as *young* ($<$ 4 years old) and *old* ($>$ 25 years old) pasture. There are no fields which have been pastures for between five and 25 years. It is reasonable, therefore, to ask whether the length of time for which a pasture has remained undisturbed (except for grazing) correlates in any way with its apparent quality as a feeding site. Figure 3.1a shows the relationship between I_c and pasture age for lapwings and golden plovers. Gulls will be discussed later. The graph shows a significant positive relationship for lapwings, but no significant trend for golden plovers. I_c, however, takes into account only the probability that a feeding flock/subflock will be present on a field. It says nothing about the bias in *numbers* towards particular fields. To see whether certain fields tend to contain a disproportionate number of birds, we calculated:

$$I_n = \frac{N_i - a}{(N_i + a) - 2((N_i - a)a)} \tag{3.2}$$

where N_i is the number of individuals of species i in field n divided by the total number of individuals of species i counted in the remainder of the study area and a is the area of field n divided by the total area of other pasture fields in the study area. I_n is similar to the index used by Patterson *et al.* (1971), Fuller

Table 3.4: I_c values for lapwings and golden plovers in fields covered by circuits

Field no. (as encountered on circuit)	I_c for Lapwing 1980/1981	Lapwing 1981/1982	Golden plover 1980/1981	Golden plover 1981/1982
1	.45	.11	.19	.11
2	.84	.61	.47	.39
3*	.88	.60	.02	.01
4	.94	.62	.72	.45
5*	.81	.45	.01	.00
6	.06	.04	.03	.04
7	.77	.50	.36	.02
8	.60	.20	.30	.10
9	.65	.26	.35	.19
10*	.76	.58	.01	.00
11*	.43	.53	.12	.00

*fields smaller than 3.0 hectares.
Data from circuits (see Table 3.1 and text).

and Youngman (1979) and Fuller and Lloyd (1981) and takes into account bias resulting from variation in field size and the number of birds feeding at the time of sampling. If I_n for lapwings and golden plovers is plotted against pasture age (Figure 3.1b), there is again a positive correlation for lapwings and a non-significant relationship for golden plovers. Taken together, therefore, the trends in Figure 3.1a,b show that the greatest concentration of lapwings occurs on old pastures, and that lapwings are seen there more regularly

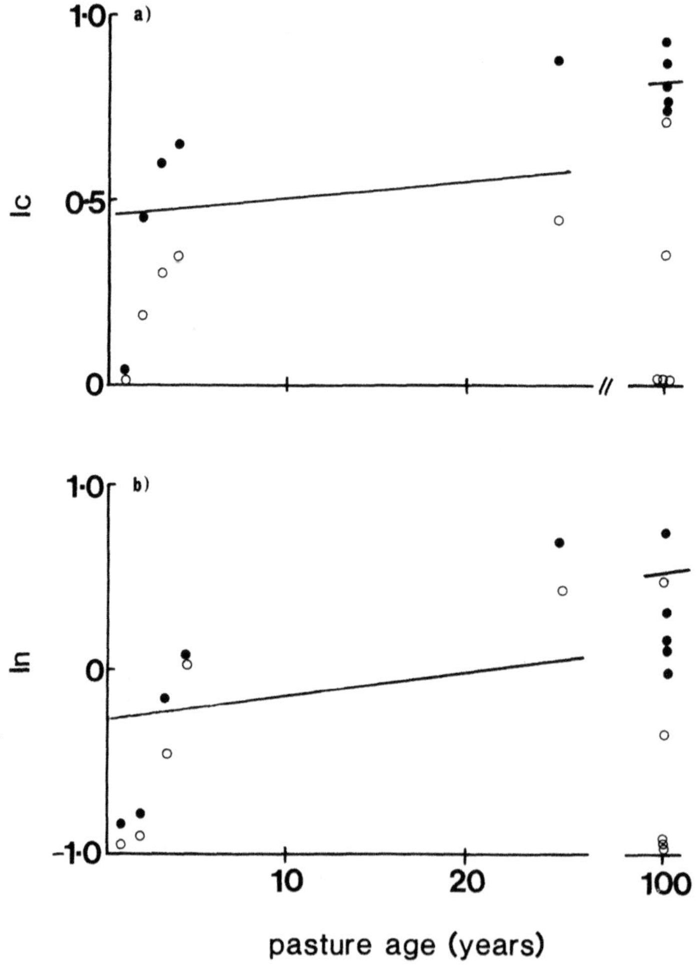

Figure 3.1: The relationships between pasture age and (a) I_c *and (b)* I_n *in lapwings (closed circles) and golden plovers (open circles). F-ratios in (a),* $F = 5.34, p < .05$ *for lapwings and 2.66, n.s. for golden plovers; ratios in (b),* $F = 10.73, p < .01$ *for lapwings and .23 n.s. for golden plovers. Data from 10 fields*

than golden plovers. This difference between the two species will become important later. First, we must consider the factors likely to account for the accumulation of foraging birds on old rather than young pastures.

Pasture age and prey availability. One obvious factor which might correlate with pasture age and bias the distribution of birds is prey availability. Plovers in the study take almost exclusively earthworms during the winter (see Chapter 2) and several studies have shown positive correlations between pasture age and earthworm density (e.g. Heppleston 1971a, Waite 1981, 1983, and see Evans and Guild 1948, Guild 1951, MacFadyen 1962). Of these, however, only Heppleston (1971a) has attempted to rank fields in terms of worm density and relate this to site preference in feeding birds (oystercatchers).

We sampled the earthworm community in pasture fields using two methods: (a) randomly-placed 0.25m-square quadrats of turf and (b) 10.2cm-diameter core samples, both to a depth of 3.0cm. Both turf and core samples were hand-sorted for invertebrates and all items found were preserved in a 10% Biofix solution for later analysis. Hand-sorting was used because it is the most effective sampling technique for surface-dwelling earthworm species (Edwards and Lofty 1977). From bill-length measurements, we judged 3.0cm to be the maximum depth to which plovers could penetrate the soil and therefore the vertical limit of worm availability. Other studies (e.g. Satchell 1971, Brown 1983) have used chemical (formalin, potassium permanganate, Biofix) extraction techniques, which result in an overestimate of deep-dwelling species such as *Lumbricus terrestris* that are forced up, and an underestimate of surface-dwellers such as *Allolobophora chlorotica*, *A. caliginosa* and *A. rosea* that tend to move down. The core samples were taken to provide estimates of the vertical distribution of earthworm density, size and species composition in different fields. An important point, however, is that no bias was apparently incurred by worms burrowing away during sampling. Owing to the heavy clay content of the soil and the relatively low temperatures which prevailed during the study periods, worms moved only very slowly through the soil. In some cases, worms were still present on the cut soil surface up to five minutes after a turf sample was removed. Using the worm samples, we compared four aspects of worm availability in young and old pasture and related them to the distribution of foraging birds.

(a) Worm density. Forty-three 0.25m-square turf and 84 core samples from a total of ten fields were examined on two consecutive days in 1980/1981 when the ambient temperature was between 8°C and 10°C and weather conditions more or less constant. Within the turf samples, there was a significant positive relationship between worm density and pasture age (up to 25 years) (Figure 3.2a). The most likely reason is that, as pastures mature, there is a gradual accumulation of the dead organic matter on which worms feed (e.g.

Nordstrom and Rundgren 1974). Furthermore, old pastures are grazed by cattle for a greater part of the year and therefore have a higher organic input from dung. Worm density, however, appears to drop in very old pasture (>100 years) (Figure 3.2a). A possible explanation is that the thick root layer in very old pastures favours large numbers of nematodes, which reduce the amount of oxygen available to earthworms (D.B.A.T. unpubl). In addition, the largest of the earthworm species, *L. terrestris*, is found in old pasture (Edwards and Lofty 1972, Brown 1983), where it can be predatory on other species. Perhaps not surprisingly, therefore, the wet weight biomass of worms increases linearly with pasture age (Figure 3.2b) owing to the increasing predominance of larger species such as *L. terrestris* and *A. longa* (see below). To see whether ploughing itself reduces worm density in any given erstwhile pasture field, we took a number of turf samples from selected fields at the same time (between 5 and 12 March when the ambient temperature was between 8.5°C and 10°C) in three successive years. Figure 3.3a,b shows worm density in two of the fields which were undisturbed in the first two years they were sampled, but ploughed up in the third. Samples taken two months after ploughing show a significant decline in worm density. In another field which was not ploughed, worm density continued to increase in the third year, as expected from Figure 3.2a. There is, therefore, a clear positive relationship between worm density in the surface-soil layer available to plovers and the length of time since a field was last ploughed.

If the abscissa in Figure 3.1a,b is now converted from pasture age to the surface 3.0cm worm density recorded in each field, significant positive relationships for both I_c and I_n emerge. This is reinforced by stepwise partial regression analysis which examines the independent effects of pasture age and worm density on I_c and I_n. In this and all other partial regression analyses referred to, we used the stepwise forward inclusion technique detailed by Nie *et al.* (1975). Data were checked for violations of the assumptions (of normality, homogeneity of variance, linearity of relationships and weak correlation between independent variables) underlying partial regression analysis (see Pedhazur 1982). Where necessary, the frequency distribution of variables was normalised using one of several transformations (e.g. $\log_{10}$ or natural log for ratio variables, arcsin for proportions and percentage variables where their distributions approached 0 and 1 (or 100) and reciprocation where the data were highly skewed: see Sokal and Rohlf 1981 for details). When transformation still failed to satisfy requirements for the analysis, data were analysed using non-parametric methods. Independent variables yielding non-significant F-ratios were not included in partial regression analysis. For the application of partial and multiple regression techniques to other charadriiform data see e.g. Bryant (1979), Goss-Custard *et al.* (1981, 1984), Pienkowski (1983b), Ens and Goss-Custard (1984).

Table 3.5 shows the results of partial regression analysis taking into account both pasture age and mean worm density. In addition, the analysis

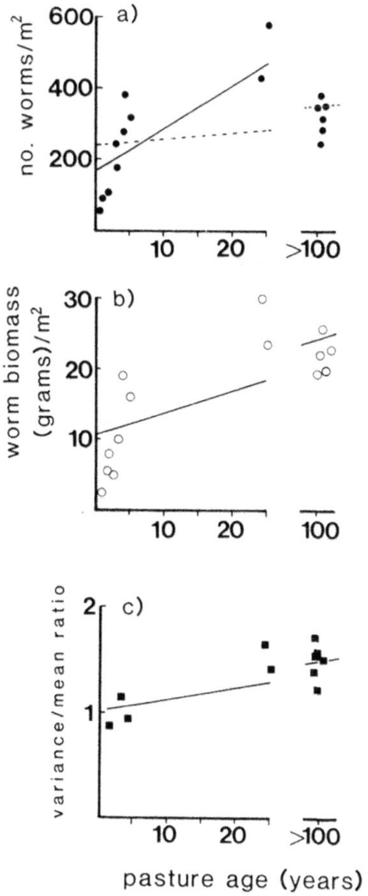

Figure 3.2: Effects of pasture age on (a) worm density, $F = 14.9$, $p < .001$ (pastures up to 25 years, solid line); the relationship is not significant when all pastures are included (broken line); (b) worm (wet weight) biomass, $F = 8.2$, $p < .05$; (c) variance: mean ratio of worm density, $F = 12.96$, $p < .01$. Data for 16 pastures

took into account field area and the number of heterospecifics present in the field; as we shall see in later chapters, both these factors have profound effects on foraging behaviour and species flock/subflock size. Table 3.5a shows that, when other factors are taken into account, the effect of pasture age on I_c for golden plovers disappears and variation in I_c is best explained in terms of variation in worm density and field area. Birds occur most regularly in large fields where worm density is high. Worm density also accounts for a significant amount of variation in I_c in lapwings, but here pasture age still exerts a significant independent effect and there is no effect of field size. When I_n is

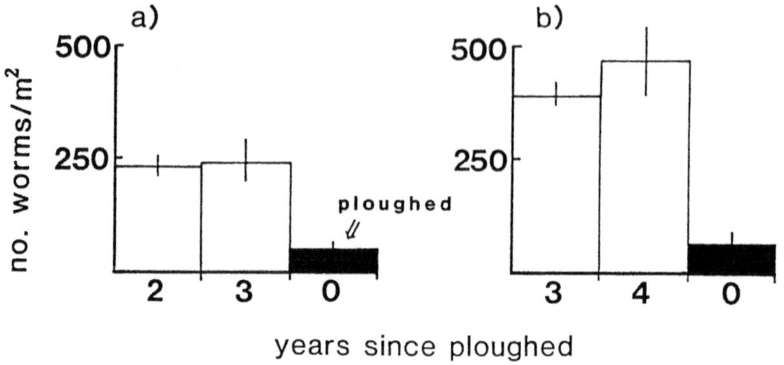

Figure 3.3: Effects of ploughing on worm density. The figure shows the relationship between mean worm density and time since ploughing in two pastures. Shaded columns show densities immediately after ploughing (see text). Bars represent standard errors

examined (Table 3.5b), the effect of pasture age disappears in both species with variation in lapwing number being explicable entirely in terms of worm density. Two important points to note here, however, are (a) I_n for golden plovers depends most on I_n for lapwings, with significant effects on field size and worm density, and (b) variation in I_c and I_n for gulls is best explained by I_c and I_n for golden plovers with no significant effect on other variables. From Table 3.5, therefore, it seems that lapwings tend to choose feeding sites partly on the basis of worm density and partly on some other correlate of pasture age, that golden plovers choose largely on the basis of intensity of use by lapwings, and that preferences in gulls are very closely correlated with those of golden plovers.

The apparent dependence of golden plovers on the distribution of lapwings suggests that they may use lapwing flocks to indicate the best fields in which to feed (see e.g. Krebs 1974, Neuchterlein 1981, Burger 1984 for similar evidence from other species). If they do, their distribution across lapwing flocks should be non-random. This appears to be the case. Significantly fewer lapwing flocks (37.1%) contain golden plovers than expected if the latter distribute themselves indiscriminately across flocks (64.6%, X^2 test comparing observed and expected, $p < .01$, $n = 569$ lapwing flocks). Golden plovers thus appear to be selective in their choice of flock. Indeed, we have observed them flying over four or more lapwing flocks before alighting. What criteria might birds use in choosing a flock? We recorded the size and density (no. birds/ha, see Chapter 6) of lapwing flocks joined or passed over by golden plovers. In each case we also measured the density of worms where lapwings were feeding. The results, summarised in Table 3.6, suggest that golden plovers are attracted to fields containing large flocks of lapwings and high worm densities. This is good evidence that the presence of lapwings acts as a guide to rich feeding areas.

Table 3.5: Beta values from stepwise partial regression analysis of the relationship between field characteristics and (I_c), and (I_n) (see text) in plovers and gulls

Dependent variables	Field age	Worm density	Field size	I_c or I_n L	GP	%
(a) I_c for:						
Lapwings	.52***	.40*	ns	—	—	53
Golden plovers	ns	.79***	.73***	ns	—	62
Gulls	ns	ns	ns	ns	.43***	77
(b) I_n for:						
Lapwings	ns	.87***	ns	—	—	76
Golden plovers	ns	.47***	.54**	.73***	ns	69
Gulls	ns	ns	ns	ns	.69*	96

Independent variables: Field age, Worm density, Field size, I_c or I_n (L, GP).

* $p < .05$, ** $p < .01$, *** $p < .001$; significance levels for F-ratio associated with beta value. ns not significant, — variable not included. % gives % variance explained.

Table 3.6: Factors affecting the attractiveness of fields to golden plovers

Mean ± se	Fields with lapwings which: attracted golden plovers	did not attract golden plovers	Mann-Whitney U test
Worms/m^2	143 ± 14.7	119 ± 15.8	7 (p = .047)
Lapwing flock size	24.2 ± 4.18	11.5 ± 3.58	6 (p = .036)
Lapwing flock density (no./ha)	4.4 ± 0.80	3.3 ± 0.92	10 (p = .120)

(b) Worm patchiness. A second aspect of prey populations which may affect predator capture rate and feeding-site preference is spatial distribution. Most natural food supplies are clumped or 'patchily' distributed rather than being scattered randomly or uniformly through the environment (e.g. Taylor 1961). There is also evidence that some predators forage more efficiently on patchy food supplies (Krebs 1979). A positive correlation with the degree of worm patchiness might help to explain the remaining effect of pasture age on feeding-site selection in lapwings. We therefore took a further series of random turf and core samples, the number of samples taken in each field being the number at which the variance in worm density levelled off (e.g.

Heppleston 1971b, and see Southwood 1978). This resulted in between 20 and 25 turf samples and between 40 and 50 core samples being taken per field. Samples were again hand-sorted.

From the labelled samples, we estimated the degree of patchiness in worm distribution using Fisher's index of dispersion (the variance:mean ratio, $s^2:\bar{x}$). Indices greater than 1.0 indicate a departure from randomness towards greater clumping. Some workers (e.g. Raw 1959), however, consider that hand-sorting underestimates the number of small worms under 1cm in length. We acknowledge that this is likely to be a difficulty in our samples because worms of this size were often adsorbed by the water film onto grass roots and easily overlooked. We therefore calculated Fisher's index both for whole samples and for samples excluding worms under 1cm. Figure 3.2c shows that in both cases there is a significant positive relationship between pasture age and the $s^2:\bar{x}$ ratio of worm distribution, with worms on old pasture being more patchily distributed than those on young pasture. The relationship is, however, more pronounced when small worms are excluded.

One aspect of prey patchiness which is particularly likely to affect predator feeding efficiency is its degree of spatial predictability. If high-density patches tend to occur regularly in particular areas, we might expect more of the predator population to aggregate there. To see whether there is any relationship between pasture age and the spatial predictability of high-density patches, we sampled earthworms at the *same* locations in an old and a young pasture on five sample days in December and January. Four 0.25m-square turf samples were taken in each of four 20m-square areas in each field and the samples hand-sorted as before. We then adopted a ranking procedure. Worm counts for the four samples from each area were summed and the totals ranked so that the area yielding the most worms was ranked 1 and that yielding the fewest was ranked 4. Table 3.7 shows the consistency of rank scores for areas within each field across sample days. The spatial predictability of the highest- and lowest-density patches is much greater in old pasture (see probabilities of rank consistency in Table 3.8). X^2 analysis comparing ranks observed on different sample days with those expected if there is no change shows significant heterogeneity on young pasture ($X^2 = 10.96$, $p < .05$), but not on old pasture ($X^2 = 0.43$, n.s.). Another important point is that variation in worm density *within* 20m-square sample areas is significantly lower in old pasture (mean $s^2:\bar{x}$ ratio in old pasture $= 0.87 \pm 0.09$, range 0.55-1.30; mean $s^2:\bar{x}$ ratio in young pasture $= 1.39 \pm 0.15$, range 0.09-2.07, $t = 2.28$, $p < .02$). In old pasture, therefore, worm density is predictable over larger areas than in young pasture. In Chapter 4, we discuss the way birds capitalise on this greater predictability.

(c) Worm size. We showed in Figure 3.2b that the wet-weight biomass of worms increases with pasture age and suggested that this may be due to an increase in the proportion of large species in the earthworm community. Item

Table 3.7: Comparison of worm distribution in old and young pastures. The rank order of 4 areas (20m-sq.) in each pasture in terms of worm density (1 highest density, 4 lowest density)

	Rank of area in each field							
Date	**Old pasture**				**Young pasture**			
11/12/80	1	2	3	4	1	2	3	4
13/12/80	1	2	3	4	4	2	1	3
18/01/81	1	3	2	4	3	4	2	1
20/01/81	2	1	3	4	1	3	2	4
24/01/81	1	3	2	4	3	2	4	1
Prob. best area always being best	.80			—	.40			—
Prob. poorest area always being poorest	—			1.0	—			.40
χ^2 test for heterogeneity between days	$\chi^2 = 0.43$, ns				$\chi^2 = 10.96$, $p < .02$			

Data for a total of 160 0.25m-sq. samples.

size is a third way in which prey availability may vary and account for the observed preferences for old pasture.

While several studies (e.g. Gerard and Hay 1979, Barnes and Ellis 1982, Edwards and Lofty 1982) have shown that drilling and ploughing reduce the density of large, deep-burrowing species such as *L. terrestris* and *A. longa* relative to the smaller surface-dwelling species such as *A. caliginosa, A. chlorotica* and *A. rosea,* they used formalin extraction which is not directly comparable with the turf sampling used here. Nevertheless, analysis of worms from turf samples taken from plough, young pasture and old pasture show that *L. terrestris* is significantly more abundant in old pasture (comparing mean numbers of individuals per m^2 in young and old pasture, $t = 3.47$, $p < .001$, $n = 100$ turf samples) but absent from ploughed fields. Surface-dwelling species such as *A. chlorotica* and *A. rosea,* however, show no significant tendency to increase in abundance with pasture age, although *A. caliginosa* is more abundant in old pasture ($t = 2.23$, $p < .05$, n as above). Comparisons were made using only mature individuals (identified from keys by Gerard 1964, Edwards and Lofty 1972), but a large proportion of samples were unidentifiable immature individuals (<10%) or broken fragments (between 17% and 32%). The reason for the high incidence of broken fragments is discussed in Chapter 5. Despite the increased abundance of *L. terrestris* in old pasture, there is no significant increase in worm size (measured as length after fixation) with pasture age when all sample fields are taken into account. Since ploughing appears to have such a pronounced negative effect on *L. terrestris* abundance, however, we repeated the analysis omitting fields which had been ploughed within a period of one year. Now, a significant positive relationship emerged between pasture age and worm size ($r = .649$, $p < .05$, $n = 14$ fields). There is some evidence, therefore, that, after an immediate post-ploughing lag, older pastures tend to contain larger worms.

(d) Surface characteristics and pasture age. A fourth way in which pasture age might influence prey availability is through variation in surface characteristics. In particular, we might expect grass density to increase with pasture age as the turf matures. Increased grass density may reduce the detectability of soil-dwelling prey to such predators as lapwings and golden plovers which appear to rely on visual cues (Pienkowski 1983a, Metcalfe 1984a, and see Chapter 6) rather than probing randomly in the soil (see Heppleston 1971a). Grass density was measured as the number of blades per m^2 and was indeed significantly higher in old pastures (mean no. blades/m^2 in young pasture = $4.8 \times 10^4 \pm 0.9 \times 10^4$, $n = 5$ 0.25m-sq. quadrat samples; mean no. in old pasture = $11.7 \times 10^4 \pm 1.2 \times 10^4$, $n = 5$ quadrat samples, $t = 8.5$, $p < .01$). This means that an average of 27.5% of the sampled ground surface in young pasture consists of bare earth or light grass cover (fewer than 1.5×10^4 blades/m^2), compared with only 13.2% in old pasture. We should therefore expect worms, or surface cues suggesting the presence of worms (e.g. casts, burrows, soil movement), to be more conspicuous to visually-foraging plovers on young pasture.

To test this, we acted as visual 'predators' by scanning the surface of ten randomly-chosen $1m^2$ quadrats on old and young pastures. Scanning was carried out on hands and knees and the numbers of worm casts and holes greater than 2mm in diameter seen were recorded. Five preliminary quadrats were sampled first to control for initial improvement in our ability to spot evidence of worm activity. The results of later quadrat counts suggest strongly that cues in bare-earth/light grass-cover areas are easier to detect visually than those in dense cover. On old pasture, 80% of recorded casts and burrows occurred in bare-earth/light-cover areas, whereas only 47% of worms recovered from turf samples occurred there. This implies that surface cues are harder to detect on old pasture, where there is denser turf. This must, however, be offset against the fact that worm density is higher on old pasture (see above) and a greater proportion of earthworms occur within 3.0cm of the surface (core samples to a depth of 0.75m show that, on average, $58 \pm 9.2\%$ of worms in old pasture occur in the 3.0cm turf layer compared with only $39 \pm 9.2\%$ in young pasture). Barnard and Stephens (1983) showed that the combination of surface-grass density and the vertical distribution of earthworms could account for differences in the range of prey sizes taken by fieldfares and redwings on old and young pastures.

Population Size and the Distribution of Foraging Flocks

The population sizes of lapwings, golden plovers and black-headed gulls within the study area vary considerably both between and within winters. Table 3.8 shows the mean and variance for the number of individuals counted on the circuits during each winter. The variation in numbers provides another means of testing for feeding-site selection. So far, we have shown that foraging birds tend to accumulate in old rather than young pastures and that this appears to be due to greater prey availability in old pastures. If birds do prefer old pasture, we should expect a positive relationship between the number of birds foraging in the study area and the number of different fields occupied, with old pasture being occupied first. To test this, we examined count data from selected days when different numbers of birds were recorded in the area, and plotted the number feeding in each field against the rank order of fields in terms of age and worm density. Figure 3.4 shows the cumulative number of lapwings and golden plovers foraging in fields of different age and worm density. The plots show that on days when fewer birds are in the area only the older, higher-worm-density pastures are occupied. As the number of birds increases, more of the lower-ranking fields are used. This supports the earlier suggestion that birds prefer to feed in older pastures. As a further test, we compared the fluctuation in the number of birds feeding on the field with the highest recorded worm density with the fluctuation in the total number of birds in the area over an 11-day period. To control for chance fluctuations in foraging-flock size, caused for example by alarm, we recorded the equilibrium flock size (EF size, see Chapters 1 and 7) over two to five hours on each day and compared this with fluctuations in daily total numbers. For reasons which will become clear in Chapter 7, we could record EF sizes only in lapwings. Figure 3.5 shows that, despite fluctuations of over 150 birds within the 12-day period, mean lapwing EF size varied by a maximum of only 20 birds. The number of birds feeding on the field with the highest worm density therefore remains relatively constant. In addition to considering flock distribution in relation to conspecific population size, we also examined the effects of variation in the sizes of heterospecific populations. Partial regression and correlation analyses failed, however, to reveal any significant relationships.

Table 3.8: Mean and variance for the numbers of plovers and gulls in the study area over two winters

	1980/81		1981/82	
	Mean	**Variance**	**Mean**	**Variance**
Lapwings	139.9	7603	162.7	9006
Golden plovers	105.2	14280	136.2	43472
Gulls	4.0	28	5.3	18

Mean and variance for daily counts. See text for details.

Foraging Preferences and Earthworm Predation

Since plovers show distinct foraging preferences in their choice of field, they are likely to have an uneven impact on their earthworm community. Such biases in their impact of predation are widespread and have important consequences for the dynamics of predator-prey relationships (e.g. Begon and Mortimer 1981). Overall depletion by predators can be substantial. Evans *et al.* (1979) suggested that waders remove some 40-50% of their prey population each winter. Summarising the results of a number of studies, Goss-Custard (1980) calculated 25-40% depletion by overwintering waders. Similarly, Zwarts and Drent (1981) estimated that oystercatchers removed about 40% of winter mussel populations. Studies of other bird groups have shown depletion rates ranging from 18% to 81% in woodpeckers (e.g. McLellan 1958, Kroll and Fleet 1979) and 5% to 97% in tits (e.g. Gibb 1960,

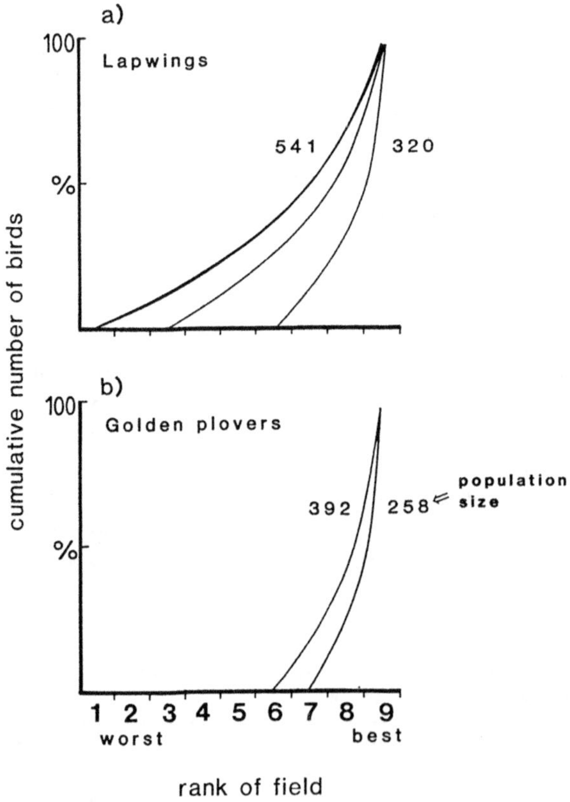

Figure 3.4: The cumulative number of birds feeding on pasture in relation to the rank order of worm density and population sizes (320-541 in lapwings, 258-392 in golden plovers) in the study area. Curves fitted by regression

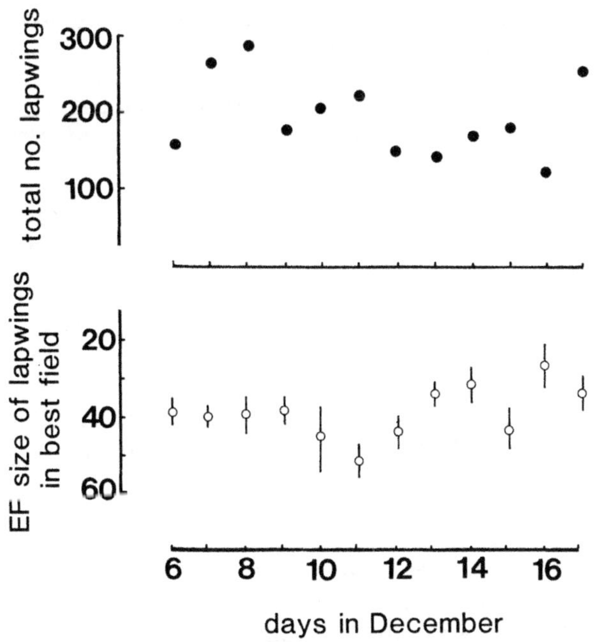

Figure 3.5: The relationship between equilibrium flock size in lapwings on the field with the highest worm density and the number of lapwings in the study area. Data taken over 11 days in December 1981. Bars represent standard errors

Nilsson 1979). To measure the impact of plovers on the earthworm community, we carried out an enclosure experiment.

We set up a number of net enclosures (2 × 1.5 × 0.2m) at random points in two fields: 16 enclosures were set up in an old pasture and 12 in a young pasture. We collected two 82-cm^2 (10.2-cm diameter) turf samples from each area to be enclosed and another two samples from an arbitrary point 1m outside. Three months later, we collected further samples as above. On both sampling occasions the ambient air temperature was between 7°C and 9°C. Soil samples were hand-sorted for invertebrates as before. Table 3.9 compares earthworm densities from the two sampling periods. In enclosed areas, worm density rose by between 10 and 12 worms/m^2/month. In exposed areas, however, it *declined* by between 40 and 45 worms/m^2/month. The decline was much greater in old pasture, where there was an apparent 71% drop in worm density attributable to predation. In young pasture, loss to predators was only 53%.

To see whether the declines tallied with observed intake by birds, we measured feeding rates in individual plovers and estimated the total amount

Table 3.9: The mean (± standard deviation) density of earthworms sampled in December (control) and March (after predation) within and outwith enclosures during the 1981/82 winter

	Density of worms (no./m^2)	
	Enclosed area	**Exposed area**
Old pasture		
December	228.8 ± 53.8	183.6 ± 69.7
March	260.5 ± 45.2	53.8 ± 15.9
Difference	+ 31.7	− 129.8
Young pasture		
December	265.4 ± 96.6	284.7 ± 92.9
March	295.9 ± 68.3	152.7 ± 34.2
Difference	+ 30.5	− 132.0

Data collected for 16 and 12 netted enclosures on old and young pastures respectively, and a total of 224 81-cm^2 samples.

of birds/hours spent feeding between the two sampling periods. Estimates suggested an average removal of some 35 ± 2.1 worms/m^2/month from old pasture and 25.3 ± 1.4 worms/m^2/month from young pasture. While the difference between pasture types is in the same direction as before, removal rates fall far short of the levels of depletion recorded in the enclosure experiment. This is almost certainly because plovers are not the only predators of earthworms on the study area. Other birds, including fieldfares, redwings, rooks and carrion crows, as well as badgers (*Meles meles*) and foxes (*Vulpes vulpes*), also deplete worms through the winter. The mismatch between estimated intake and measured depletion is in the opposite direction to that recorded on an immediate time scale when plovers are the only predators (see Chapter 6). Further evidence that predation was responsible for the decline came from the fact that, while the plots with the highest worm density remained constant in enclosed areas, they changed in exposed samples (Thompson 1984). This could be due to birds and other predators concentrating their attention where prey is locally abundant (see Chapter 4).

While other factors, such as the presence of livestock and changing grass length/density, could conceivably have contributed to the decline in worm density (Bengtson *et al.* 1976, and see Edwards and Lofty 1972, Waite 1983), these can be dismissed in this case. No livestock was present in the fields in question and grass length and density remained constant over the three-month period (also H.A. Stephens unpubl. data).

Environmental Conditions and the Distribution of Foraging Flocks

In the previous section, we considered the effects of variation in species population size on the distribution of foraging birds across pastures. We said nothing, however, about the factors which brought about *day-to-day* fluctuations in population size in the first place. Many studies of charadriiform and other birds which feed on subterranean prey in open and exposed habitats have pointed to the importance of climatic and other physical environmental conditions on habitat distribution, feeding behaviour and survivorship (e.g. King and Farner 1974, Beuchat *et al.* 1979, Barnard 1980b, Pienkowski 1982, 1983a,b, Walsberg 1983). Climatic factors appear to influence feeding behaviour in at least two ways: firstly, by altering food demand as more or less energy is channelled into body maintenance, and, secondly, by affecting prey availability. There is evidence for both effects in plovers.

Dugan *et al.* (1981) studied the effects of temperature and windspeed (chill factor) on fat deposition (a good indicator of energy demand) in grey plovers (*Pluvialis squatarola*) feeding on an estuary. From catches of grey plovers through several winters, they found that, during periods when the weather was relatively mild, body weight increased through the autumn to mid-winter. Birds tended to lose weight quite rapidly, however, when weather became severe. By matching up periods of weight loss with the sequence of weather changes, Dugan *et al.* concluded that the main factor in weight loss was windspeed, not only because it is likely to increase heat loss but also because it inhibits, and when very high prevents, foraging. If windy periods coincide with, or are followed by, periods of low temperature, as is likely in exposed habitats during winter, then birds will lose weight and be unable to regain it because of their high energy demand.

The effects of temperature on food requirement were also examined by Pienkowski (1982), this time in grey and ringed (*Charadrius hiaticula*) plovers. From calculations of daily energy requirements (see Chapter 6) and expected intake from polychaete and crustacean prey, Pienkowski estimated the minimum densities of prey needed by birds at different temperatures. For worms, the curves for the two species decreased exponentially with increasing sand temperature so that the minimum acceptable density to meet requirement doubled with less than 1°C drop when sand temperature was below 5°C, but doubled over 5°C when it was between 5°C and 10°C. The exceptionally high densities needed at low temperatures are likely to be rare. As Pienkowski points out, this may have accounted for the build-up of large plover flocks on the few high-density areas in his study site during cold days.

Although birds need more food when it is cold, low temperatures often correlate with a reduction in prey availability, thus compounding the problem. Pienkowski (1983b) found that capture rate by grey and ringed plovers was markedly reduced at low temperatures. The reduction was associated with decreased activity in their prey (again polychaetes and crustaceans). Prey activity tended to bring animals within reach of a bird's bill and, in some cases, led to tell-tale disturbance at the surface. Depressed activity at low temperatures is a general phenomenon among ectotherms. In addition to reducing their activity, burrowing species tend to move deeper into the substrate in winter (e.g. Edwards and Lofty 1972, Reading and McGrorty 1978). The problems are probably less acute for predators such as sandpipers (Scolopacidae) which detect prey by tactile means and can probe the substrate. Plovers, however, being primarily visual foragers, rely heavily on surface cues of food availability. In cold weather, plovers feeding intertidally tend to spread out, with individuals being separated by anything from 0.5m to 50m (Pienkowski 1983b). Interestingly, Pienkowski found that plovers 'foot-tremble' more when it is cold. In this context, foot-trembling may mimic the mechanical action of tide flow and thus stimulate activity in invertebrate prey.

Temperature and Daylength Effects

An immediately apparent effect of climate on the formation and distribution of foraging flocks is the disappearance of large numbers of birds from the study area during very cold periods. When the maximum daylight temperature falls below 0°C, both lapwings and golden plovers tend to depart, probably to southern England and Ireland (e.g. Sharrock 1979, Hale 1980, Cramp and Simmons 1983). Characteristically, birds remain at roost sites without foraging for up to 18 hours before leaving. When temperatures rise again, the numbers of birds of both species return to their pre-cold spell levels within days. Fuller and Youngman (1979) noted similar reductions in the number of lapwings and golden plovers during cold spells in their study area near Oxford, and a similar increase back to previous levels when temperatures rose.

There is also a marked reduction in the number of foraging birds on very mild days when maximum daylight temperature passed 6-7°C. In this case, however, the reduction is due not to emigration from the study area but to birds remaining either at roosts or in non-foraging flocks on sown fields. The reduction in foraging at the extremes of recorded temperatures means that most foraging took place when daily maximum temperatures are between 2°C and 6°C. This is shown in Figure 3.6a,b, where the observed distribution of foraging flocks/subflocks in relation to daily maximum temperature is compared with the distributions expected if there was no temperature effect. In both lapwings and golden plovers the observed distribution differs significantly from that expected, with a pronounced peak at intermediate temperatures. Emigration during cold spells appears to be due to reduced food availability as a result of frozen soil and downward migration of surface worms. The reasons for decreased foraging during mild weather, however, may have been more complex. Firstly, higher temperatures are likely to reduce energy demand and therefore the amount of time birds need to spend foraging, and, secondly, more feeding can potentially be carried out at *night* because worms are more abundant at the surface then (Ralph 1957, C.A.J. Brown pers. comm.) and birds are safe from aerial predators.

One interesting factor which is suspected to influence night feeding in plovers is moonphase. There is some evidence that earthworms, particularly *L. terrestris*, are more active on the grass surface (and hence easier to detect and capture) at night during the full-moon period (Ralph 1956, and see Edwards and Lofty 1972). Furthermore, lapwings have been recorded foraging up to one hour before dawn on arable fields which normally contain only pre- or post-foraging flocks (Spencer 1953, Hale 1980). Hale (1980) presents data showing that, during full-moon periods, lapwings often slept

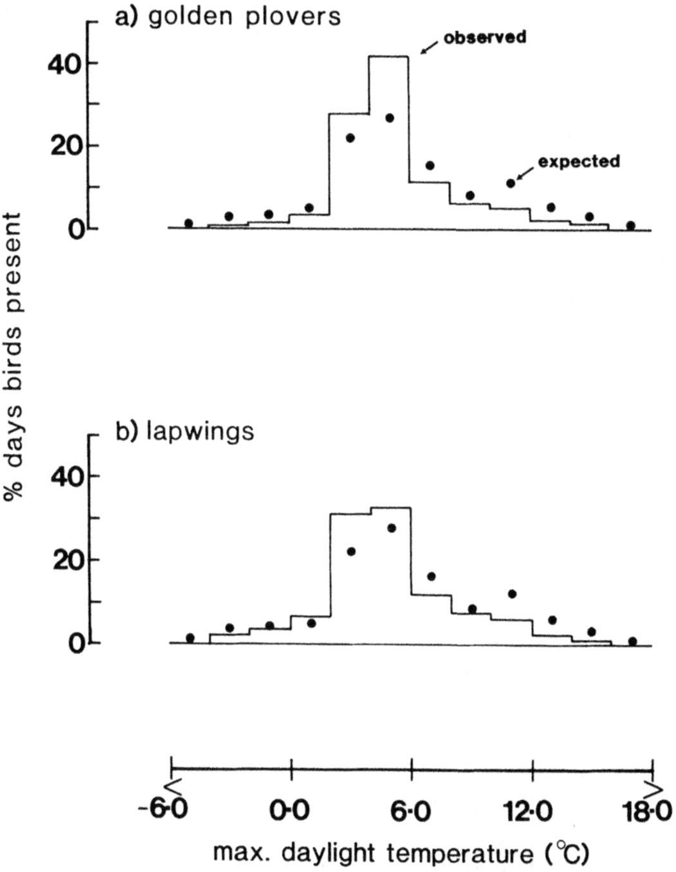

Figure 3.6: The observed (histograms) and expected (circles) occurrence of plovers on pastureland in relation to maximum daylight temperature. Observed and expected distributions are significantly different (Dmax = .23, p < .05 for lapwings, Dmax = .21, p < .05 for golden plovers; Kolmogorov-Smirnov one-sample test). Data for 97 days

and preened during the day, apparently because they had been feeding at night. Recently, Milsom (1984) has provided further support for an influence of moonphase on lapwing foraging behaviour. Data from two lunar cycles showed that the amount of time birds spent feeding within days correlated positively with the number of days from the nearest full moon. Minimum overnight temperature, however, had an important modifying influence on the effect of moonphase. Interestingly, Hale (1980), presenting data collected by A.W. Boyd, also shows that, in early spring, curlews (*Numenius arquata*) were absent from roost sites only during full-moon periods. At these times, birds fed some 30km away on the coast.

We carried out circuits of the study area on 11 nights when there was a full moon and searched for foraging birds with a 'nightcat' image intensifier. No foraging birds, however, were observed. Despite this, there was a significant negative correlation between moonphase and the number of fields recorded as occupied by lapwings during daytime circuits ($r_s = -.441$, $p < .01$, suggesting that some night-feeding may have been occurring. An important consequence of the reduction in numbers of lapwing and golden plover flocks/subflocks during cold, mild and full-moon periods is an increase in the tendency for gulls to feed for themselves. When we observed gulls foraging for themselves there were fewer (< 10-15%) plovers recorded foraging in the flock. When plovers are available, gulls obtain most of their prey by klepto-parasitism (see Chapter 8). These broad effects of changes in temperature suggest that climatic factors might be an important influence on the day-to-day formation and species composition of foraging flocks.

To test this, we carried out a series of partial regression analyses. Before we discuss these, however, an important point needs to be made. Clearly, some of the environmental factors we consider here (e.g. temperature variables) are likely to influence the foraging behaviour of birds indirectly through their effects on earthworm abundance (see earlier). When earthworm abundance is controlled for, however, environmental factors still account for between 7% and 12% of the variation in the number and distribution of foraging flocks, and therefore merit separate discussion.

In the regression analyses, we took into account ambient air temperature at the time flocks were recorded, the minimum temperature over the preceding 24 hours, and the maximum temperature for the day on which the circuit was made. Several other variables were also included: these were the circuit day (because the number of both plovers and gulls tend to decline slightly through the winter); daylength (because lower temperatures are associated with shorter days and daylength has been shown to be an important determinant of flock-size distribution in other species (e.g. Barnard 1980a)); moonphase (see above); and the number of lapwings and golden plovers, including ratios of lapwings: golden plovers, and the total number of lapwings plus golden plovers (because con- and heterospecific numbers had turned out to be important predictors of golden plover and gull distribution in earlier analyses

(Table 3.5)). In the first analysis, we used the total number of each species recorded in the study area as dependent variables to see whether environmental factors have a directional day-to-day effect. No significant relationships emerged, however, and it appears that temperature acts in an all-or-nothing way to determine the presence or absence of birds (see above). In the second analysis, we tested for the effects of environmental variables on the distribution of foraging birds.

Table 3.10 shows the results for the total number of foraging lapwings, golden plover and gulls recorded on each circuit day. Only independent variables yielding F-ratios with associated p values of $<.05$ are presented. The table shows that the number of lapwings foraging correlates best with maximum daylight temperature and daylength. Fewer birds forage on warm or long days. While the number of foraging golden plovers also correlates negatively with maximum temperature and daylength, it correlates best with the number of lapwings present; more golden plovers feed in flocks containing large numbers of lapwings. The number of black-headed gulls appears to depend mainly on the total number of lapwings and golden plovers, with more gulls in larger flocks. The number of gulls correlates negatively with daylength, but appears to be independent of variations in temperature. Once again, therefore, lapwing numbers correlate best with physical environmental variables, while golden plovers and gull numbers depend more on the number of heterospecifics (lapwings and total number of plovers respectively, see also Table 3.5). As might be expected from the effects of temperature on the number of birds foraging, birds are spread over fewer fields when temperatures are high and days long (Table 3.10). The table also shows the effect of con- and heterospecific population sizes (within the study area) on the distribution of birds. The number of fields occupied by lapwings and golden plovers is best predicted by conspecific population sizes. While the number of fields occupied by gulls correlates with conspecific population size, it also correlates with the number of lapwings.

Since climatic factors influence both numbers of birds foraging and the number of fields they occupy, they must also influence flock/subflock sizes. The analysis in Table 3.11 shows the relationship between flock/subflock *size* in each species and the social and environmental factors considered above. It also includes measures of food availability which are likely to influence choice of feeding sites (see earlier). As expected from the earlier analyses, flock/subflock size of lapwings and golden plovers in a field correlates strongly with worm density (so also does bird density, $p<.001$ for each species). Golden plover numbers, however, tend to drop as worm patchiness increases. The number of birds decreases with increasing temperature and, in lapwings, time of day as expected feeding priority drops (Chapter 6). Environmental factors have little effect on the number of gulls, although there is a weak tendency for more to be present on longer days. Instead, gull numbers are most strongly related to the overall density of plovers. We shall see why this is so in Chapter 8.

Table 3.10: Beta values for the interrelationships between the total number of lapwings, golden plovers and gulls foraging, the number of fields occupied and environmental factors

	Independent variables						
					Total no. of		
Dependent variables	Min °C	Max °C	Daylength	L	GP	GP+L	B–HG
no. lapwings	ns	−.39***	−.31**	—	—	—	—
no. fields occupied	−.28*	ns	−.36***	.41***	ns	ns	ns
no. golden plovers	ns	−.25*	−.22*	.44***	—	—	—
no. fields occupied	−.19*	ns	−.24*	ns	.53***	ns	ns
no. gulls	ns	ns	.21*	ns	ns	.37***	—
no. fields occupied	ns	ns	−.18*	.35**	ns	ns	.20*

* $p < .05$, ** p. $< .01$, *** $< .001$; significance levels for F-ratios associated with beta values. ns not significant ($p > .05$) relationship, — variable not included, $n = 79$ circuit days.

Table 3.11: Beta values for the relationship between lapwing, golden plover and gull flock/subflock sizes and environmental variables

	Independent variables						
Subflock size of:	Worms/m²	$S^2/\bar{X}$	Time of day	Max °C	Daylength	GP/ha	(L+GP)/ha
Lapwings	.75***	ns	−.21*	−.26*	ns	—	—
Golden plovers	.44**	−.25*	ns	.34*	ns	—	—
Gulls	ns	ns	ns	ns	.19*	−.27*	.93***

Data for 43 lapwing flocks/subflocks, 35 golden plover subflocks and 317 0.25m-sq. turf samples.

$S^2/\bar{X}$, the variance:mean ratio of earthworm density.

Other independent variables included in the analysis: worm biomass (grams/m^2), minimum overnight temperature, ambient temperature, time of day, moonphase, field size, no. lapwings/hectare; and for gulls the subflock sizes of lapwings and golden plovers, and the total number of plovers. See Table 3.10 for other details.

Effects on the formation of mixed flocks. Earlier work has suggested that climatic and other physical environmental factors are important in the formation of mixed-species flocks both in lapwings and golden plovers (see Fuller and Youngman 1979) and other species (Barnard and Stephens 1983). In fieldfares and redwings, Barnard and Stephens (1983) found that single-species flocks tended to form at the beginning of winter, when they build up on different types of pasture: fieldfare flocks on young pasture and redwing flocks on old pasture ('young' and 'old' as defined earlier). Barnard and Stephens argued that this was due to differences in the availability of worms on the two pasture types during the relatively mild early-winter weather. The formation of mixed flocks appeared to be a consequence of reduced worm density in the turf layer of young pasture as temperature decreased. Worm densities on old pasture remained relatively constant through the winter. To see whether temperature or other environmental variables affect the formation of mixed charadriiform flocks, we carried out a partial regression analysis with the proportion of flocks which were mixed (contained at least lapwings and golden plovers) as the dependent variables. The analysis was carried out separately for old and young pastures.

The results suggest that the effects of climate and species population size differ between pasture types. The tendency to form mixed flocks increases with decreasing temperature (minimum overnight) on old pasture but decreases on young pasture. Mixed flocks are, however, more likely to form on both types of pasture on short days. Not surprisingly, the proportion of flocks which are mixed increases with the total number of plovers in the area and the relationship is strongest on young pasture which, as we have seen, acts as an overspill area when competition for preferred old pasture is high.

Choosing Where to Feed: Résumé

Plovers and gulls appear to have clear preferences when choosing where to feed. Lapwings usually arrive first and tend to land in old pastures which they use throughout the winter. There are several reasons why old pasture might be preferred. The first, and apparently most important, is that worm density increases with pasture age. In addition, however, worms are more patchily distributed in old pasture with greater spatial and temporal predictability in their local density. By preferring old pasture, lapwings concentrate where food availability is highest. The resulting positive correlation between lapwing flock size and worm density provides golden plovers with an easy means of finding good feeding sites. Golden plovers seldom feed in single-species flocks, but instead associate with lapwings, discriminating between flocks so that most end up where worm density is high. The cues they use to do this are discussed in the next chapter. Gulls prefer to land where the density of plovers is high and are unaffected in their choice by climatic or habitat factors. As we shall see later, feeding efficiency in gulls depends on a particular size range of worms being procured by plovers, and this does not necessarily relate closely to the overall density of worms. Golden plovers and gulls therefore capitalise on lapwings in different ways and for different reasons.

Not all birds, however, forage in fields containing the highest worm densities. Numbers in old pasture eventually reach a ceiling capacity and birds are forced to spill over onto lower-quality fields. During days when the total number of plovers in the area is high, a greater number of less desirable fields are occupied. The number of birds in rich, old pasture tends to remain constant as plover population sizes vary. Dispersion in relation to feeding-site quality is also influenced by temperature. Plovers occur on young pasture more frequently and in greater numbers on mild days (early and late in the winter). This possibly reflects the greater degree of exposure and vulnerability of worms to variation in temperature in young pasture, where grass cover is thinner. As might be expected, the effects of temperature on the distribution of birds influence the formation of different types of foraging flock (single- or mixed-species) across fields. Predation over the winter period results in a marked depletion of the worm supply. Enclosure experiments revealed the disproportionate effect of concentrated predation in the preferred old pasture, highlighting the important consequences of the aggregative response in predators for prey communities.

Summary

1. Lapwings regularly forage in both single- and mixed-species flocks. Golden plovers and gulls occur almost exclusively in mixed flocks.
2. Lapwings selectively forage in old pasture where worm availability is greater and more predictable.
3. Golden plovers and gulls use birds which are already feeding as indicators of the best fields in which to land.
4. Population sizes and temperature variation affect the distribution of birds across fields.
5. Predation through the winter reduces worm density and depletion is most pronounced in old pasture.

Chapter 4

Choosing Where to Feed: Choice within Fields

Most natural food resources are *patchily* distributed, either as discrete entities, such as rotting tree stumps and bunches of berries, or as statistical heterogeneities in an otherwise continuous distribution, such as the occurrence of earthworms in a field. Whichever form it takes, patchiness in food resources has important consequences for the foraging strategies we should expect predators to adopt (e.g. Pyke *et al.* 1977, Krebs 1978, 1979, Krebs and McCleery 1984). At a simple level, food patchiness should affect the way predators spend time in different parts of their environment. If they are to maximise their feeding efficiency, we should expect them to spend most time where food is most abundant, i.e. in the richest patches. In socially-feeding predators, the amount of time spent in a given area by different individuals translates into the number of individuals we might expect to encounter in that area at any given time (e.g. Barnard 1980b). In the last chapter, we saw that gulls and plovers prefer to feed in old pasture, where earthworms are most abundant. Choice of field on the basis of worm availability is a response to patchiness at one level. As we have seen, however, differences in patchiness can also be measured *within* fields. In this chapter we shall look at the way birds respond to prey patchiness at this second level by examining their distribution within fields.

Choosing Where to Land

One major difference between old and young pasture lies in the spatial and temporal predictability of worm availability. The locations of the richest and poorest areas are less changeable in old pasture, and this may be one of the reasons it is preferred by plovers. A bird's choice of field may therefore depend, at least partly, on the ease with which it expects to find good feeding areas. Is there any evidence that birds tend to land where worm density is locally high?

We recorded arrivals and departures of birds in selected old and young pastures for 36 days spread over two winters (1980/1981 and 1981/1982), using the techniques described in Chapter 3; we sampled surface (3cm-deep) worm densities in the immediate area in which birds landed and elsewhere in the same and adjacent fields where they did not land. Samples were always taken within 2.5m of where the *first* bird landed within a field. Sampled worms were allocated to one of six size classes on the basis of length (class 1 = < 16mm; 2 = 16-32mm; 3 = 33-48mm; 4 = 49-54mm; 5 = 55-80mm; 6 = >80mm). These same size classes form the basis of the diet-selection and feeding-efficiency analyses in the following two chapters. For each sampling time we also recorded minimum overnight, maximum daylight and ambient temperatures, time of day and daylength, all of which might influence both worm density and food demand in birds (see especially Chapter 6).

Table 4.1 shows the density and size-class distribution of worms in the turf layer in areas where plovers do and do not land. The table shows that golden

Table 4.1: Density (mean ± se/m²) and size-class distribution of earthworms available to lapwings (L) and golden plovers (GP) within and between fields

Sample location	Earthworm density Total	Size class 1	2	3	4, 5, 6
(a) where L land (n=95)	216.2±8.8*	106.6±13#$$	69.4±7	29.2±4**	11.7±2#
(b) where GP land (n=93)	246.1±9.3#$$	146.0±21$$	73.8±9	17.3±2	8.3±1
(c) outside flocks area in same field (n=68)	197.9±10.8	114.6±15	55.7±8	17.2±4	9.0±2
(d) outside flock area in adjacent field (n=86)	174.6±20.4	73.8±6	69.6±7	24.8±2	5.3±1

One way ANOVA shows a significant effect of sample location on worm density ($F = 7.4$, $p < .01$, d.f. = 3,338) allowing the following comparions:
* $p < .05$, ** $p < .01$; t-test comparing worm density between (a) and (b).
\# $p < .05$; t-test comparing (a) or (b) and (c).
$$ p < .01; t-test comparing (a) or (b) and (d).

plovers tend to land in areas where local worm density is significantly higher than in other areas sampled elsewhere in the same field and in an adjacent field where birds do not land. Lapwings also land where worm density is higher, but comparisons with other areas are significant only between fields. Furthermore, the density of worms where golden plovers land is significantly higher than for lapwings. A reason for this difference between the two species will be discussed later. Clearly, however, neither plover species is perfect. In many instances, birds land in areas which do not contain the highest local worm density. While it is possible to explain this in adaptive terms — for example, it may pay birds to sample areas other than those they think are currently the best because worm distribution changes — it is stretching the point. Birds are unlikely to assess local worm density directly. They are more likely to use simple approximate indicators (see Barnard 1984b) such as local grass quality, density of dung pats and mole hills or the location of other birds. They can therefore be expected to make mistakes. Nevertheless, relative worm density does change within fields, even in old pastures, and birds do not end up on food areas simply by landing in the same place every day. In fact there is a probability of only 0.2 and 0.28 that lapwings and golden plovers respectively will return to the same spot first on consecutive days (judged from arrivals within 20m-square cells, see later). Two-way analysis of variance (with replication) of worm density in ten fields over five days (Table 4.2) shows that density varies significantly between fields (as we might expect from Chapter 3), but that there is also a significant interaction between field and time of day. In some fields worm density decreases through the day; in others it increases. Field/time effects such as these may account for some of the variation in flock distribution within and between days.

Although there is a tendency for plovers to alight on locally dense worm supplies, it is not easy to see how they do it. As we intimated above, they probably use some approximate guide. From our previous discussion (Chapter 3), the most likely candidate is some aspect of the variation in flock composition within fields.

Table 4.2: Analysis of variance summary table for the effects of time of day and field on earthworm density

Source	Sum of squares	d.f.	Mean squares	F-ratio	P
Time of day	1748.5	1	1748.5	0.48	ns
Field	48438.7	4	12109.7	3.32	< .05
Interaction	54784.3	4	13696.1	3.76	< .05
Error	36429.5	10			
Total	141400.9	19			

Analysis for 268 core samples from the upper 3.0cm soil layer. (Morning samples 10.00-11.00 GMT, afternoon samples 14.00-15.00 GMT)

Local Flock Composition and Worm Density

In common with several other species which feed in single- and mixed-species flocks (Chapter 1), plovers appear to use the presence of other foragers in deciding where to start feeding (Chapter 3). Since lapwing flock/subflock size correlates strongly with worm density, birds may use the distribution of lapwings within a field as a guide to local worm density.

To quantify the distribution of birds within a flock, we plotted the position of each individual on a duplicate scale map of the field. Positions were plotted using a Ranging 1200 distance finder (calibrated for distances between 46m and 1000m) and vegetational landmarks. Altogether, 114 flocks were mapped in this way and examples of one flock at three different stages are shown in Figure 4.1. We then superimposed a grid of scaled-down 20m-square cells over the maps to measure local flock composition and bird density. This is similar to techniques used with shorebirds (e.g. Bryant and Leng 1975, Bryant 1979, Thompson 1981, Rands and Barkham 1981, Halliday *et al.* 1982).

Since lapwings are virtually always the first birds to begin feeding in a field, we looked at where lapwings and golden plovers landed in relation to the local density of already-foraging lapwings. Figure 4.2a compares the mean density of lapwings in the immediate areas (20m-square cells) in which birds landed with the mean density taken over the whole flock. Both lapwings and golden plovers tend to land in areas where the density is higher than average for the flock, but the trend is significant only in golden plovers. Once golden plovers have landed, however, the picture changes. Samples taken between three and eight minutes after the arrival of golden plovers reveal an inverse relationship between the local densities of the two plover species ($r = -.39$, $p<.01$, d.f. = 49 20m-square cell counts); in established flocks there tends to be a much lower than average concentration of lapwings in those parts of the flock containing golden plovers and *vice versa* (see Figure 4.1b,c).

Although lapwings and golden plovers associate in the same flocks, there are some marked differences in their pattern of distribution within fields. Firstly, there is a tendency for golden plovers to feed in higher densities than lapwings ($t = 4.56$, $p<.001$, d.f. = 48, see Figure 4.3). This may, in part, be due to golden plovers being more numerous than lapwings in most mixed flocks (mean number of golden plovers per flock = 80.7 ± 21.8, mean number of lapwings = 32.3 ± 5.58; $d = 2.17$, $p<.01$, d.f. = 40 mapped flocks). Secondly, golden plovers arrive and depart in variable-sized, often very large, groups. This has important consequences for flock dynamics and we shall return to it again in Chapter 7. Thirdly, there are differences between the two species in the relationship between subflock size and density. In lapwings, density increases steadily with subflock size, but begins to level off at around 150 birds. In golden plovers, it increases more rapidly so that high densities are reached at much smaller subflock sizes (Figure 4.4).

Is there any evidence that variation in plover density within flocks corre-

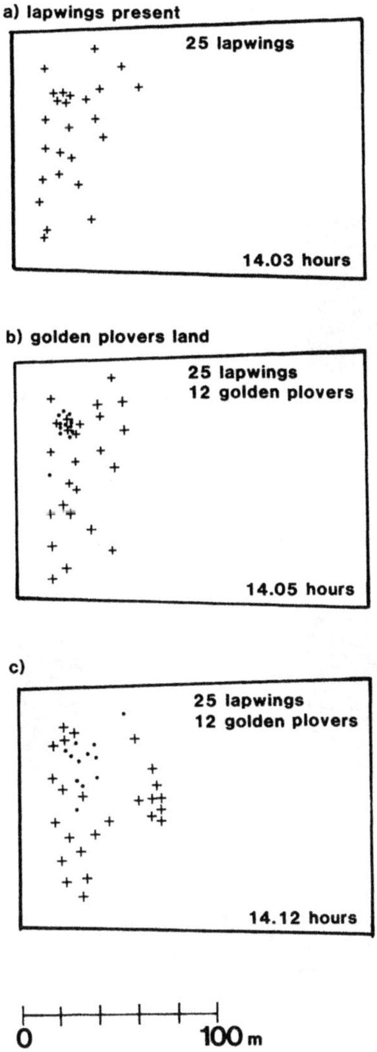

Figure 4.1: Examples of spatial distribution within (a) a single-species lapwing flock, (b) a mixed flock in which golden plovers have just arrived and (c) an established mixed flock. Data for a young pasture recorded on 8/12/82

lates with local worm density? Figure 4.3a shows that there is when cells containing only one species are considered. In both species, bird and worm density increase together. The data for old and young pastures are shown separately, because of the pronounced preference of plovers for the former.

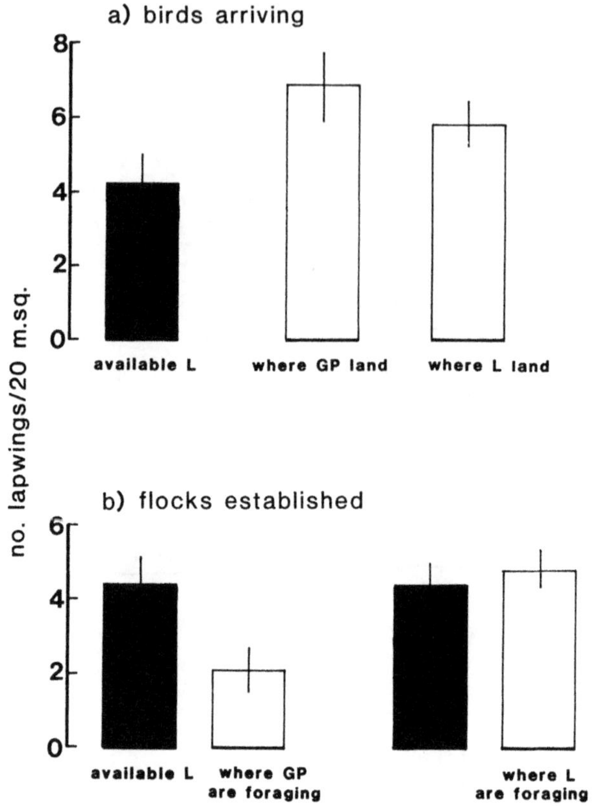

Figure 4.2: The mean (± se) density of lapwings in parts of a flock where (a) lapwings and golden plovers land and (b) where birds of each species are feeding 3-8 minutes after landing. Shaded columns, average density of lapwings in the flock

The relationships are significant for both species in both types of pasture, but they are stronger in golden plovers. This is as we might expect if lapwings tend to locate high-density patches by sampling and are then used by golden plovers as a guide to good feeding areas. In golden plovers, the relationships are linear, in lapwings they are best described by second- and third-order curves. Interestingly, there tend to be higher densities of birds for any given worm density on old pasture. This is particularly pronounced in lapwings on low worm densities. A possible explanation is that the location of high-density patches is more predictable on old pasture (Chapter 3), so birds spend less time sampling other, lower-density, areas.

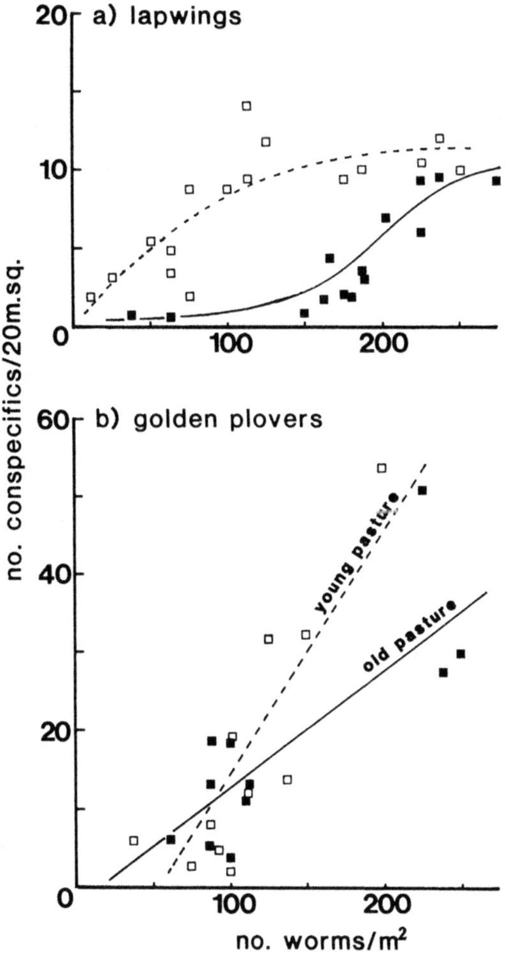

Figure 4.3: The relationship between local bird and worm densities where individuals of each plover species are feeding. Open squares and broken lines, young pasture; solid squares and lines, old pasture. (a) Relationships for lapwings; on old pasture $F = 8.26$, $p < .05$, on young pasture $F = 13.41$, $p < .01$. (b) Relationships for golden plovers; on old pasture $F = 17.6$, $p < .001$, on young pasture $F = 24.8$, $p < .001$. Each point represents estimates for four 0.25m.sq turf samples. Lines fitted by polynomial regression

Local Enhancement and Worm Density: experiments with models

We have seen that plovers use the presence of others in deciding where to feed. They also use variation in local flock density as a guide to good areas within fields. Nevertheless, it is still possible that cues other than the presence and density of birds are used in locating good feeding areas. Obvious candidates are indicators of local soil quality such as the length, colour and density

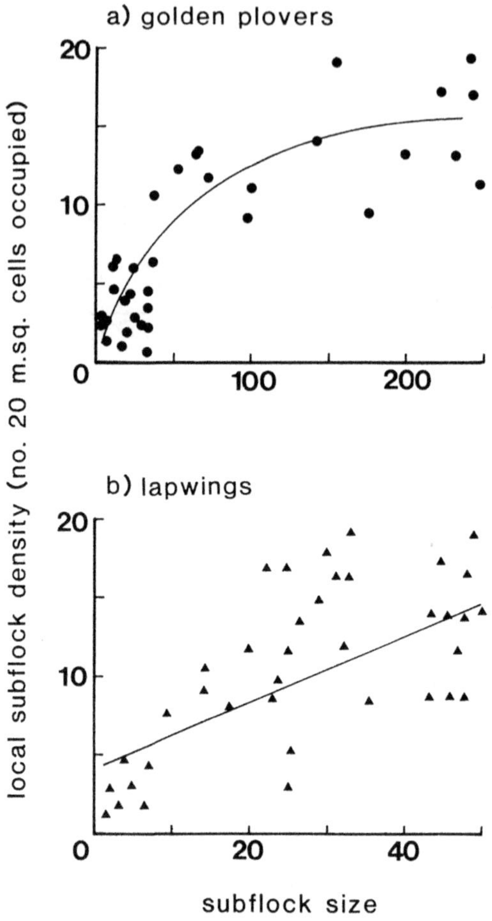

Figure 4.4: The relationship between plover flock/subflock size and the number of 20m.sq cells occupied by birds. (a) Relationship for golden plovers, $F = 37.8$, $p < .001$; (b) relationship for lapwings, $r = .62$, $p < .001$. Data for 34 mixed flocks. Lines fitted by regression. See text for details

of grass, or the presence of fertilising agents such as dung. One way to test for this is to use model birds. By setting up artificial flocks on areas of known high or low worm density, we should be able to distinguish between the effect of the presence of birds and the effects of other potential indicators of high worm density. We carried out some simple experiments using model lapwings.

We chose two adjacent pastures (one young, one new) which were approximately equal in size but which differed in mean worm density. The 9.11-ha young pasture had last been ploughed two years previously and had a mean worm density of 148 ± 13.5 worms/m^2; the new pasture was 8.7 ha, had been ploughed four months previously and had an average of 62 ± 6.3 worms/m^2. Despite not having been pasture for long, the latter was used regularly by lapwings and golden plovers, particularly during late winter (February to March) when the experiment was performed. Tests were carried out between 08.00 and 10.00 GMT when birds were seeking the first feeding sites of the day. Five model lapwings were set out in an ellipse of 20m maximum length and 10m maximum breadth in the following treatments:

Treatment 1 — models absent from both fields.
Treatment 2 — models present in the new pasture only.
Treatment 3 — models present in the young pasture only.

We distinguished two measures of interest in a field. The probability of birds wheeling over the fields (an almost invariable prelude to landing, in itself suggesting some kind of local habitat assessment) and the probability of landing. Table 4.3 shows that, when models were absent from both fields (Treatment 1), neither species wheeled over or landed in the new pasture. Both showed a tendency to wheel over the young pasture, but only lapwings landed. By contrast, both species were likely to wheel over the new pasture when it contained the model 'flock' (Treatment 2). The increase in the probability of wheeling was significant in both cases (Fisher's exact probability for lapwings = 0.054 and for golden plovers = 0.021) and it is noteworthy that no birds wheeled over the young pasture (cf. Treatment 1). In Treatment 3, when models were present in the young pasture, both species were likely to wheel over and land in the young pasture. The differences between Treatments 2 and 3 in young and new pasture were significant for landing in both species (see Table 4.3), but significant for wheeling only in lapwings ($p = 0.041$ for lapwings and 0.145 for golden plovers).

Table 4.3: The effect of model lapwings on the probability of lapwings (L) and golden plovers (GP) landing in or wheeling over fields

		Young pasture			New pasture		
Experimental treatment	Plover sp.	prob. land	prob. wheel	n	prob. land	prob. wheel	n
1. Models and birds absent from both	L	.13	.25	8	0	0	9
	GP	0	.13	6	0	0	9
2. Models in NEW pasture only	L	.20	0	10	.10	.40	10
	GP	0	0	5	0	.60	5
3. Models in YOUNG pasture only	L	.88	.78	9	0	0	9
	GP	.66	.66	3	0	0	3

Data for 10 observations per treatment (see text). n = the number of flocks that flew over the field.

When birds did land in one of the fields, they landed significantly closer ($p < .001$) to the models than expected by chance. Indeed 71.4% of lapwings and 50% of golden plovers landed *amongst* the models. Comparisons with chance expectation were calculated on the basis of the model 'flock' occupying approximately 2% of the surface area of each field. If plovers alighted at random with respect to the 'flock', then only 2% of birds should have landed near (<10m from an edge bird) or within the 'flock'.

Together, the results of the model experiment suggest that the presence of other birds influences the degree of interest in a field in both lapwings and golden plovers. It appears, however, that other factors are necessary to induce birds actually to land. These presumably correlate in some way with food availability. In this case, the relative density of grass cover may have been used, since this is a good guide to worm density and detectability (see Chapter 3). There is also evidence that the density and distribution of dung pats influences the local distribution of birds within fields (Waite 1983, 1984a, K.R. Futter unpubl.). That other factors appeared to be important in these experiments may have been a consequence of the small model 'flock' and the postural uniformity of the model birds. The attractiveness of a flock might be expected to depend on its size, species composition and the spacing between birds (see also Chapter 7). There is also evidence that birds take account of the posture of individuals within flocks. Drent and Swierstra (1977), for instance, found that barnacle geese (*Branta leucopsis*) were more likely to join model birds set in a head-down position ('feeding') than those set head-up ('scanning for predators'). Our model lapwings were all head-up. Combined with their small number, they are therefore likely to provide a very conservative estimate of the effect of flocks on birds' choice of feeding site.

Local Bird Density and Pasture Age

As we might expect from Chapter 3, the local density of plovers within fields varies with pasture age. The increased density of birds on old compared with young pasture in Figure 4.3 is significant (for lapwings $t = 5.35$; $p < .001$, d.f. $= 34$; for golden plovers $t = 2.63$; $p < .05$, d.f. $= 34$). This is not simply because birds occur in larger numbers in old pasture. In fact, there is no significant difference between mean flock/subflock sizes on old and young pasture (mean number of lapwings on old pasture $= 30.2 \pm 2.97$, mean number of golden plovers $= 82 \pm 14.5$; mean number of lapwings on young pasture $= 25.9 \pm 5.9$, mean number of golden plovers $= 80.3 \pm 26.1$). Instead, the differences appear to be due to the way birds forage on the two types of field. Figure 4.5 shows a plot of the number of 20m-square cells occupied by lapwings or golden plovers in old and young pasture over a 16-day period (27 November-12 December 1980). Birds of both species tend to move through a greater number of cells during a given period and are more dispersed on young pasture. The difference is particularly pronounced between days, where birds are regularly seen in the same cells on old pasture but not on young. Figure 4.6a shows the mean probability of plovers being present in ten randomly-picked cells on consecutive days in the two types of pasture. Lapwings are significantly more likely to be present in the same cells on old pasture than on young ($d = 2.16$; $p < .05$, d.f. $= 16$), but there is no difference in golden plovers. This fits with our previous observation that local worm density tends to be less variable on old pasture. Figure 4.6b shows that the probability of the same 20m-square cell containing the highest densities of both worms *and* birds is much greater on old pasture. Since plover subflock size varies with worm density, the greater variation in worm density in young pasture might result in greater variation in the number of plovers. This appears to be the case. Both lapwing ($F = 2.71$, $p < .01$) and golden plover ($F = 14.6$, $p < .001$) subflock sizes are more variable on young pasture.

Bird Feeding Hours and Pasture Age

A more conventional measure of spatial and temporal variation in bird density at a feeding site is *bird feeding hours* (*BFHs*) per unit area (e.g. Bryant and Leng 1975, Bryant 1979, Buxton 1981, Thompson 1981, Prater 1982, Halliday *et al.* 1982). BFHs are calculated by averaging the number of birds observed feeding in a given area at hourly intervals through the period of observation.

Figure 4.7 shows BFHs/20m-square cell for lapwings and golden plovers in the same old and young pastures as above. BFHs are summed over the 17-day period. From the figure, it appears that lapwings tend to forage intensively within certain cells on old pasture, whereas golden plovers forage more intensively on young pasture. This may, however, be a reflection of both the greater variation in local worm density on young pasture and the constraint imposed on the movement of golden plovers around rich feeding areas by lapwings. The significant negative correlation between BFHs in lapwings and golden plovers in both pastures bears this out ($r = -.312$, $p < .05$ on old pasture, and $-.423$, $p < .01$ on young). It is also interesting that the cell used most intensively by lapwings on the old pasture contained a higher density of worms than those used most intensively by golden plovers on either pasture. With the exception of golden plovers in young pasture, there are significant positive correlations between *daily* BFHs and the density of worms in given cells (Table 4.4). In other words, birds spend more time in those parts of the field containing the highest worm densities, even though they may change from day to day.

Prey Depletion and Return Times

Although we can account for the difference in the distribution of feeding effort between old and young pasture in terms of worm density, it might arise

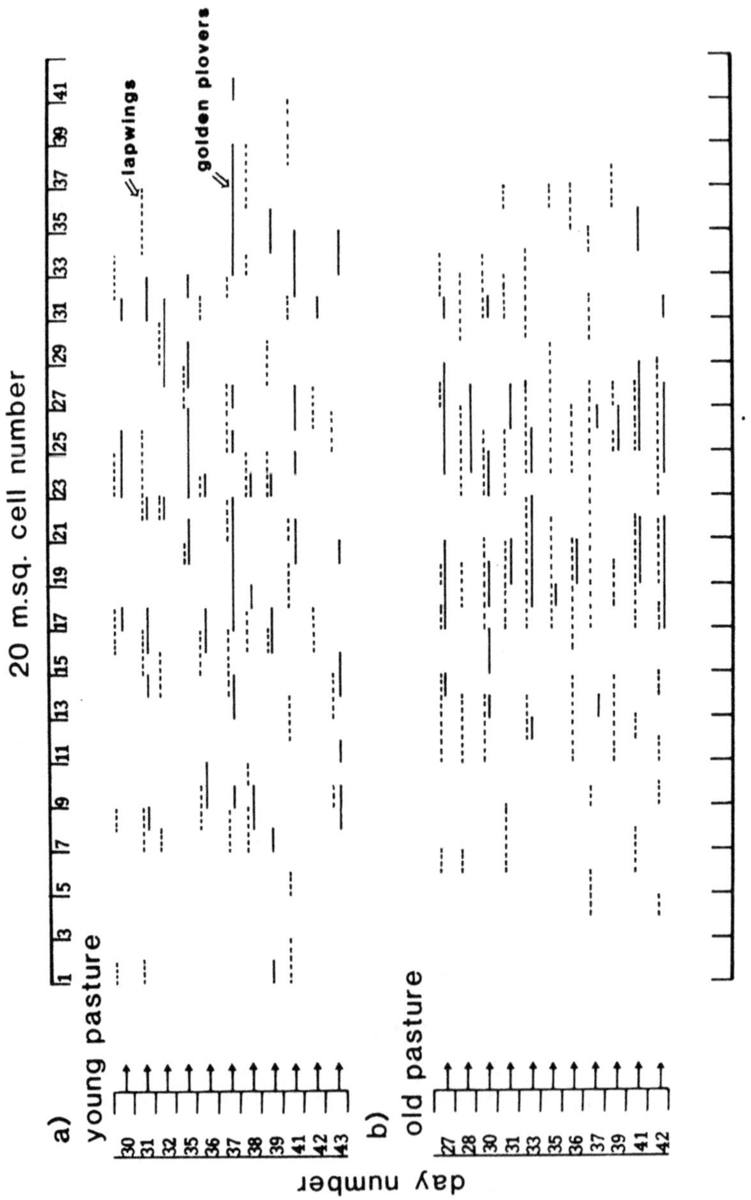

Figure 4.5: The presence or absence of lapwings and golden plovers in 20m.sq cells in young and old pasture through the day. Data for 17 days. See text

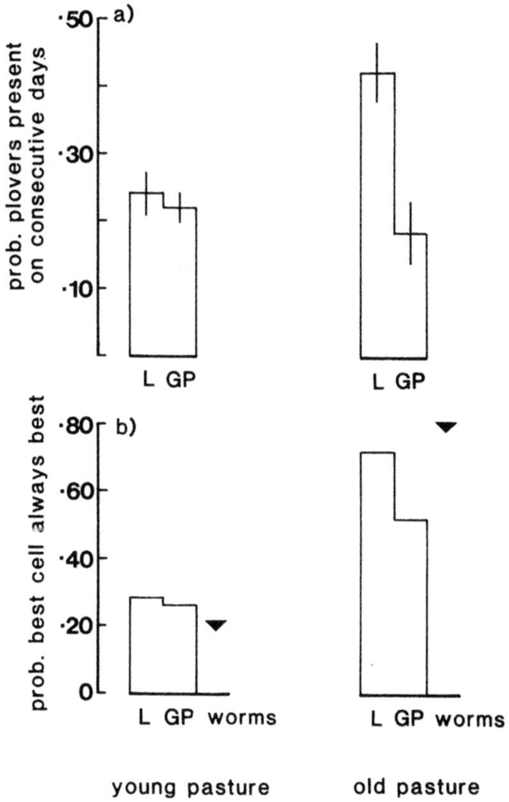

Figure 4.6: The consistency with which plovers forage in given 20m.sq cells in young and old pasture. (a) The mean (± se) probability of a cell containing plovers on consecutive days, (b) the probability of the same cell containing the greatest number of plovers or worms (arrow) on consecutive days. See text

in other ways. In particular, it might reflect differences in the patterns of prey depletion and renewal. Young pasture has sparser grass cover than old pasture (Chapter 3, Barnard and Stephens 1983). This has two effects on earthworm availability: it renders the fewer worms that are present more detectable (at least to a human observer) and it increases exposure to climatic factors such as temperature fluctuations and frost. Barnard and Stephens (1983) found a disproportionate decrease in the number of worms in young pasture compared with old pasture in cold weather and attributed this to the greater degree of exposure. If birds feeding on young pasture also inflict a disproportionately higher mortality (because worms are spotted more easily), then the rate of prey depletion and time taken for densities to recover may be greater. In fact, this appears to be the case. We observed a number of plovers

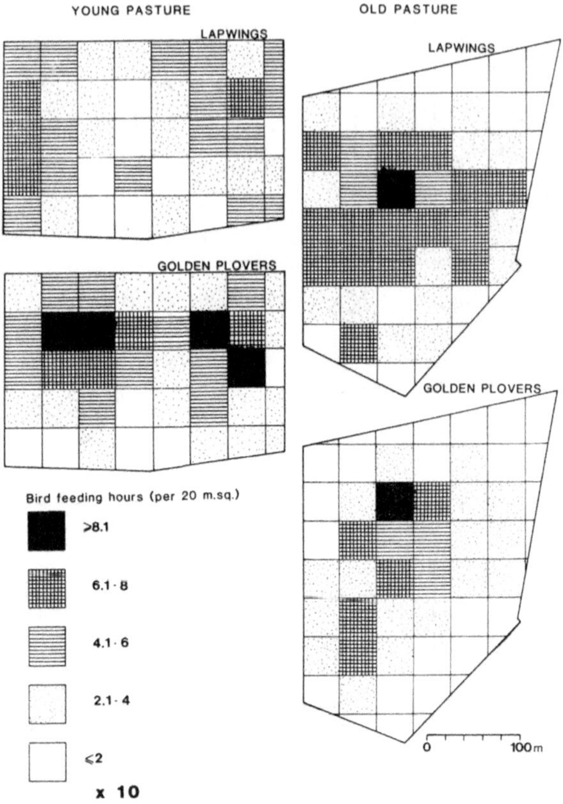

Figure 4.7: The distribution of bird feeding hours on young and old pasture. Data for hourly counts at two pastures over 17 days. See text

Table 4.4: Beta values and F-ratios from partial regression analysis of the relationships between number of 'bird feeding hours'/20m-sq. cell/day and local worm density

Bird feeding hours:	Beta value	F-ratio	nth order relationship	No. 20m-sq. cells sampled	p
LAPWINGS					
Old pasture	.61	8.73	2nd	14	$< .01$
Young pasture	.67	8.21	2nd	16	$< .01$
GOLDEN PLOVERS					
Old pasture	.47	5.31	1st	19	$< .05$
Young pasture	.39	3.23	1st	12	ns

Data for a total of 109 0.25m-sq. turf samples.

foraging on young and old pasture and sampled worms along their search paths (for details of locating search paths and sampling, see Chapter 5). Worms were sampled three times after birds had begun to feed: before 30 minutes had elapsed, and 30 to 60 and 60 to 90 minutes after the start of feeding. Table 4.5 shows that worm density decreased from the first to second sample in both types of pasture. The decrease, however, was greater in young pasture (22.7%) than in old (15.2%). Furthermore, while worm densities recovered to some extent by the 60-90 minute sample in both pasture types, the recovery was slightly less marked in young pasture (86.6% of the initial sample versus 92.1% in old pasture). The impact of predation does seem to be greater on young pasture. Since plover flocks tend to be concentrated in a small part of the field at any given time (see Figure 4.1) and work their way around the field during the day, we might, therefore, expect birds' *return times* to given areas to be greater on young pasture, perhaps resulting in birds feeding in more localities within young pastures than within old during any given period.

Optimal Return Times

As Krebs (1978) points out, where feeding sites replenish themselves rapidly enough to make them worth revisiting, we can consider how rapidly a predator ought to return to a particular place. An obvious possibility is that predators will travel around their foraging area in a way which maximises their net rate of food intake from particular patches at each visit. Davies and Houston's (1981) study of pied wagtails (*Motacilla alba*) shows clearly how the timing of revisits to a replenishing patch critically determines the rate of food intake. In their case, territorial wagtails were revisiting small bays along a river bank where insects and other edible debris periodically accumulated. When they left intervals of about 20 minutes between successive visits to the same bay, birds found roughly twice as much food per visit as when they left about eight minutes.

Of course, in order to be able to optimise return times, a predator must have exclusive use of its foraging area (Charnov *et al.* 1976). If competitors

Table 4.5: Mean worm density (± se) along the search path of lapwings and golden plovers at different times since the flock landed

	Worm density		
	< 31 mins	31-60 mins	61-90 mins
Old pastures n = 64, 33, 50	151 ± 11.3	128 ± 14.2	139 ± 6.6
Young pasture n = 21, 36, 19	97 ±18.4	75 ± 11.1	84 ± 19.5

n = number of 0.25m-sq. turf samples for each time period.

are also able to harvest food supplies, their rate of renewal becomes unpredictable. Intruders in the territories of Davies and Houston's pied wagtails halved the effective return times of territory owners by depleting bays while the latter were only half-way around the rest of their foraging circuit. Interference can be reduced if predators defend territories (e.g. Charnov *et al.* 1976, Kamil 1978, Davies and Houston 1981) or forage in groups (e.g. Cody 1971, Prins *et al.* 1980). Kamil (1978) studied pairs of Hawaiian honeycreepers (*Loxops virens*) defending breeding territories around blossoming trees. By marking clusters of blossom in a number of territories, Kamil was able to monitor the visitation pattern of known birds. He found that territory owners paid a second visit to particular clusters only after an appreciable period since their first visit. Moreover, their return time was very close to that expected if they were maximising their net rate of nectar intake from flowers (in fact they made just one more visit per cluster than expected). Intruders, on the other hand, which did not know the sequence in which clusters had been visited, were much more likely to visit a recently-depleted cluster. Consequently, they obtained only about two-thirds as much nectar as residents. It turned out that members of a resident pair avoided interfering with each other's foraging circuit by feeding in different parts of their territory.

Feeding in a group provides another way of regulating visits to renewing food supplies. If all the predators within a given area forage together, they can avoid visiting recently-depleted feeding sites. Of course, this depends on the degree of group cohesion and can operate only at the level of coarse-grained patches (areas of locally high food abundance within patches are likely to be depleted by other group members). Cody (1971) considers the optimisation of return times to be an important factor selecting for aggregation in predators which exploit renewing food supplies. He supports his argument with examples from mixed-species finch flocks. One set of species fed close to a mountain range, where it was relatively wet and ripening seeds were abundant but renewed slowly. Another set fed further away in a desert, where it was dry and seeds were sparse but renewed more rapidly than those near the mountains. Cody found that flocks in the desert moved faster and took bigger turns (bringing them back to the same place more quickly) than flocks near the mountains. This is as we might expect from the differences in food-renewal rate. Unfortunately, as Krebs (1978) points out, Cody's data confound food density and renewal rate. It is therefore difficult to tell whether birds were really reducing their return times in the desert habitat or simply responding to the lower food density. An excellent experimental field study of return times in socially-feeding predators, however, has been carried out by Prins *et al.* (1980) who looked at brent geese (*Branta bernicla*) feeding on sea plantain. It is worth describing their study in some detail.

Geese were studied on an island in the Wadden Sea where they were staging during their spring migration. The sea plantain is one of their principal food plants during the stopover. Prins *et al.* noticed that the geese tended to forage in large flocks and regrazed the same area of the island about every four days. An obvious possibility was that this reflected the amount of time needed for plants to regenerate after the previous grazing. To test this, Prins *et al.* performed a simple experiment. By marking plants and clipping them every four or eight days, they showed that the growth zone of each leaf was mainly at its base. Light clipping (removing 10mm of leaves over 10mm in length) resulted in an increase in total herbage accumulation compared with no-clipping and heavy-clipping (clipping leaves over 5mm back to 5mm) treatments. Clipping also had a profound effect on the production of new leaf tissue, which the geese preferred, and its distribution within a rosette. In unclipped rosettes, the production of new tissue was more or less evenly distributed over the leaves. An increased degree of clipping, however, resulted in a shift of production from the bottom towards the top. Light clipping caused an increase in the production of new tissue, heavy clipping caused a decrease.

It turned out that light clipping, removing about 32% of the leaf material, was roughly equivalent to the amount removed by geese. Light clipping (= grazing) therefore increased the productivity of the plantain crop if it was carried out at four-day intervals. The clipping experiment thus suggests that the foraging cycle of the brent geese maximises the growth rate of the new leaves on which they feed, and that the geese return to harvest regenerated areas at an interval which allows for the appropriate type and amount of growth. As a further test, van Eerden (cited in Prins *et al.* 1980) applied organic fertiliser to test areas of plantain. Since areas treated with fertiliser showed an accelerated rate of growth, we might expect geese to regraze the same localities after a shorter interval than in non-fertilised areas. Indeed, van Eerden found that return times were greatly shortened following fertilisation.

Return Times in Plovers

To see whether return times in plovers were related to the rate of earthworm replenishment, we calculated the time taken for foraging birds to revisit previously-sampled 20m-square cells in old and young pasture. Return times were scored in the following way. The daytime period was taken as 08.00 to 17.00 GMT. The median hourly period for this was thus 12.00-13.00 hours. When a cell was occupied only during the median period on two consecutive days, a return time of 1 was recorded. When it was occupied during the median period for the second time on the next but one day, we recorded a return time of 2, and so on. If birds were present in the cell at any other time, return times were scored as fractions. Thus birds initially present in a cell between 12.00 and 13.00 hours on Day 1 and which were next recorded there between 09.00 and 10.00 hours on Day 2 scored 0.67 (6/9 hours). Birds present next at 10.00-12.00 hours on Day 2 scored 0.89 (8/9 hours). This, of course, assumes that birds did not occupy cells during the night (see Chapter 6).

Return times for lapwings and golden plovers to five arbitrarily-selected

Table 4.6: Mean (± se) return time (days) to 20m-sq. cells by plovers in old and young pastures

	Return time in		
	Old pasture	Young pasture	t-test
Lapwings	1.20 ± .091 (range = 0.5-5.7)	2.68 ± .423 (range = 0.6-6.4)	3.42**
Golden plovers	1.87 ± .231 (range = 0.6-7.0)	3.50 ± .446 (range = 0.7-11.2)	3.67*
t-test	2.69*	1.70	

* $p < .05$, ** $p < .01$.

cells in two fields (young and old pasture) are shown in Table 4.6. There is a significant tendency in both species for return times to be longer in young pasture. Golden plovers also tend to have longer return times than lapwings, but the difference is significant only on old pasture. Were these differences in return time due to worms taking longer to recover in young pasture? We have seen already (above and Table 4.5) that predation per unit area appeared to be comparatively heavier in young pasture. To relate this to return times, we sampled worms in two 20m-square cells in each pasture occupied by plovers on consecutive days, and compared them with samples from cells not occupied on any day. Figure 4.8 shows that worm density declines significantly between days in young pasture, but not in old pasture. The trend is the same for both plover species. There is no significant decline in worm density between days in control areas on either type of pasture (Mann-Whitney U = 29, d.f. = 6,6, n.s. for old pasture; U = 11, d.f. = 6,6, n.s. for young pasture). The decline in areas where plovers feed (in young pasture), therefore, cannot be attributed to other factors such as temperature variation or weather differences which might confound the comparison. The results in Figure 4.8 thus support the idea that longer return times in young pasture correlate with longer recovery periods for worm density.

An obvious prediction from the above is that the rate at which birds return to feed in a given area should correlate positively with the rate at which worm density recovers. We tested this by looking at the number of 20m-square cells containing birds before and after worm densities in the cells had recovered to their levels prior to previous predation. We collected five core samples (see Chapter 3) from each of 25 cells where birds (of either plover species) were observed feeding. To measure worm densities immediately after birds had been feeding, we took samples from cells just as birds moved out of them. We then sampled them again when birds were next seen in the cells. Birds were scared off as soon as they entered cells for the second time so that little or no predation took place before samples were taken.

From our samples, it appears that worm densities take, on average, about

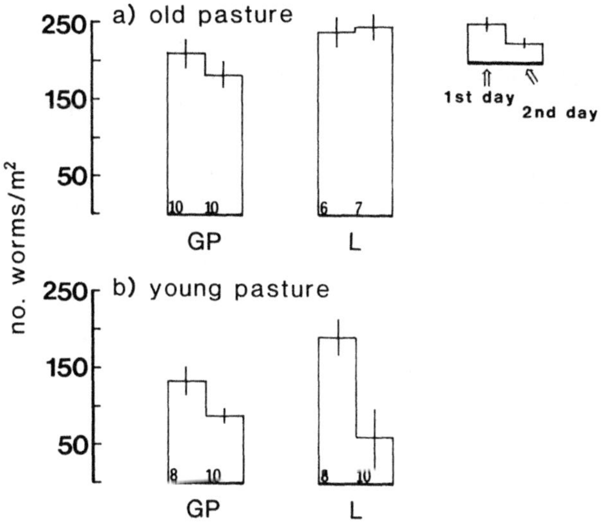

Figure 4.8: Changes in worm density between days in areas occupied by lapwings and golden plovers on consecutive days. Mann-Whitney U comparing densities on consecutive days = 3, $p < .001$ and 8.5, $p < .02$ for lapwings and golden plovers respectively on young pasture. U = 24 and 18, n.s. on old pasture. Sample sizes in figure

1.7 days (0.3-6.1 days) to recover (cf. bird return times in Table 4.6). Mean (before and after) worm densities which were within 20 worms/m^2 of each other were considered to show no change. This provided a rough control for the effects of temperature on worm density. Lapwings were seen next in nine cells which had recovered to their previous levels, in two which had lower worm densities on the second sampling and in five which remained the same. The corresponding figures for golden plovers were seven, two and one. Both species, therefore, tend to revisit areas within fields after sufficient time has elapsed for food availability to recover, although the trend is significant only in lapwings ($X^2 = 6.40$, $p < .02$; for golden plovers $X^2 = 2.77$, n.s.). It is worth emphasising, however, that our estimates of renewal times are minima. In some cases, worm density may have replenished itself in a cell before we next recorded birds there. It is also possible (though unlikely, see Chapter 3), that birds visited cells during the night and hence were not recorded on their very next visit. From what we know of the foraging strategies of the two plovers (Chapters 3 and 5-7), it may not be surprising that lapwings appear to come closer to optimising their return times than do golden plovers. Since lapwings are usually the first to begin feeding in a field, they rely on detecting differences in worm density directly. We might therefore expect them to be more efficient discriminators and hence time revisits more accurately.

Resettlement Flights

Foraging behaviour in plovers is frequently punctuated by short flights within the flock. These take two forms: *resettlement* flights in which one or more individuals, or the entire flock, take off in alarm and resettle in the same field; and *internal* flights where birds move within flocks through displacement or to attack or area-copy (see Chapter 6) another bird (see also Waite 1981, 1983a). The cause of resettlement flights is obscure. They do not appear to occur in response to any identifiable alarm stimulus (see Chapter 9 for a discussion of alarm flights). This may, however, reflect movement with respect to local worm density. In mixed flocks of rooks (*Corvus frugilegus*), carrion crows (*C. corone*), jackdaws (*C. monedula*) and magpies (*Pica pica*), Waite (1983) found that birds took flight more often and were less likely to return to the spot where they had just been feeding when local worm density was low. Is this the case in plovers?

Figure 4.9 suggests that it is. Figure 4.9a shows an inverse relationship between the frequency of resettlement flights in plover subflocks and local worm density. The plots take into account only those instances where birds landed again within one minute. For each subflock, we took eight 3cm-deep core samples from the area where birds had been feeding prior to take-off. Taken together, the frequency with which the two species take flight declines in a significant second-order manner with worm density. Another factor which might influence resettlement flight is, of course, worm distribution. If worms are very patchily distributed, birds may need to move a considerable distance to the next good feeding area. We might therefore expect a positive correlation between flight frequency and worm patchiness. In fact, there is only a weak relationship. Figure 4.9b shows a non-significant increase with the variance:mean ratio of worm density in the area of take-off.

In some cases, birds take flight because of disturbance by potential predators rather than to move to a new feeding area. Once in the air, however, it may pay birds to seek a new area if the benefit of doing so exceeds the cost of returning and searching for prey where it is already partly depleted. If this is the case, birds should be more likely to move on if worm density in the current area (from which they are disturbed) is low and the variance:mean ratio high. The latter prediction arises because the patchier the distribution of worms within a given field, the smaller each patch is likely to be and the greater the probability that the current patch will have been depleted to a level where an alternative patch is more profitable. To test this, we recorded the proportion of birds returning to the area from which they had been disturbed and sampled worm densities in the area immediately after they returned. Figure 4.10a shows that a greater proportion of both lapwings and golden plovers return to the same spot when current worm density is high. Birds are also more likely to return when the variance:mean ratio of worm density is low (Figure 4.10b) and local depletion less likely. There is no tendency, however, for birds to spend less time in the air when they are disturbed from a high-density area. This is not surprising if birds use time in the air to assess the source of alarm and simultaneously select another feeding area before deciding to move on.

Choosing Where to Feed: Résumé

Food availability appears to be the main factor influencing the distribution and movement of plovers within fields, just as it influences choice between

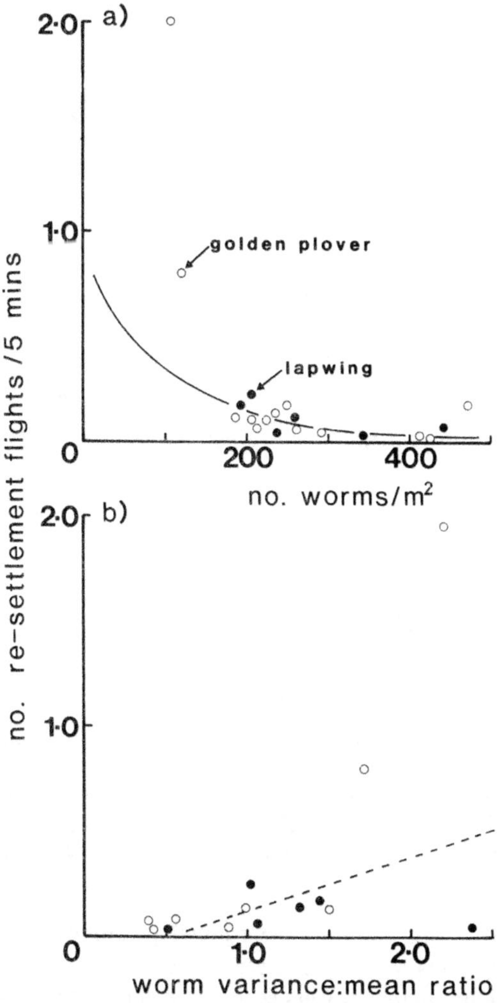

Figure 4.9: The relationship between the frequency with which plovers take off and (a) worm density, $F = 7.4$, $p < .001$, and (b) the variance: mean ratio of worm density, $F = 4.3$, n.s., in the area where they have just been feeding. Data for 17 resettlement flights. Lines fitted by regression

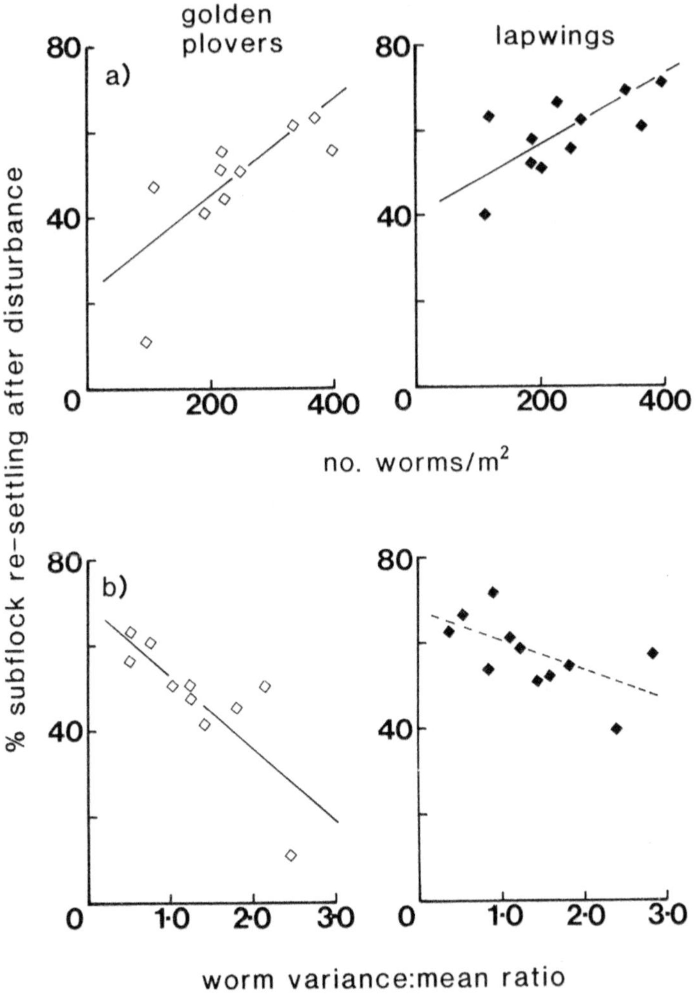

Figure 4.10: The relationship between the percentage of plovers returning to their feeding area after disturbance and (a) local worm density, r = .83, p < .01 for lapwings, .86, p < .001, for golden plovers and (b) the variance:mean ratio of local worm density, r = −.69, p < .05 for lapwings, and −.92, p < .001, for golden plovers. Data for 11 resettlement flights in lapwings and 10 in golden plovers and 232 core samples. Lines fitted by regression

fields. Intriguingly, both newly-arriving lapwings and golden plovers tend to land in parts of fields where earthworm density is higher than average, though, at first sight, golden plovers seem to be better at it than lapwings. The reason is almost certainly that lapwings arrive first and therefore have fewer cues available to them than golden plovers, which can use the presence of other birds to indicate locally-rich areas. While cues such as grass quality and the density of dung pats may provide a guide to local worm availability, lapwings may depend to some extent on spatial memory. The preference for old pasture seems to be based partly on greater temporal and spatial predictability in the surface (<3cm-deep) worm distribution. Thus, the richest and poorest areas tend to remain in the same place. There is evidence that birds of other species use spatial memory during foraging, at the levels of both multi-prey patches (e.g. Smith and Sweatman 1974) and single prey items (e.g. Sherry 1984). If lapwings do remember where good feeding areas are from one visit to the next, however, they appear to do so (or to use immediate experience) only within days.

The use made of birds which are already foraging is shown by the experiment with models. Both lapwings and golden plovers are more likely to wheel over or to land in a flock when lapwings are apparently present, but the effect is most pronounced in golden plovers. Similarly, when landing in real (rather than dummy) flocks, golden plovers are more likely to land where lapwing density is high. Clearly, however, factors other than local bird density are also used.

Several aspects of foraging behaviour in plovers suggest that their movement within fields is geared to the rate of worm replenishment. The distribution of birds is more variable in young pasture where worm density and its rate of replenishment are low. Unfortunately, the relative rate of replenishment in young and old pastures is confounded by differences in the spatial and temporal predictability of worm density. Nevertheless, plovers tend not to turn up in areas a second time before worm densities have recovered, at least to their pre-predation levels.

Worm density also influences where birds land after disturbance or other causes of longer-range movement within fields. Birds are more likely to return to the same spot if local worm density is high. Both when first arriving at a field and when feeding, therefore, lapwings and golden plovers choose where they go on the basis of local food availability.

Summary

1. Plovers tend to land within fields where worm density is higher than average and may use local bird density as a guide to good feeding areas.

2. Worm densities recover after predation and plovers appear to allow sufficient time for recovery before revisiting a given area. Return times are longer in young pasture because recovery rates are slower.

3. If birds are disturbed or take-off for other reasons, they are more likely to return to the same area if it contains a high worm density.

Chapter 5

Choosing What to Eat

In the last two chapters, we discussed the way feeding flocks build up on rich food supplies. Birds use the presence of conspecifics or birds of other species to select fields with high surface densities and predictable spatial distributions of earthworms, and areas within fields where worms are locally abundant. Locating rich feeding areas, however, is only one problem facing hungry birds. Earthworms occur at different depths in the soil, in different sizes and so on, making it easier or more difficult for birds to detect and capture them. Is there any way we can predict how (or, indeed, whether) birds should choose between the different types of worm they encounter?

Optimal Foraging Theory

An approach which has proved remarkably powerful, in terms of generating precise and subsequently well-supported hypotheses about foraging behaviour is optimal foraging theory (OFT). As Krebs *et al.* (1983) put it, OFT is an attempt to find out if there are any general rules about where animals go to feed, how they search for food and what they feed on. We can think about these questions in terms of 'decision rules' employed by hungry animals: an animal in a given situation decides to eat this item, but not that, it decides to hunt here, but not there, and so on. OFT assumes that these basic decision rules will have been shaped by natural selection to enable the animal to feed as efficiently as is feasible. Models are developed which make an educated guess at what 'efficiency' means in a particular case and what the constraints might be that limit it. For instance, some OFT models assume that efficiency can be measured as net rate of energy intake and that the constraints on maximising this include such things as handling time (the time taken to capture and mandibulate prey), the time needed to recognise cryptic prey and the requirement for certain nutrients (see e.g. Belovsky 1978, Hughes 1979, Krebs *et al.* 1983). It is important to emphasise that OFT is not limited to models based simply on energy intake. It is reasonable in some cases to formulate models based on, say, trade-offs between foraging and territorial defence (e.g. Kacelnik *et al.* 1981, Barnard and Brown 1981) or foraging and vigilance for predators (Milinski and Heller 1978, Lendrem 1982) (see Krebs and McCleery 1984 for a good review).

The essence of OFT is that every decision a foraging animal makes is likely to have both costs and benefits in terms of the animal's probability of survival and reproductive potential. The animal may benefit in that the decision results in food being discovered and eaten, but it may also be at a disadvantage by having to expose itself to predators to get the food. 'Benefits' therefore refer to the object of the animal's goal-seeking behaviour (energy, nutrients etc), while 'costs' refer to the deleterious consequences of deciding to obtain the goal in a particular way. OFT assumes that, in the past, natural selection has acted on variation in the efficiency of foraging decision rules so that extant predators will tend to maximise the benefit:cost ratio in the decisions they make. Given that we can accurately identify the goal an animal is seeking and the costs inherent in its goal-seeking behaviour, this assumption allows us to predict how the animal should respond to any given foraging problem. Of course, OFT is just one application of optimality theory to behavioural ecology. As Krebs *et al.* (1983) point out, the distinctions between optimality models of foraging (e.g. Krebs 1978), life-history strategies (e.g. Horn and Rubenstein 1984), mating behaviour (e.g. Bateson 1983), territorial defence (e.g. Davies and Houston 1984), social aggregation (e.g. Caraco and Pulliam 1984) and so on will disappear as the principles of OFT are gradually extended.

Optimal-diet Selection

All prey items confer a benefit in terms of nutritional value and inflict a cost in terms of, say, the time, energy and risk of predation needed to find and ingest them. The profitability of a prey item can therefore be calculated as its net nutritional value (gross value minus the metabolic costs of handling, assimilation and so on) divided by its searching and handling time. OFT assumes that predators will have evolved to distinguish between prey of different profitabilities and to choose the most profitable. We can therefore make some simple predictions about which of a range of prey items a predator should take. These are summarised in Figure 5.1. The graphs model the effects of varying prey abundance (and therefore search time between items) on the number of prey types a predator should include in its diet. If prey are abundant (rich environment), it will pay the predator to ignore all but the most profitable types (Figure 5.1a). Inclusion of less profitable types will reduce the predator's overall rate of nutrition intake ($V/h+s$ in Figure 5.1). If prey are scarce (poor environment), however, the predator will do better by including some of the less profitable items in its diet (Figure 5.1b). As prey abundance decreases, therefore, optimal-diet breadth is likely to increase and predators become more generalist. Three broad predictions emerge from this simple optimal-diet model: (i) Predators should prefer profitable prey; (ii) Predators should be more selective when profitable prey are abundant; (iii) Predators should ignore unprofitable prey, no matter how common they are, when profitable prey are abundant.

A number of laboratory experiments and field observations have attempted to test these predictions (see e.g. Pyke *et al.* 1977, Krebs *et al.* 1983, Krebs and McCleery 1984 for reviews), but many have tested them only qualitatively (e.g. tested whether animals take a greater range of prey in poor environments) rather than quantitatively (tested whether animals take prey types 1 and 2, but not 3, 4 and 5, at encounter rate i with profitable prey and 1, 2 and 3, but not 4 and 5, at encounter rate j). Clearly, the more precise the predictions, the more rigorous the test of the model and the surer we can be that our initial assumptions about the predator's decision rules were correct. Lack of support for the predictions in a particular case cannot, however, be taken as evidence that a predator is not foraging optimally. Diet selection and other OFT models cannot be used to test whether animals are or are not optimal. They merely test whether a particular hypothesis (e.g. maximising net rate of energy intake subject to handling, recognition and nutrient constraints) accurately describes a predator's foraging behaviour. If an energy-maximisation model, say, is not borne out, it does not mean there is no plausible maximisation model which would account for observed behaviour.

Early OFT models of diet selection were the simplest, and generally took into account only the energy content and searching and handling time costs of different prey types (e.g. Krebs *et al.* 1977, Goss-Custard 1977a). Later models catered for more complex costs and benefits, such as requirements for specific nutrients (e.g. Belovsky 1978, Owen-Smith and Novellie 1982), difficulties in identifying prey (Erichsen *et al.* 1980, Rechten *et al.* 1983), the threat of competition (Barnard and Brown 1981), and the need to return to a fixed point with prey (e.g. Orians and Pearson 1979, Bryant and Turner 1982). The better tailored the model to a particular problem, the more precise the predictions it can make, although, of course, there is a consequent restriction in its applicability to other foraging problems. Most rigorous tests of optimal-diet models have been in the laboratory. For obvious reasons, the factors influencing prey profitability are more difficult to measure in the field. Nevertheless, we shall see below that plausible optimal-diet models can be tested in the field.

Optimal-diet selection in waders. As it happens, the field studies which

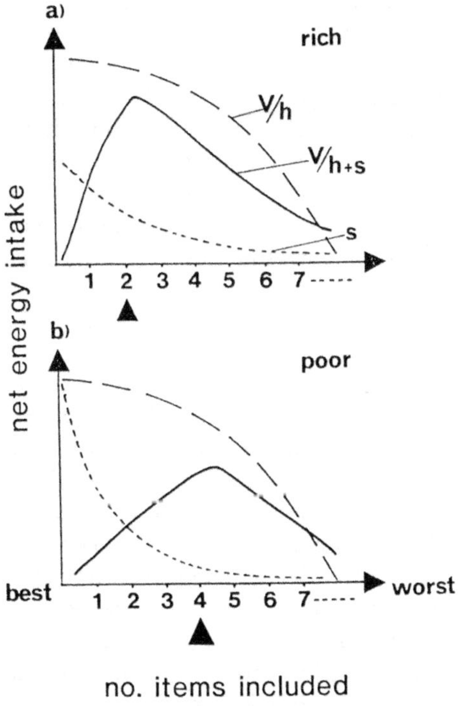

Figure 5.1: Optimal-diet selection in rich (a) and poor (b) environments. V *is the gross energy content of each item,* h *the handling time and* s *the searching time. More items should be included in the predator's diet in a poor environment (the peak of* V/h + s *occurs further to the right than in a rich environment. Arrows indicate optimal-diet breadth*

Source: modified after Krebs (1978)

have so far provided some of the best tests of optimal-diet models have been done on wading birds. The oldest and best known is Goss-Custard's (1977a) study of redshank taking different-sized worms (*Nereis* and *Nephthys* spp.) on mudflats.

Goss-Custard compared the response of redshanks to different size categories of worm in a variety of situations to see whether variations in responsiveness were consistent with the predictions of optimal-diet selection. The model could not be tested directly by comparing the food intake of birds selecting different sizes of worm because birds feeding under similar conditions tended to take the same type of prey. Instead, Goss-Custard compared observed ingestion rates with those expected if birds had selected particular worm sizes from the mud.

Using dry weight rather than energy as a measure of intake, the results showed that, as predicted by the optimal-diet model, redshanks selected the combination of worm sizes that provided higher rates of food intake than any alternative combination. The rate of taking large, profitable worms was positively correlated with their presumed encounter rate in the mud (measured as worms available per metre of a bird's search path). There was no correlation between the intake rate of small, unprofitable worms and their presumed availability. The crucial observation, however, was that the risk of taking small worms (the number taken per encounter) was negatively correlated with the rate of taking large worms; that is birds were less likely to take unprofitable worms, however common they were, if their encounter rate with profitable worms was high.

Sutherland (1982) studied oystercatchers (*Haematopus ostralegus*) selecting cockles (*Cerastoderma edule*) on mudflats. Like Goss-Custard (1977a), Sutherland used dry weight as an index of the food value of different-sized cockles. Although both dry weight and handling time increased with cockle length, longer cockles were disproportionately heavier so profitability (dry weight/unit handling time) increased with cockle size. If oystercatchers behaved according to the predictions of OFT, therefore, they should have preferred the largest cockles. Comparison of the sizes taken by oystercatchers with those present in the mud showed that birds did take a disproportionate number of large cockles. Furthermore, the capture rate of large cockles was positively correlated with their density in the mud, but the capture rate of small cockles was independent of theirs. Unfortunately, Sutherland was unable to show conclusively that the probability of taking small cockles was negatively related to the capture rate of large cockles. The weak correlation coefficients for small cockles, however, suggest that factors other than their own density influenced their capture rate by birds.

Optimal Prey Selection in Lapwings and Golden Plovers

Like redshanks and oystercatchers, lapwings and golden plovers searching for earthworms on pasture face the problem of choosing between prey items which are usually invisible. Whatever the benefit criterion birds may use to select prey, the costs of selection are likely to be diverse. In this chapter, we develop a simple optimal-diet-breadth model based on energy intake and a variety of costs measured directly in the field. The model is then used to predict the number of different sizes of worm that plovers should take.

Foraging flocks were located opportunistically and the behaviour of arbitrarily-chosen lapwings and golden plovers ('focal' birds) dictated as a series of discrete activities: stepping, scanning, crouching and pecking. *Stepping* is more or less self-explanatory: birds move horizontally in a series of short, rapid-stepping actions with the head held erect and each step measuring about 3.5cm in lapwings and 5cm in golden plovers. *Scanning* (Plate 3a) was assumed to occur when birds stopped after a burst of rapid stepping and the head was held in an extreme erect position. Scanning appears to reflect vigilance for predators and is similar in context and posture to vigilant behaviour in other species (e.g. Lazarus 1972, 1978, Lazarus and Inglis 1978, Barnard 1980a, Elgar and Catterall 1981, Lendrem 1984). Predation risk and anti-predator responses are discussed in detail in Chapter 9. *Crouching* (Plate 3b) refers to a posture sometimes adopted by a bird immediately before pecking, although pecking does not always follow. The bird orientates so that the head points downwards and the tail up, with the long axis of the body held at approximately 20° to the ground. This is in marked contrast to the erect posture otherwise adopted between pecks. During crouching, birds frequently change the orientation of the long axis of the body before pecking or moving on in the normal stepping posture. *Head turning* (rotating the head so that one eye looks up and the other down) is also observed during crouching. *Pecking* describes all the actions associated with catching and handling worms once a bird first pecks at the ground. Pecking therefore includes pulling worms out of the ground and mandibulation. Details of pecking success and handling (extraction and mandibulation) times were also recorded. *Handling time* was measured directly in the field as the time elapsing between the bill contacting the ground and the cessation of swallowing movements. The size of worm caught in each case was measured as length relative to bill length (taken from the tip to the end of the gape). This proved a reliable method as both lapwings and golden plovers usually pull worms clear of the ground before mandibulating. There is therefore enough time to estimate worm length and we were unable to do so in fewer than 1% of cases. All observations of focal birds were made through a 15-50x zoom telescope.

At the same time, counts of the number of lapwings, golden plovers and gulls in the flock were made as well as recordings of date, time of day, field, ambient temperature and the maximum and minimum temperatures for the 24-hour period. These data and the behaviour categories defined above will be referred to more extensively in the next chapter, when we discuss time budgeting. In addition, we sampled the earthworm community occurring in the top 3cm of soil (see Chapter 3) at arbitrarily-chosen points along the search path followed by focal birds during observation. Sample points were determined using a distance finder (Chapter 3) and vegetational landmarks. Between four and eight 0.25 x 0.25m quadrat samples were taken per focal bird and the turf hand-sorted for worms. As before, worms from each sample were preserved in 10% Biofix solution for later analysis. In this way, the range of worm sizes taken and available along the bird's search path could be compared.

Worm Size and Energy Content

Since plovers appear to take exclusively earthworms when feeding on pasture (Chapter 3), the problem of deciding on a currency for prey selection is

(a)

(b)

Plate 3. Lapwing (a) scanning and (b) crouching and golden plover (c) scanning and (d) crouching (see text). Photographs courtesy of the copyright holder Dr Tony Holley (lapwings) and Professor Richard Vaughan and Terence Dalton, publishers (golden plovers)

(c)

(d)

simplified. Although there are some instances in which earthworm species or size classes may differ in nutrient content or other aspects (for instance, Piearce (1972) found that *Lumbricus rubellus* contained up to twice as much calcium in the crop and gizzard as *Allolobophora chlorotica* and *A. caliginosa*), these were not a problem in the community we studied (e.g. *L. rubellus* made up only 1.5% of the analysed worm samples). Possible exceptions which we shall discuss later are size-related parasite load and pollutant concentration. We have therefore assumed that birds used energy content as a measure of prey quality. While it is true that nutrient and other qualities may correlate positively with energy content and therefore confound the assumption (see also Sutherland 1982, Turner 1982, Vickery 1984), it is most parsimonious to assume an energy currency in the first instance because energy expenditure per unit time is likely to be greater than that for particular nutrients.

Worms taken by birds were classified into 22 different sizes (ranging from <0.5 to 6 bill lengths for each bird species). A number of worms of approximately equal length and wet weight were selected from different parts of the size range and prepared for bomb calorimetry (see e.g. Southwood 1978). The relationship between worm length and energy content was then calculated after first establishing the relationship between length and dry weight. An important correction had, however, to be made to the calorific values of each size class before they could be used to calculate the relative profitability of different-length worms.

The reason for the correction was that a substantial proportion (between 24% and 30%) of the worms recovered from turf samples were broken pieces. Three things suggest that breakage was a result of bird feeding activity. Firstly, great care was taken by three independent samplers to ensure that worms were not broken during hand-sorting and that fragments were therefore not an artefact of the sampling technique; secondly, several birds were observed to snap worms as they extracted them and not to retrieve the remaining fragments; and, thirdly, 80% of the broken worms were 'tail-ends' and had thus lost a number of anterior segments, which might be caused by the birds' tendency to pull worms out of their burrows by their anterior ends (e.g. Heppner 1965). Supporting this, Burton (1974) and S-A Bengtson (pers. comm.) found that a large proportion of earthworms in golden plover guts were broken anterior ends of larger worms. Some of the worms observed to be taken by birds were therefore likely to be broken pieces of larger worms. Since weight per unit length increases with worm length (longer worms are disproportionately fatter), a 16mm worm, say, which was really part of a 60mm worm would be fatter and contain more calories than a complete 16mm worm. It is reasonable to expect that the risk and position of breakage would vary between worm sizes. To estimate the risk of breakage, we extrapolated the original length of a broken worm from the fragment obtained from the turf. Each unbroken worm was allocated to one of six 16mm size classes (see Chapter 4) by dividing the weight of the fragment by its length and using the regression equation for the relationship between the length of unbroken worms and wet weight per mm to find the length which yielded the same ratio. We then calculated the amount of each worm which was missing, assumed taken by a bird, by dividing the percentage of all sampled worms broken which fell into each class by the percentage of all worms in each class that were observed to be taken by birds. This gives the relative likelihood of a worm in each size class breaking. The probability of breakage and the proportion lost to the bird in relation to worm size class are shown for lapwings in Figure 5.2a,b and for golden plovers in Figure 5.3a,b.

From estimates of the probability of worms of given observed length taken by birds really being broken pieces of larger worms and the approximate length of worm which was left behind in the soil, we calculated the energy content of each worm size as:

$$E_i = (l_i / l_r) E_r \quad (5.1)$$

where E_i is the corrected energetic value (in calories) of taken worm i, l_i the observed length of worm i, l_r the estimated original length of worm i and E_r the energetic value of the original-length worm calculated from the regression equation for the relationship between worm length and energy content. E_i is shown as the mean value for each size class in Figure 5.4a,b (lapwings) and Figure 5.5a,b (golden plovers). We have calculated E_i separately for the two species and for birds in the presence and absence of black-headed gulls because the risks of breakage differ between species and conditions. The presence or absence of gulls has an important influence on several aspects of lapwing and golden plover foraging behaviour (see also Chapters 6, 7 and 8), and throughout this chapter data are separated accordingly. While the correction detailed above spreads the risk of breakage over all worm sizes, it is probably a more accurate reflection of what birds actually obtain for two reasons. Firstly, breakage occurs in worms of all size classes and, secondly, the correction makes very little difference to energetic value until observed worm size reaches 55-60mm. This reflects the situation in the field, where it is the larger worms which are most likely to break.

Costs of Worm-size Selection

Having calculated the absolute energy content of different-sized worms, we now consider the costs involved in worm-size selection. We have already taken one cost (the risk of breakage) into account in calculating a corrected energy value. A number of other size-specific costs, however, must also be accounted for.

Handling costs. Larger worms take longer to extract from the ground and swallow. On average, 65% of handling time is taken up with extraction and 35% with mandibulation and swallowing. The relationship between handling time and worm size for the two species is shown in Figures 5.2c and 5.3c. In

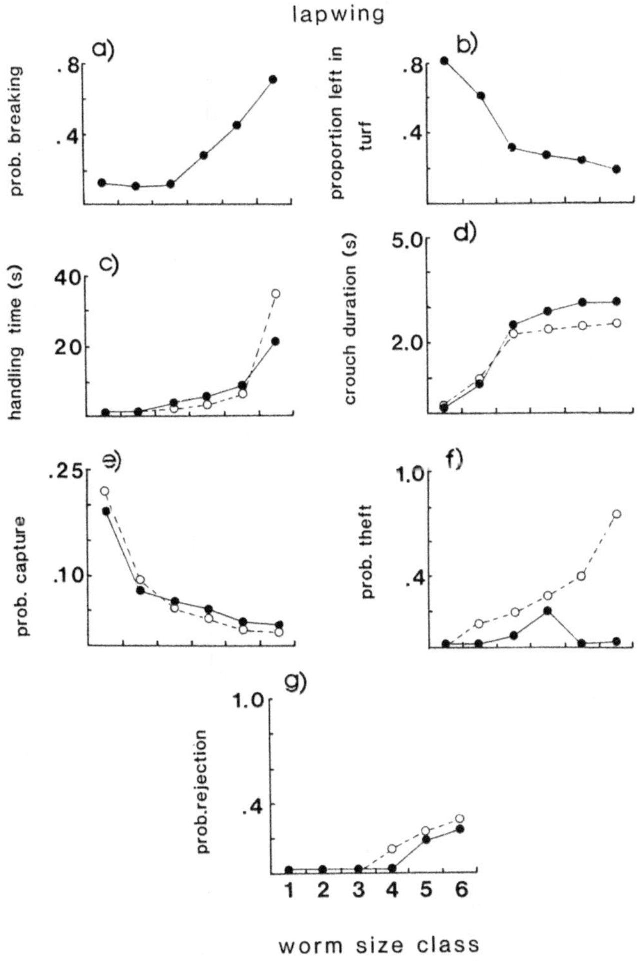

Figure 5.2: Time and probability costs associated with taking different worm sizes in lapwings: ——— flocks without gulls, - - - - flocks with gulls. (a) The probability of a worm snapping (n = 1,586 worms), (b) the proportion of a broken worm left in the ground (n = 1,586 worms), (c) the handling time (s) of the average-sized worm in each class (n = 129 worms), (d) the length of crouch (s) preceding the capture of an average-sized worm in each class (n = 119 crouches), (e) the probability of the average-sized worm in each class being captured during a peck, (f) the probability of the average-sized worm in each class being stolen by plovers (solid line) or gulls (broken line) (data for 463 worms), (g) the probability of the average-sized worm in each class being discarded (n as in (f)). See text for details of calculations

Source: Thompson and Barnard (1984)

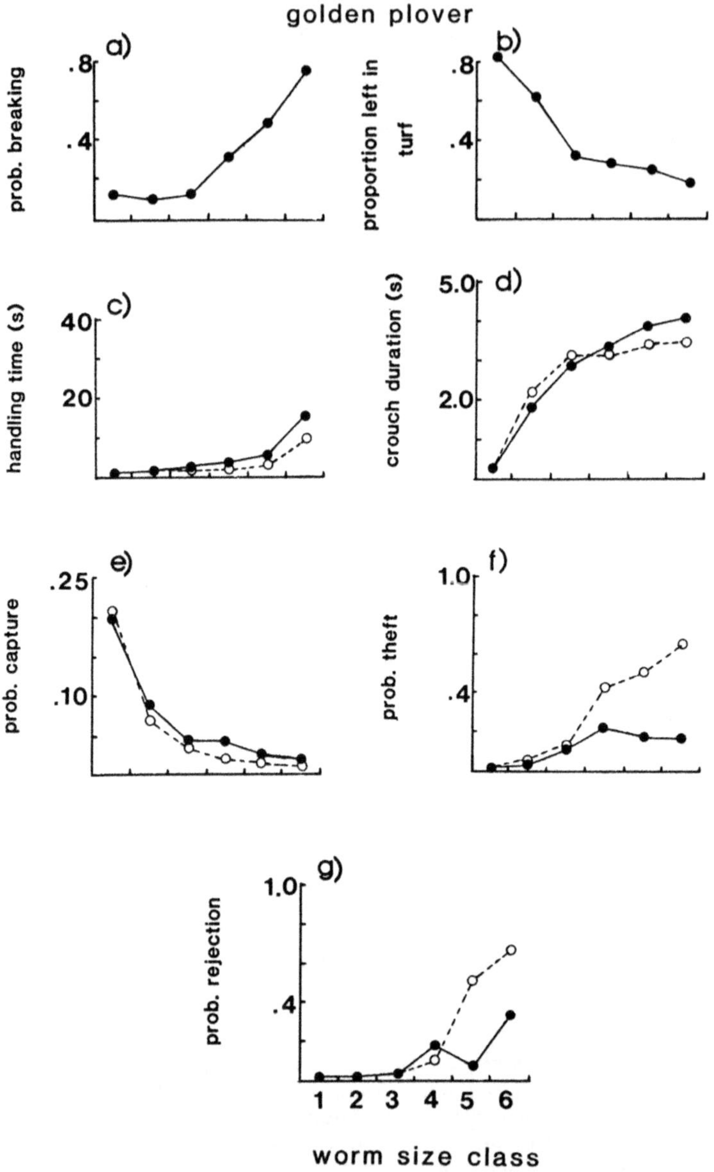

Figure 5.3: Time and probability costs of worm-size selection in golden plovers. See Figure 5.2

Source: Thompson and Barnard (1984)

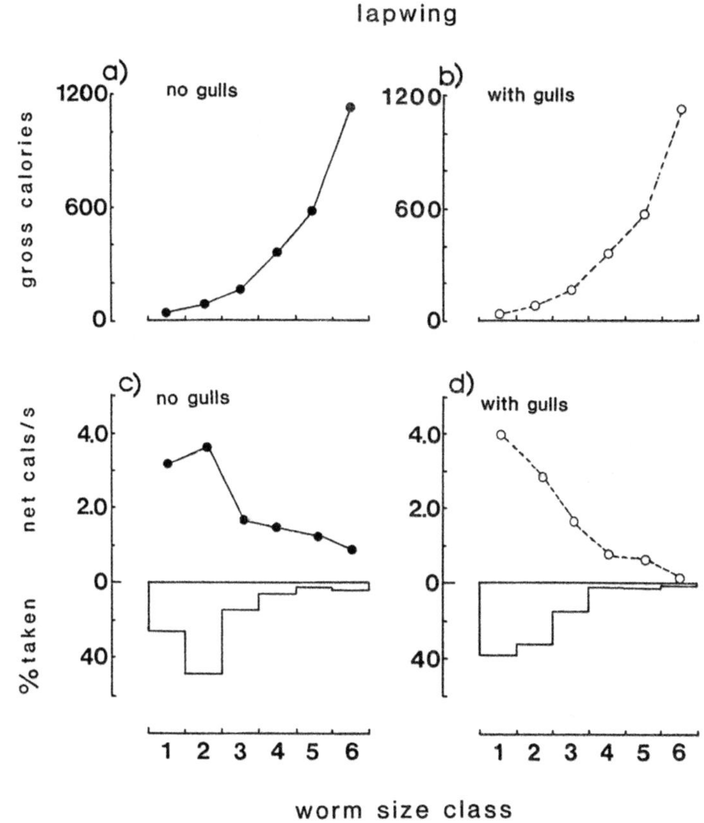

Figure 5.4: The gross energy content and profitability of different worm sizes and their representation in the diet of lapwings. (a) The gross energy content of the average-sized worm in each class without gulls; (b) as (a) but with gulls; (c) the profitability of different-sized worms (curves) and their contribution to the diet (histogram) without gulls (n = 289 worms); (d) as (c) but with gulls (n = 174 worms)

Source: Thompson and Barnard (1984)

both species there is a disproportionate increase in handling time with increasing worm length. In the presence of gulls, however, the mean handling time per unit length of worm is significantly shorter in both species when compared across the range of worm sizes taken (mean handling time (s/mm) for lapwings without gulls = 0.87 ± 0.006, with gulls = 0.74 ± 0.003, t = 2.23, $p < .05$; mean for golden plovers without gulls = 0.069 ± 0.007, with gulls = 0.054 ± 0.002, t = 2.05, $p < .05$). The sharp increase in handling time with

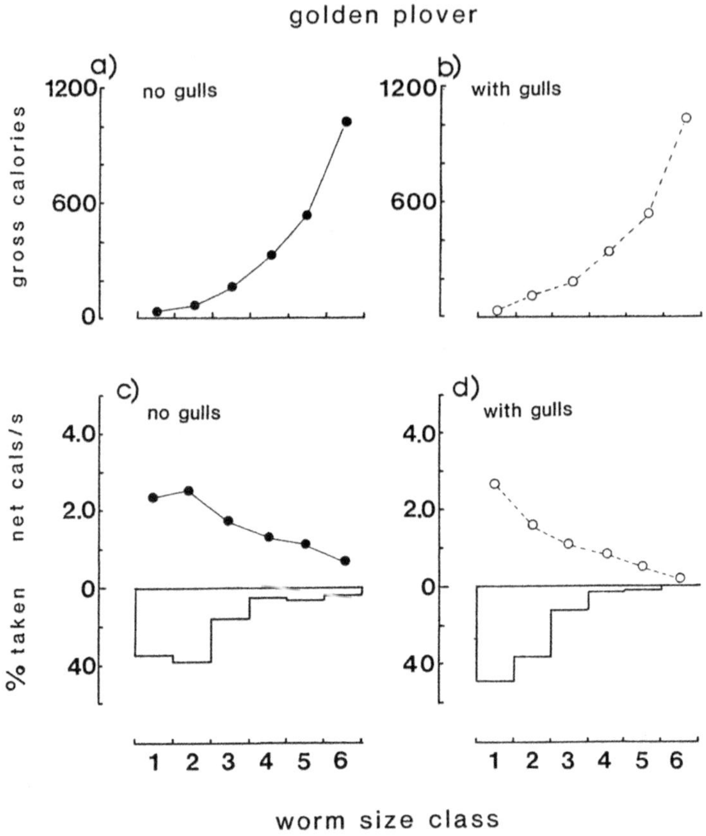

Figure 5.5: The gross energy content and profitability of different worm sizes and their representation in the diet of golden plovers. Sample sizes for (c) and (d) 283 and 482 worms respectively. See Figure 5.4

Source: Thompson and Barnard (1984)

worm size is due firstly to long worms being disproportionately fatter (see above) and secondly to long worms being deeper down (see later) and harder to pull out. The extraction times for very large worms may be as long as 70s (see Plate 4).

Assessment costs. There is a positive correlation between worm length and depth in the surface-soil layer (see Chapter 6) so that deeper pecks tend to procure larger worms. Table 5.1a compares worm sizes obtained from shallow (up to 1.5cm depth) and deep (1.5-3cm depth) pecks (where 'shallow' and 'deep' were measured against bill length: see Chapter 6). The crouching posture which is often adopted by lapwings and golden plovers prior to pecking appears to be a means of locating deep worms. This is discussed in

Plate 4: Large worms may take a long time to extract. Photograph courtesy of Professor Richard Vaughan and Terence Dalton, publishers

detail in Chapter 6. We might therefore expect a positive correlation between the length of worm taken during a peck and the duration of the crouch preceding the peck. Figures 5.2d and 5.3d and Table 5.1b show that crouch duration increases significantly with worm size and is disproportionately high for very large worms. Because they are deeper down, large worms appear to require more orientation time to ensure their capture. This needs to be taken into account in calculating profitability.

If deeper worms are more difficult to locate, we might expect a higher failure rate for deep pecks. A comparison of deep and shallow pecks (Table 5.1) shows that deep pecks are indeed much less successful. Since worm size and depth are positively correlated, there is likely to be a different risk of failure for different-sized worms. This can be calculated as:

$$P_{capi} = (p_s \cdot p_c)S + (p_d \cdot p_c)(1 - S) \qquad (5.2)$$

where P_{capi} is the probability of obtaining a target worm of size class i, p_s the

Table 5.1a: Mean (± se) worm size (mm) taken and probability of capture, from shallow and deep pecks

	Lapwing		Golden plover	
Pecks	**Worm size captured**	**prob. of capture**	**Worm size captured**	**prob. of capture**
Shallow	17.0 ± 2.42	0.56	16.8 ± 1.53	0.51
Deep	31.9 ± 2.71	0.44	26.3 ± 2.50	0.29
t-test	4.10***		2.45**	
X^2 test		3.50*		4.03*

* p < .10, ** p < .05, *** p < .001.
Data for 51 observations of lapwings and 109 of golden plovers, see text.

Table 5.1b: The mean (± se) crouch duration (s) preceding shallow and deep pecks

	Crouch duration		
	Lapwing	**Golden plover**	**t-test**
Shallow	0.65 ± .08	1.09 ± .19	2.13**
Deep	1.17 ± .13	1.81 ± .14	3.47***
t-test	4.07***	3.07***	

** p < .01, *** p < .001.
Data for 84 observations of lapwings and 71 of golden plovers. See Table 5.1a for other details.

probability of a worm of class i occurring in the top 1.5cm of soil, p_d the probability of a worm of class i occurring in the lower 1.5cm, p_c the probability that the target worm will be caught, and S the probability that the bird will peck at the upper rather than lower soil layer. Figure 5.2e and 5.3e show that, in both species, the probability of success decreases sharply with increasing worm size.

Costs of kleptoparasitism. A major potential cost of feeding in a group is the loss of prey through competition or theft (e.g. Brockmann and Barnard 1979, Goss-Custard 1980, Barnard 1984a, Ens and Goss-Custard 1984, see Chapter 1). While both lapwings and golden plovers probably lose prey to other birds through area-copying (Barnard and Sibly 1981, and Chapters 1 and 6), it was not possible to quantify the impact of this on the intake rate of copied birds. Kleptoparasitism, however, is a much more serious and easily measurable form of food loss. On rare occasions, plovers steal from each other (both within and between species, see also Wallace 1983), but most loss is to gulls. As might be expected, worm sizes differ in their risk of being stolen. Figures 5.2f and 5.3f show the probability of losing worm to gulls. Probabilities were calculated from direct observations of attacks against focal birds during the recording of time budgets. In both species, the risk of loss increased with worm size. With very large worms (size class 6), over 60% of items is lost to gulls. Also shown in Figures 5.2f and 5.3f is the combined probability of losing worms to other plovers (both species). As before, golden plovers are more likely to lose large worms, but the effect is not so pronounced in lapwings.

Theft of food items is a direct cost of kleptoparasitism. In addition, however, birds sometimes discard worms they have extracted (see also Kushlan 1978, Sutherland 1982, Hulscher 1982, Curtis *et al*1985). Discarding occurs whether or not a bird is being or is about to be attacked. Figures 5.2g and 5.3g show that, as with the risk of theft, the likelihood of a worm being discarded increases with size. Large worms are also much more likely to be discarded if gulls are present, suggesting that the risk of attack is an important factor selecting for rejection. Birds which discard worms, of course, pay the cost of extraction with no energetic returns.

Other potential costs. Besides the factors considered here, it is possible that others also affect birds' preferences for different types of worms. One important factor might be the risk of parasitic infection. British earthworms act as hosts for a large number of endoparasites, including monocystid gregarines, astome ciliates and nematodes (see Cox 1968 for a good review). Among the nematode parasites are the third-stage larvae of *Syngamus trachea*, for which birds are among the definitive hosts. In *S. trachea*, the egg is ingested by the earthworm and the third-stage larva, which develops within the shell, is liberated to encyst in the worm until it is eaten by the definitive host (Cox 1968). Adult *S. trachea* occur in the respiratory tract of birds and, in some cases, may be a serious cause of disease. Syngamiasis has been recorded in several species, including pheasants (*Phasianus colchicus*), grey partridges (*Perdix perdix*), red-legged partridges (*Alectoris rufa*), blackbirds (*Turdus merula*), starlings and fieldfares (Lee 1958, Keymer *et al.* 1962, Wynne-Owen and Pemberton 1962). Parasite species do not have equal prevalence in all species of earthworm (Cox 1968) and the risk of infection to birds is likely to vary with earthworm species and age (and therefore size). Evidence suggesting that birds can discriminate between prey on the basis of parasite load comes from oystercatchers feeding on bivalves (Swennen *et al.* 1979, Hulscher 1982). Hulscher (1982) found that oystercatchers frequently rejected bivalve (*Macoma balthica*) prey after opening them, and that rejection correlated with infection by the trematode parasite *Parvatrema affinis*.

Another factor, which is likely to have only local importance, is the tendency for earthworms to accumulate pollutants. Studies (see Ireland 1977 for a review) have shown that heavy metals (including lead, cadmium, copper and arsenic) may occur at up to four times the soil concentration in certain worm species. Pesticides may also accumulate in earthworm tissues (see Edwards and Lofty 1972, Satchell 1983). Worms can live in soil containing large amounts of some persistent organochlorine insecticides which are lipophilic and become absorbed into earthworm tissues as soil is passed through the gut. Dieldrin and DDT appear to be the most easily assimilated. Worms also assimilate radioactive soil contaminants, though to a much lesser degree than heavy metals and pesticides. *L. terrestris* and *Octolasion* species, for instance, appear to assimilate about 12% of the ^{137}Cs and ^{59}Fe they consume.

Worm Size and Profitability

Having quantified the potential costs involved in taking different-sized worms, we are now in a position to make some predictions about prey selection. Expected rate of energy intake can be calculated as:

$$E_{ri} = \frac{P_{capi}(E_i) - (p_{li} \cdot (P_{capi}(E_i)))}{h_i + c_i} \tag{5.3}$$

where E_{ri} is the expected rate of energy intake from worms of size class i, E_i is as calculated in equation (5.1), P_{capi} is as in equation (5.2), p_{li} is the probability of losing a worm of class i by theft or rejection, h_i the handling time for the average-sized worm in class i and c_i the crouching time to take the same. E_{ri} was calculated for each size class for the two species in the presence and absence of gulls and is plotted in Figures 5.4c,d and 5.5c,d. Plotted below E_{ri} values are the proportions of worms taken by birds which fell into each class.

Taking flocks without gulls (Figures 5.4c, 5.5c) first, the most profitable worms for both species are those in class 2 (17-32mm) with a sharp decrease in profitability as size increases. In golden plovers, however, the profitability of classes 1 and 2 are closely similar. We should therefore expect birds to prefer class 2 and avoid large worms, but golden plovers to show more similar preferences for classes 1 and 2. The histograms in Figures 5.4c, 5.5c show a pronounced preference for class 2 worms in lapwings and a similar preference, but less pronounced in relation to class 1, in golden plovers. When gulls are present (Figures 5.4d, 5.5d), the predictions for both species change. Now class 1 worms are the most profitable and their representation in the diet increases accordingly.

So far, we have shown that large worms with a high gross-energy content are reduced in profitability because of handling, assessment, theft and rejection costs. Small to intermediate-sized worms are therefore more profitable. As would then be expected, birds prefer the most profitable worm sizes. This, however, provides only qualitative support for optimal-prey selection. To provide quantitative support we must show that selectivity depends on birds' encounter rates with the most profitable worm sizes. OFT predicts that predators will be unselective when their encounter rate with profitable prey is low, and selective, regardless of encounter rate with unprofitable prey, when it is high (i.e. high enough for the inclusion of less profitable items to reduce net intake rate over the foraging bout). We tested this initially by examining the correlation between rate of intake of the most profitable worm sizes and encounter rates with different size classes.

Table 5.2 shows the results for both species in the presence and absence of gulls. In lapwings without gulls, capture rate of class 2 worms correlates significantly with their density (assumed to be proportional to encounter rate) in the turf samples but not with the density of any other size class. Furthermore, the capture rate of other size classes does not correlate with the density of any class, even their own. Partial regression analysis shows that, even when variation in the density of other size classes is controlled for, capture rate of class 2 worms is still best predicted by its own density. Similar results emerge for golden plovers, except that there is also a significant negative correlation between capture rate of class 2 and the density of class 4 worms. This second correlation is difficult to explain because, at least in the absence of gulls, golden plovers take very few class 4 worms.

As expected from Figures 5.4d and 5.5d, the correlations change when gulls are present. Now class 1 worms are the most profitable. In golden plovers, the capture rate of class 1 worms correlates with their density and there is no longer a significant correlation for class 2 (Table 5.2). While there is not a significant correlation for class 1 in lapwings, the strong correlation for class 2 has also disappeared. These changes are also borne out by partial regression analysis and thus provide some quantitative support for optimal-prey selection. Birds prefer the most profitable worms independently of the availability of less profitable worms and size preferences shift appropriately when there is a change in relative profitability. The fit between prediction and

Table 5.2: Correlation coefficients (r) for the relationship between capture rate and density of each worm size class (see text)

	Worm size class						
	1	2	3	4	5	6	n
LAPWINGS							
without gulls	.04	.56***	.08	−.05	−.10	.02	26
with gulls	.11	.29	.18	−.25	.01	.02	29
GOLDEN PLOVERS							
without gulls	.24	.61***	.28	.02	−.15	.01	34
with gulls	.49***	.18	.05	.11	.16	−.01	55

*** $p < .001$; r coefficients.
Most profitable size classes in each case underlined.

results, however, is not perfect. One reason may be that we have considered responses to individual size classes, whereas birds might do better by taking a mixture of sizes.

Optimal-diet breadth. To see whether birds should takc a mixture of size classes, we used an optimal-diet-breadth model similar to that in Figure 5.1. This involved a modified version of an equation proposed by Charnov (1976) which incorporated encounter rate with worms and the size-specific costs included in equation (5.3). We calculated the rate of energy intake from taking one or more size classes as:

$$E_{r1\ldots n} = \frac{\sum \lambda_{1\ldots n} \cdot E_{1\ldots n}}{1 + \sum \lambda_{1\ldots n}(h_{1\ldots n} + c_{1\ldots n})} \qquad (5.4)$$

where $E_{r1\ldots n}$ is the expected rate of energy intake from taking worms of the first to nth most profitable size (n is any number from 1 to 6), $E_{1\ldots n}$ is the corrected energetic value for worms of the first to nth most profitable sizes (given by the numerator of equation (5.3) calculated for classes 1 to n and therefore including the probabilities of capture and subsequent loss for target worms, $h_{1\ldots n}$ and $c_{1\ldots n}$ are the handling and crouching times respectively for worms of the first to nth classes, and $\lambda_{1\ldots n}$ is the encounter rate with the same. Each term in the equation except λ was calculated on the basis of the proportion of each size class available in the turf. $\lambda_{1\ldots n}$ was estimated as the reciprocal of the expected travel time between consecutive worms in the turf; that is as:

$$\lambda_{1\ldots n} = \frac{1}{100((1/d_{1\ldots n})/1.71)} \qquad (5.5a)$$

for lapwings and:

$$\lambda_{1\ldots n} = \frac{1}{100((1/d_{1\ldots n})/2.70)} \tag{5.5b}$$

for golden plovers, where $d_{1\ldots n}$ is the number of worms of the first to nth classes per m^2 and 1.71 and 2.70 are the mean rates of movement (cm/s) of lapwings and golden plovers respectively (calculated from the mean observed stepping rate and the approximate distance covered by a step in each species).

We then ranked worm size classes in order of profitability and, using equation (5.4), calculated the rate of energy intake expected from taking just the most profitable class, then the two most profitable classes, then the best three and so on until all six classes were taken into account. The results are shown in Figure 5.6a-d.

In all cases, expected profitability increases as more size classes are added, but peaks after the three most profitable classes. Birds appear to do better by taking these three classes than by taking only the most profitable class or all six classes. OFT therefore predicts that birds will take the three most profitable size classes exclusively when their encounter rate with them is high. This seems to be the case.

Figure 5.6a-d shows the tendency to take different-sized worms as a cumulative percentage of the diet. Birds include the three most profitable classes but seldom take the less profitable ones (the slope of cumulative intake drops off rapidly after class 3). Table 5.3 shows the relationship between the probabilities of taking profitable (classes 1-3) and unprofitable (classes 4-6) worms and their densities in the soil. The probabilities were calculated by dividing the proportion of each type of worm in the diet by its proportion in the turf samples. Three important points relating to the predictions of the optimal-diet model emerge. Firstly, in the absence of gulls, birds take a greater proportion of profitable worms when their density is high. Worm patchiness (see Chapter 3) increases with density ($r = .34$, $p < .05$) so that birds are likely to encounter a greater proportion of worms at high densities. Capture rate in both species increases with worm patchiness, and similar increases have been found in other predators (e.g. Krebs 1979). While there is no significant correlation in the presence of gulls, worms are more patchily distributed in fields containing gulls ($t = 2.37$, $p < .05$, Table 6.8) and it is noteworthy that the probabilities are in the higher range of those recorded without gulls (see intercept values in parentheses in Table 5.3). Secondly, similar analyses for unprofitable worms show that, except for golden plovers in the absence of gulls, there are no significant correlations between the probability of being taken and density in the soil. In most cases, therefore, the intake of unprofitable worms is independent of their assumed encounter rate. Thirdly, and most importantly, the probability of taking unprofitable worms is inversely related to that of taking profitable worms. When birds are taking a lot of profitable worms,

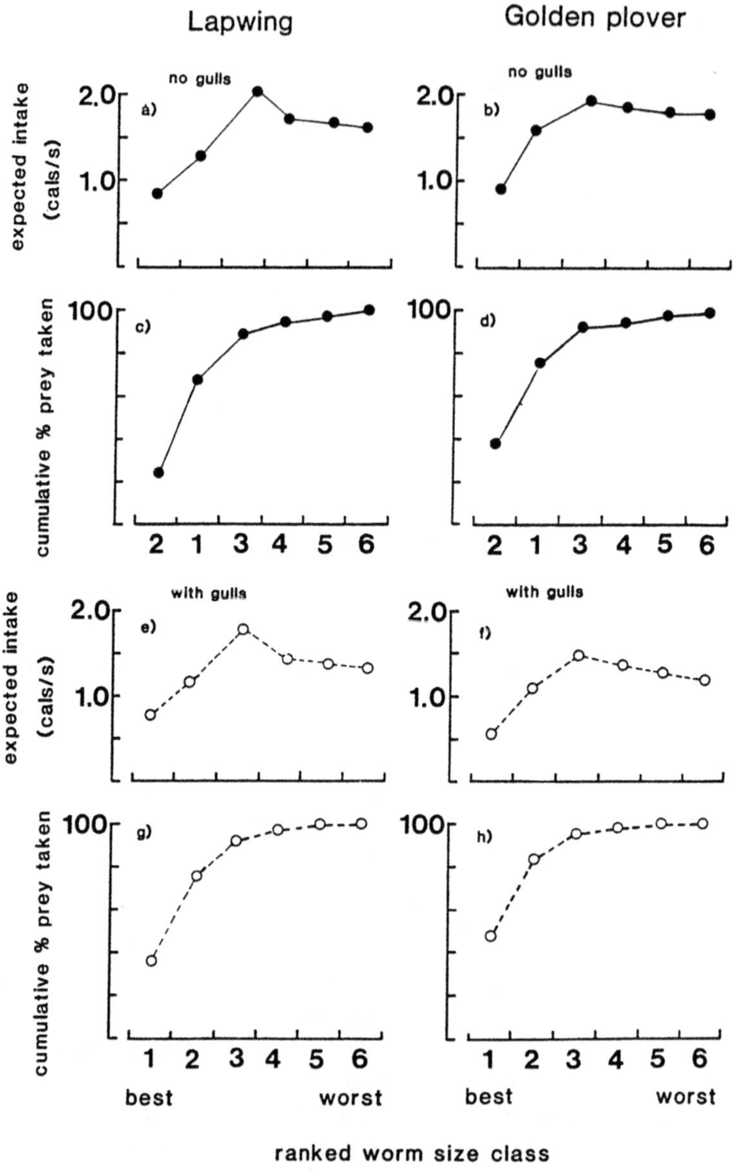

Figure 5.6: The rate of energy intake expected from taking different mixtures of worm sizes in the absence of gulls (a, b) and with gulls (e, f) and the contribution of each mixture to the diet of lapwings (c, g) and golden plovers (d, h). Worm sizes ranked in order of profitability (see text). Data for 55 observations of lapwings and 89 of golden plovers

Source: Thompson and Barnard (1984)

they are unlikely to take unprofitable ones. The results therefore agree very strongly with the predictions of OFT.

While the plovers provide good quantitative support for optimal-diet selection, it is important to show that their apparent selectivity results in greater feeding efficiency. The bias towards classes 1-3 worms may be due simply to these being the most abundant. Using equation 5.4, we compared the rate of energy intake from the mixture of worm sizes taken with that expected if birds simply take worms as encountered. Table 5.4 shows that, even though birds do not take exclusively profitable worms, they do better in every case than they would by feeding unselectively. Further, independent, support for the selection of profitable worms comes from the results of the enclosure experiments discussed in Chapter 3. Here, nets were set out to prevent birds feeding in certain parts of fields. By comparing the abundance of different worm sizes in enclosed and unenclosed areas before birds fed and then again after they had been allowed to feed in the latter areas, we could measure the changes in relative abundance attributable to the birds' feeding activity. Comparisons revealed a disproportionate decrease in the density of size classes 1-3 (Figure 5.7) in areas where birds were able to feed.

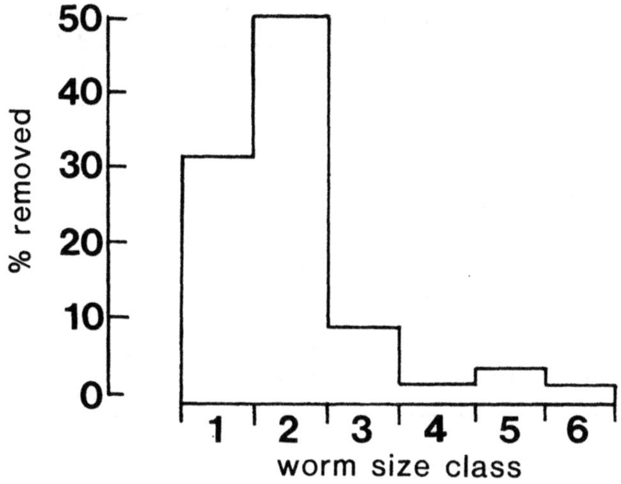

Figure 5.7: The percentage of worms of each size class removed from the turf over winter. Comparison between enclosed and exposed turf samples (see text). Data for 2240.25m.sq turf samples

Source: Thompson and Barnard (1984)

Table 5.3: Correlation coefficients (r) for relationships between the probability of plovers taking profitable or unprofitable worms and the prevalence of each in the soil or diet (see text)

		Lapwings without gulls	Lapwings with gulls	Golden plovers without gulls	Golden plovers with gulls
(a) prob. taking PROFITABLE worms x density of PROFITABLE worms	r	+.41* (−.18)	−.16 (+.12)	+.44** (−.25)	−.13 (+.16)
(b) prob. taking UNPROFITABLE worms x density of UNPROFITABLE worms	r	+.11	+.01	+.37*	+.10
(c) prob. taking UNPROFITABLE worms x prob. taking PROFITABLE worms	r	−.40 *	−.65***	−.98***	−.57***
d.f.		26	29	34	55

* $p < .05$, ** $p < .01$, *** $p < .001$.
Variables normalised where necessary using log or log (1 + variable) transformations. The bracketed values are y-axis intercepts for fitted regression lines.
Source: Thompson and Barnard (1984)

Table 5.4: Mean (± se) rate of energy intake (calories/foraging second) from the mixture of worms taken by birds and that expected if birds were unselective

	cals/s from mixture of worms taken by birds	cals/s from random sample of worms in turf	t-test	d.f.
LAPWINGS				
without gulls	1.44 ± .09	1.19 ± .07	2.19*	26
with gulls	1.21 ± .06	1.01 ± .05	2.56*	29
GOLDEN PLOVERS				
without gulls	1.03 ± .04	0.84 ± .04	2.12*	34
with gulls	0.97 ± .07	0.78 ± .05	2.71**	55

* $p < .05$, ** $p < .01$.
Source: Thompson and Barnard (1984)

Prey Selection by Plovers: Résumé

In this chapter, we have measured some of the costs and benefits accruing to plovers from taking different-sized prey items. We have assumed that birds take worms for their energy content because the rate of energy expenditure is likely to exceed that of other nutritional components. Nutrient, trace element etc content, however, is likely to correlate with energy content and may be a confounding factor.

Although large worms obviously contain more energy, several factors reduce their profitability to birds. The most important time factors are those of searching, handling and assessment. Large worms are rarer, deeper down, more difficult to locate and take longer to extract and mandibulate. The apparent difficulty of locating worms results in a reduced success rate for deep pecks and constitutes an additional cost to selecting for large worms. In addition, large worms are more likely to be stolen by gulls and, to a lesser extent, other plovers. At least in the case of gull attacks, there is probably also a risk of injury and possibly (but see Chapter 6) an additional time and energy cost (of escaping) to the victim. Apparently in connection with the increased risk of attack, large worms are sometimes discarded once extracted.

When time and other costs are taken into account, small to intermediate-sized worms turn out to be the most profitable. As predicted by OFT, both plover species bias their intake towards profitable worms. Capture rates of profitable worms correlate with their availability but are independent of the availability of other worms. While capture rates of unprofitable worms are not related to their availability, they do correlate (negatively) with the rate of intake of profitable worms. Both lapwings and golden plovers shift their intake appropriately when the presence of gulls alters the relative profitability of different worm sizes.

For the birds studied here, therefore, prey energy content (or some factor closely correlated with it) and the time and risk costs measured in the field appear to be good predictors of prey selection. The fact that the predictions of the optimal-diet model are not borne out perfectly (e.g. birds still take some unprofitable worms even when profitable worms are abundant, and the probability of taking unprofitable worms is not wholly independent of their availability) can be explained in a number of ways (see e.g. Krebs and McCleery 1984). The most likely explanation is the problem of prey recognition. Birds are taking concealed prey at different depths in the soil and appear to use depth as an approximate guide to worm size. Mistakes are therefore likely. This is supported by the fact that birds frequently reject certain worms during or after extraction. Also, of course, our categories of worm size are very broad and it is quite likely that partial preferences (e.g. Krebs and McCleery 1984) confound comparisons of selectivity between classes.

Summary

1. Prey selection by plovers appears to be predictable from prey energy content and a number of time and probability costs.

2. The profitability of different prey sizes is affected not only by searching and handling costs, but also by the difficulty of locating and catching concealed prey, time spent targeting prey and direct and indirect loss due to kleptoparasites.
3. The presence of gulls causes shifts in the relative profitability of prey sizes and plovers alter their preference accordingly.
4. Birds do better by feeding selectively than by taking prey at random.

Chapter 6

Time Budgeting and Feeding Efficiency

Our treatment of diet selection by lapwings and golden plovers has so far been taken across birds feeding in very different types of flock. It is well known, however, that the size and species composition of a flock can influence individual foraging behaviour and feeding efficiency (see Chapters 1 and 5). Effects of flock size and composition may be another reason why plovers did not follow the predictions of our simple optimal-diet model exactly. We have seen that, in lapwings and golden plovers, both characteristics of flocks may correlate with feeding efficiency because of their relationship with the spatial and temporal availability of food (for instance, flocks tend to be bigger and denser where the density of earthworms is greatest). Flocking can also influence individual feeding efficiency by allowing birds to devote more time to foraging activities. The various anti-predator effects of flocking may allow birds to spend less time scanning for predators and thus more time doing other things. The relationship between flocking and time budgeting has been especially well studied in single-species flocks (e.g. Lazarus 1972, 1978, Powell 1974, Caraco 1979a,b, Caraco *et al.* 1980a,b, Barnard 1980a, Draulans 1981). In mixed flocks, however, attention has tended to focus on the form of interspecific association and interaction (e.g. Morse 1970, 1978, Rubenstein *et al.* 1977, Greig-Smith 1978, Balph and Balph 1979, Powell 1980) rather than on time budgeting and, except for a few studies (e.g. Barnard and Stephens 1983, Waite 1983, Sullivan 1984), the effect of mixed flocking on time allocation has been ignored.

Many studies have also failed to take into account the important effects of climatic factors, especially temperature and daylength, on individual foraging behaviour and its relationship with flocking. As we saw in Chapter 3, these can influence the behaviour of predators through effects on either prey availability or food requirement. Pienkowski (1983a) showed that environmental factors had an important effect on time budgeting in grey and ringed plovers feeding intertidally. Pienkowski distinguished a similar series of behaviour categories to that used here. With our equivalent terminology (see Chapter 5) in brackets, these were: 'up' (scan), 'run' (step), 'down' (crouch) and 'peck' — also including details of any prey caught (peck). When birds were foraging for any given type of prey, e.g. *Arenicola*, inter-bird distance tended to decrease with rising temperature as prey activity increased. Several aspects of feeding behaviour also changed with temperature. When feeding low on the shore, ringed plovers took longer to handle a particular category of prey ('thin' worms) when it was warm. Similarly, grey plovers higher on the shore took longer to handle *Arenicola* as temperature rose. Increased windspeed (see also Chapter 3) also increased the handling time of 'thin' worms in grey plovers, perhaps because birds found it difficult to make directed movements in strong wind. Temperature affected a number of correlations between other aspects of foraging behaviour and both environmental factors and flock composition. For instance, giving-up times (the time spent in the 'up' position before moving to the next 'up' position) for ringed plovers on the low shore correlated negatively with the capture rate of *Arenicola*, mud temperature and cloud cover and positively with windspeed and time after high water when air temperature was below 6°C, but with none of these variables when it was above 6°C. Similarly, waiting time (time spent in the 'up' position before pecking at prey) correlated negatively with mud temperature and positively with rainfall, time after high water and distance to nearest neighbour (same or other species) below 6°C, but only (positively) with mud temperatures and time after high water above. The importance of flock composition and environmental factors varied not only with temperature, but also with species and position on the shore. In general, flock size and density had most effect on ringed plovers feeding on the high shore when it was cold.

The rate at which birds performed 'down' postures increased under conditions which usually led to greater selection for large 'thin' worms — increased temperature, reduced windspeed, increased flock density and, in grey plovers, later stages in the tide cycle and decreased rate of taking *Arenicola* (which was usually detected during 'up'). Pienkowski (1983a) suggested that 'down' was a means of locating a cue first seen from a distance, or indicated waiting for a cue to appear again. It was more characteristic of grey than ringed plovers. As we shall see, there are close parallels between 'down' in intertidal birds and crouching in lapwings and golden plovers on pasture. In this chapter, we shall look at the influence of both flock size and composition and environmental/climatic factors on time budgeting and feeding efficiency in lapwings and golden plovers. We shall also examine the effects of single- and mixed-species flocking on the birds' ability to meet their daily energy requirement.

Feeding Efficiency: Effects of Flock Composition and Environmental Conditions

We recorded sequences of behaviour in plovers, and related feeding efficiency and the time spent in different activities to flock/subflock size and density, earthworm availability, temperature and daylength. Following Barnard *et al.* (1982), we estimated feeding efficiency in two ways:

$$E_f = \frac{n_i \cdot E_i}{t_t + t_c + t_h} \tag{6.1a}$$

and

$$E_b = \frac{n_i \cdot E_i}{T} \tag{6.1b}$$

where n_i is the number of worms of size i (see below) taken by a focal bird within a recorded sequence of behaviour, E_i is the corrected (see Chapter 5) gross energy content of a worm of size i, t_t the time spent moving between captures (stepping), t_c the amount of time spent crouching, t_h the amount of time spent handling worms and T the total time over which the sequence was recorded. E_f is therefore energy intake per unit foraging time (the components of the denominator in equation 6.1a are all directly related to searching for and dealing with food), while E_b is the rate of intake taken over the total time for the sequence and therefore including time spent scanning. Worms were measured relative to bill length as detailed in Chapter 5.

Flocks were located and focal birds chosen for observation as in Chapter 5. Because birds often departed, became obscured or performed non-foraging behaviours such as preening or resting, recordings tended to last only 3-4 minutes (mean duration = 183 ± 18.6s, range = 51.2-848.5s). All the environmental variables mentioned above were recorded for each focal bird (except worm availability, which was estimated from turf samples for a smaller number of birds — see Chapter 5). Most are self-evident, but temperature was recorded as ambient, minimum overnight and maximum and mean daylight temperatures.

The density of birds was measured in three ways: firstly, as the number of birds per hectare (flock size/field area); secondly, as the number of birds within a ten-lapwing-length (about 3m) radius of the focal bird; and, thirdly, by measuring (with a distance finder) the maximum length and breadth of flocks and subflocks. In the second case, counts of surrounding birds were made every 20s through recordings, using a calibrated electronic tone as a prompt. In the third case, the measurements were used to calibrate average inter-neighbour distance. Since flocks tended to be elliptical, we calculated inter-neighbour distance as:

$$D = \sqrt{\frac{1}{\pi(m_l \cdot m_b)N}} \qquad (6.2)$$

where D is the mean inter-neighbour distance in metres, m_l the radius at the maximum length of the flock/subflock, m_b the radius at the maximum breadth and N the number of birds in the flock/subflock. Distances calculated from equation (6.2) were confirmed by direct measurement of distances between given birds and between wooden models placed at known distances from each other and an observer. Measurements were accurate up to about 150m from the observer (X^2 test). Data for birds further than 150m away were therefore discarded when analysing the effects of inter-neighbour distance.

Time Budgets and Feeding Efficiency in Lapwings

Feeding efficiency. Table 6.1a shows the results of partial regression analysis relating variation in environmental factors and E_f and E_b (equation (6.1)) in lapwings when no gulls are present. Details of the type of analysis are given in Chapter 2. The table shows that, of the variables considered, the density (number of birds per hectare), rather than the number, of birds has the most pronounced effect on feeding efficiency. The more tightly lapwings are packed, the better they do. Golden plovers and the total size of the flock appear to have no effect. In part at least, birds seem to do better in large subflocks because they take larger worms. As we have already seen, however, feeding efficiency is only partly a function of prey size. This explains why the number of golden plovers relative to lapwings reduces the size of worm taken but not feeding efficiency.

Quite apart from the effects of golden plovers and other lapwings, feeding efficiency is also influenced by environmental changes, increasing when days are warmer (E_f) and longer (E_b). The effect of daylength appears to be due to increased capture rate and size of worm taken when worm availability is high at the beginning and end of winter.

When gulls are present, the picture changes (Table 6.1b). The positive effects of subflock size and density are now reduced and feeding efficiency is depressed by increasing numbers of golden plovers. Feeding efficiency is also reduced when there are a lot of gulls (E_f) and the ratio of golden plovers to gulls is high (E_f and E_b). The negative effect of gulls is caused by lapwings taking smaller, less profitable worms (see Chapter 5) as a result of the gulls' kleptoparasitic activities. Although E_f still increases on warmer days (based on maximum daylight temperature), the effects of climatic factors are reduced when gulls are present.

Table 6.1a: Beta values from stepwise partial regression analysis of the relationship between feeding efficiency and flock composition and environmental variables for lapwings without gulls

Dependent variable	Independent variables: No. L	L/ha	GP:L ratio	Daylength	Ambient t°
E_f	ns	.57**	ns	ns	.28*
E_b	ns	.60**	ns	.24*	ns
Capture rate	ns	ns	ns	.27*	ns
Mean worm length taken	.27*	ns	−.47*	.26*	ns
% pecks successful	ns	ns	ns	ns	ns

L = lapwings, GP = golden plovers, ha = hectare, t° = temperature in degrees Celsius.
Data for 39 flocks.
* $p < .05$, ** $p < .01$, *** $p < .001$; significance level of F-ratio associated with Beta value. ns, not significant.
The following independent variables were originally included in the analysis: subflock sizes of lapwings and golden plovers, total flock size of plovers, golden plover:lapwing ratio, no. lapwings/ha, no. golden plovers/ha, no. plovers/ha, the local density (average no. birds within 2-3m of focal bird) of lapwings, golden plovers and all plovers, daylength, time of day, and temperature (maximum daylight (Max t°), minimum over preceding 24 hours (Min t°), and ambient (Amb t°)).

Table 6.1b: As Table 6.1a but for lapwings with gulls

Dependent variable	Independent variables: No. L	No. GP	GP:gull ratio	No. gulls	Max. t°
E_f	.24*	−.26*	−.36*	−.27*	.25*
E_b	ns	−.23*	−.43**	ns	ns
Capture rate	ns	ns	ns	ns	ns
Mean worm length taken	ns	ns	−.35*	−.32*	.32*
% pecks successful	ns	ns	ns	ns	ns

Data for 43 flocks.

The effects of gulls are drawn from data for focal birds in a number of different flocks. As a means of double-checking them, we can compare the feeding efficiency of lapwings within a given flock before and after the arrival of one or more gulls. To do this, we selected data from flocks in which gulls arrived during observation but in which there was no large (>3 birds) change in the number of lapwings or golden plovers. This narrowed the sample size quite considerably, because gulls seldom joined small flocks of plovers and were almost permanently present in large flocks. They also frequently arrived with additional plovers (see Chapter 7). Nevertheless, in the cases where we can make comparisons, the arrival of gulls results in a decrease in lapwing feeding efficiency. Figure 6.1a(i) shows the change in E_b (a similarly significant drop occurred in E_f). We also compared the effects of two or more

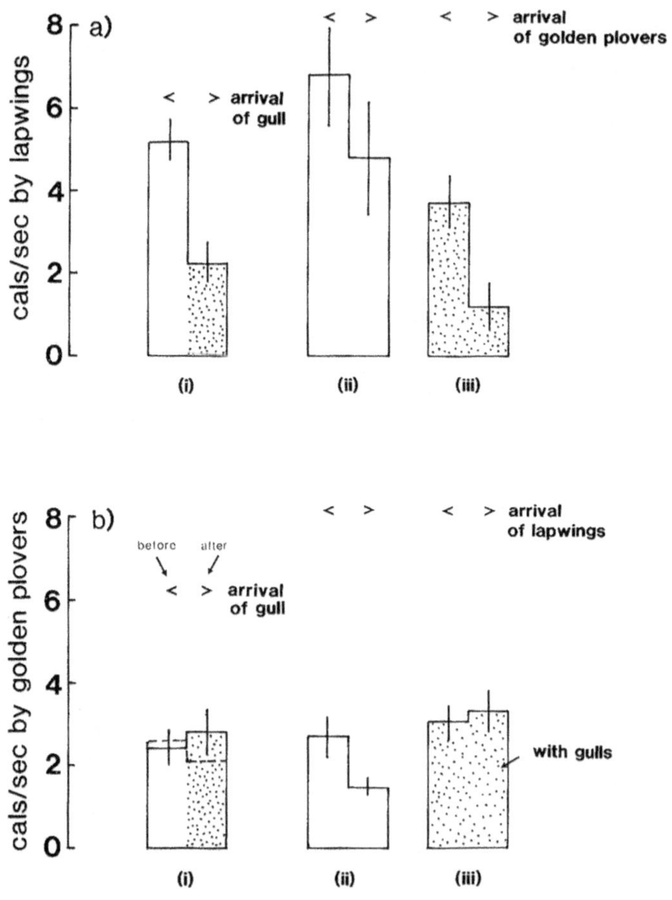

Figure 6.1: The effects of gulls, lapwings and golden plovers arriving at a feeding site on feeding efficiency (E_b) in plovers. E_b before arrival, ($<$) E_b after arrival ($>$) of additional birds. (a) Effect on E_b of lapwings: (i) arrival of gulls(s) causes a significant decline ($t = 3.3$, $p < .005$, d.f. 28), (ii) arrival of golden plovers has no effect in flocks without gulls ($t = 1.23$, n.s.), (iii) arrival of golden plovers causes a decline in flocks with gulls ($t = 2.72$, $p < .05$, d.f. 13). (b) Effect on E_b of golden plovers: (i) arrival of gull(s) has no significant effect ($t = 0.98$, n.s.), (ii) arrival of lapwings causes a decline in flocks without gulls ($t = 3.16$, $p < .01$, d.f. 17), (iii) arrival of lapwings has no effect in flocks with gulls ($t = 0.72$, n.s.). Bars indicate standard errors. Cals = calories

golden plovers arriving. Although lapwing E_b tends to decrease whether or not gulls are present, the drop is significant only when they are present (Figure 6.1a(iii) — as we might expect from Table 6.1). Golden plovers therefore seem to bother lapwings only when the latter are already under pressure from gulls.

Time budgeting. To what extent might these changes in feeding efficiency be attributable to the sort of re-allocation of time between activities found in other flocking species? Table 6.2 shows the relationship between time spent in different activities and the same independent variables used in Table 6.1. We distinguished the same activities as in Chapter 5, and stepping, crouching, pecking, scanning and head-turning are as defined there.

In the absence of gulls (Table 6.2a), lapwings spend a greater proportion of their foraging time crouching as their subflock size increases. Both the percentage time spent crouching and the proportion of pecks preceded by a crouch increase with the number of lapwings. We shall discuss the relationship between crouching and feeding efficiency again later. Interestingly, time spent pecking does not vary with flock composition. This suggests that the effects on feeding efficiency are brought about through greater investment in assessing prey prior to pecking rather than increased sampling. Indeed, Barnard and Stephens (1981) found that birds crouch more when worm density is high and they can afford to feed selectively. Scanning decreases, however, with total flock size and correlates negatively with crouching, so there may be a direct trade-off between time spent scanning and crouching which is related to flock composition. Scanning is influenced by temperature and daylength, which suggests a relationship with feeding priority; lapwings scan more often when it is mild (high maximum temperature) and for a greater proportion of their time on longer days (Table 6.2a). Under these conditions, feeding priority is likely to be low and worm availability high, so birds can spend more time looking out for predators.

When gulls are present, however, the effects of conspecific subflock size disappear (Table 6.2b). Instead, the amount of time spent crouching and pecking declines as a function of the number of gulls. This explains why the number of gulls correlates with a reduction in both the size of worm taken and feeding efficiency (Table 6.1b). Related to this is the positive correlation between crouch rate and the total number of plovers. As we shall see in Chapter 8, the efficiency with which gulls attack lapwings is reduced when they feed with golden plovers. Nevertheless, when the ratio of golden plovers to lapwings is high, there is an overall reduction in lapwing crouch rate which may account for the negative effect of golden plovers on lapwing feeding efficiency (Table 6.1b). Not surprisingly, the proportion of time spent crouching and pecking increases with temperature as worms become more abundant near the surface. Stepping rate declines with increasing (maximum) temperature, presumably because more worms are encountered within a given searching distance.

Table 6.2a: Beta values for the relationship between time budgeting and flock composition and environmental variables for lapwings without gulls

Dependent variables	Independent variables GP/ha	L	(GP+L)/ ha	Max t°	Ambient t°	Daylength
Step rate	.36*	ns	ns	ns	ns	ns
Scan rate	ns	ns	ns	ns	−.53**	ns
Head-turning rate	ns	−.52**	ns	ns	ns	ns
% time crouching	ns	.30*	ns	ns	.28*	ns
% time scanning	ns	ns	−.39*	.31*	ns	ns
% pecks + crouch	ns	.29*	ns	ns	−.36*	ns

% pecks + crouch = proportion of pecks preceded by a crouch.
Data for 39 flocks, other details as in Table 6.1a. Time budget variables also included but not yielding significant results were crouch rate, peck rate, % time pecking and % time head-turning.

Table 6.2b: As in Table 6.1a but for lapwings with gulls

Dependent variables	Independent variables Gulls	(GP+L)/ ha	GP/L	Max t°	Ambient t°
Step rate	ns	ns	ns	−.57**	ns
Crouch rate	ns	.79**	−.43*	ns	ns
Peck rate	ns	−.36*	ns	ns	ns
% time crouching	−.27*	ns	ns	ns	.42*
% time pecking	−.28*	ns	ns	ns	.30*
% pecks + crouch	−.38*	ns	ns	ns	ns

Data for 43 flocks.

Time Budgets and Feeding Efficiency in Golden Plovers

Feeding efficiency. Similar analyses for golden plovers show that, in the absence of gulls, conspecifics have little effect on feeding efficiency (Table 6.3a). Capture rate, pecking success and both measures (E_f and E_b) of feeding efficiency increase, however, with the overall density of plovers in the flock. E_b declines with the number of lapwings, apparently as the result of fewer pecks being successful. As in lapwings, feeding efficiency increases on warmer (high minimum temperature) and longer days, when capture rates are higher and larger worms taken.

Both the negative effect of lapwing subflock size and the positive effect of increasing temperature disappear when gulls are present (Table 6.3b). Plover density, however, remains a good predictor of E_f. Overall, the presence of gulls has a much less marked effect on the feeding efficiency of golden plovers compared with lapwings. We shall return to this in Chapter 8. The only direct effect of gulls is to decrease capture rate and pecking success as their density in the flock increases. Golden plovers still fare better on longer days, for the same reasons as in flocks without gulls. The reduced effect of gulls on golden plover feeding efficiency appears, at first sight, to be underlined by Figure 6.1b(i), which compares E_b for golden plovers within flocks before and after the arrival of one or more gulls. As we mentioned above, however, gulls often arrived with groups of plovers. The trend shown by the solid histogram bins therefore potentially confounds the effects of arriving gulls and additional plovers. Interestingly, when we consider cases when a gull arrives without plovers, there is a small but significant drop in E_b (dashed bins). As expected from Table 6.3, the arrival of more lapwings has a deleterious effect on

Table 6.3a: As Table 6.1a but for golden plovers without gulls

Dependent variable	Independent variables: No. L	L+GP/ ha	GP:L ratio	Daylength	Ambient t°	Min t°
E_f	ns	.16*	ns	.23***	ns	.26***
E_b	−.18**	.17**	ns	.35***	ns	.24***
Capture rate	ns	.35***	ns	.27***	ns	ns
Mean worm length taken	ns	ns	ns	.16**	ns	.17**
% pecks successful	−.24**	.16*	.19**	ns	ns	ns

Data for 298 flocks.

Table 6.3b: As Table 6.1a but for golden plovers with gulls

Dependent variable	Independent variables: L/ha	(GP+L)/ ha	Gull/ ha	GP:gull ratio	(GP+L):gull ratio	Daylength
E_f	ns	.16*	ns	ns	ns	.33***
E_b	ns	ns	ns	ns	.15*	.37***
Capture rate	.32**	ns	−.20*	ns	ns	.28**
Mean worm length taken	ns	ns	ns	.16*	ns	.22*
% pecks successful	ns	ns	−.17*	ns	ns	ns

Data for 257 flocks.

golden plover feeding efficiency only when there are no gulls (Figure 6.1b(ii), (iii)).

Time budgeting. Table 6.4a shows the relationship between the same independent variables and time budgeting as in lapwings for golden plovers in the absence of gulls. The effect of the overall density of plovers on feeding efficiency appears to be due to the increased time spent crouching and pecking when density is high. Similarly, golden plovers spend more time crouching and pecking when the minimum overnight temperature has been high and days are long. Time spent scanning decreases as both flock density and daylength increase. The negative effect of lapwing subflock size on golden plover feeding efficiency appears to be brought about by reduced rates of pecking and stepping, although lapwing density rather than number turns out to be the best predictor. Increased lapwing density, however, also correlates with an increased number of pecks preceded by a crouch. While golden plovers locate sites with high food availability by alighting with lapwings (Chapter 3) and therefore crouch frequently, lapwings still depress feeding rates compared with what might be expected in their absence.

In flocks with gulls (Table 6.4b), golden plovers spend more time scanning as the number of gulls increases (although they scan at a lower rate). While there is no direct effect of gull number or density on crouching, birds crouch more often when there are fewer gulls per plover. The proportion of pecks preceded by a crouch also increases with the number of lapwings per gull, presumably because lapwings are the gulls' preferred targets and golden plovers are less likely to be attacked if there is a large number of lapwings to choose from. Perhaps for this reason, lapwing density has a positive (cf. Table 6.4a) effect on crouching, pecking and stepping rates and plover density a positive effect on scanning rate when gulls are present. It must, however, be remembered that worm distribution may be a confounding factor here. Worms tend to be more patchily distributed in fields where gulls are present (see Chapter 5 and later). The effects of daylength on crouching, pecking and scanning disappear and there is an interesting reversal of the effects of temperature; now maximum temperature correlates negatively with rates of behaviour.

It is clear from the above that flock composition and environmental factors influence feeding efficiency in plovers and that this is related to changes in the amount of time allocated to different behaviours. One of the most important changes appears to be in the amount of time devoted to assessing prey prior to capture.

Crouching and Prey Selection

Throughout Chapter 5 and the discussion of time budgeting above, we have emphasised the relationship between feeding efficiency in plovers and the amount of time spent apparently assessing prey before pecking. Birds appear

Table 6.4a: As Table 6.2a but for golden plovers without gulls

Dependent variables	Independent variables								
	GP/ha	L/ha	(GP+L)/ha	GP+L	GP/L	Max t_o	Min t_o	Ambient t_o	Daylength
Step rate	ns	−.18***	ns	ns	ns	ns	ns	−.37***	ns
Crouch rate	ns	ns	.30***	ns	ns	ns	.22***	ns	.26***
Peck rate	ns	−.16**	.31**	ns	ns	ns	.25***	ns	ns
Scan rate	.17**	ns	ns	ns	ns	−.24***	ns	ns	ns
Head-turning rate	ns	−.14*	−.16**	ns	ns	ns	ns	ns	ns
% time crouching	ns	ns	.14*	ns	ns	ns	ns	.23***	.15**
% time pecking	ns	ns	ns	ns	ns	ns	.24***	ns	.16**
% time scanning	ns	ns	−.12*	ns	ns	ns	ns	ns	−.18**
% time head-turning	ns	ns	ns	−.11*	−.10*	ns	ns	ns	ns
% pecks + crouch	ns	.15*	ns	ns	ns	ns	ns	+.13*	ns

Data for 298 flocks.

Table 6.4b: As Table 6.2a but for golden plovers with gulls

Dependent variables	Independent variables							
	L/ha	GP	(GP+L)/ ha	GP/L	Gull	L/gull	(GP+L)/ gull	Max t_o
Step rate	.27**	−.21**	ns	ns	ns	ns	ns	−.20**
Crouch rate	.13*	ns	ns	ns	ns	ns	.22**	−.14*
Peck rate	.29*	ns	ns	.17*	ns	ns	ns	−.12*
Scan rate	ns	ns	.39***	ns	−.26**	ns	ns	−.32***
Head-turning rate	−.20*	ns	ns	ns	ns	ns	ns	ns
% time crouching	ns	ns	ns	ns	ns	ns	ns	.11*
% time pecking	ns	ns	.18*	ns	ns	ns	ns	ns
% time scanning	−.52***	ns	ns	ns	.34*	.18*	ns	ns
% time head-turning	ns	−.12*	ns	ns	ns	ns	ns	ns
% pecks + crouch	−.17*	ns	ns	ns	ns	.27**	ns	ns

Data for 257 flocks.

to assess prey during a characteristic posture which we have referred to as crouching. In Chapter 5, we suggested that birds do not assess the size of worms directly but use depth as an index of size. Crouching is therefore assumed to reflect the difficulty of locating deep worms prior to pecking. So far, we have taken only a superficial look at the relationship between crouching and feeding efficiency. We shall now examine it in more detail.

The depth and profitability of prey. If plovers use depth as a guide to prey quality, deeper worms should be more profitable. We have already shown (Table 5.1) that worm size increases with depth; what we want to know now is whether worms at the limited depth (3cm) to which plovers can peck are more profitable than those at the surface. To test this, we compared the rate of energy intake birds could have expected by taking 'surface' (<1.5cm deep) and 'deep' (1.5-3cm deep) worms. Rates of intake from each type of worm were calculated as:

$$E_{r1 \ldots n} = \frac{(\sum (F_i P_i)) p_c}{t + (c_i p_i) + (h_i P_i)} \tag{6.3}$$

where $E_{r1 \ldots n}$ is the rate of energy intake expected from taking worms of size classes 1 to n (see equation (5.5)), E_i is as in equation (6.1), P_i is the proportion of intake consisting of worm size class i, c_i and h_i as in equation (5.5), and t the estimated travel time between worms presumed encountered at the two depths. t was calculated as $1/\lambda_{1 \ldots n}$ using equation (5.7a) and (5.7b) respectively for lapwings and golden plovers. Pecks made at the surface layer were more likely to be successful than those made deeper down (see Table 5.1a). p_c is therefore as in equation (5.4). Figure 6.2 compares the rate of energy intake expected from the vertical distribution of worm sizes in the soil (calculated from the core samples described in Chapter 3) with that calculated from the worms taken by birds. Both species do better by taking deep worms. Figure 6.2 shows, however, that neither species does as well as it could from the deep layer, given the size range and density of worms available. The reason for this is not clear, but it is most probably due to a smaller proportion of deep worms being detected and therefore available. There is little difference in the extent to which the two species exploit the surface and deep layers and the size range of worms available. Following Hurlbert (1978), an index of foraging-niche overlap (α_{lg} was calculated as:

$$\alpha_{lg} = 1 - 0.5(P_{xl} - P_{xg}) \tag{6.4}$$

where p_x is the probability of lapwings (l) and golden plovers (g) exploiting x (here a given layer or size of worm). As overlap increases, α_{lg} tends to 1.0, at which point overlap is total. Comparing the tendencies to exploit different layers of soil and size range of worms, α_{lg} = 0.9 and 0.84 respectively with lapwings tending to take slightly larger worms from the deep layer.

Effects of crouching on the size of prey taken. Crouching results in larger worms being taken by both species (mean length captured by lapwings with crouch = 34.9 ± 2.25mm, without crouch = 27.2 ± 1.49mm; by golden plovers with crouch = 23.9 ± .94mm, without crouch = 15.5 ± 1.26mm). The two species differ, however, in the proportion of pecks preceded by a crouch and the average duration of crouching. Table 6.5 compares crouching behaviour in flocks with and without gulls. Golden plovers crouch significantly more often prior to pecking than lapwings, but both species crouch less often when gulls are present. Golden plovers also crouch for longer than lapwings when there are no gulls, but the difference is reduced in the presence of gulls. The reduction in the tendency to crouch results in smaller worms being taken when gulls are present, although the difference is significant only in golden plovers. Interestingly, these changes result in a reduction in the correlation between crouch duration and worm size taken in golden plovers but not in lapwings (Figure 6.3).

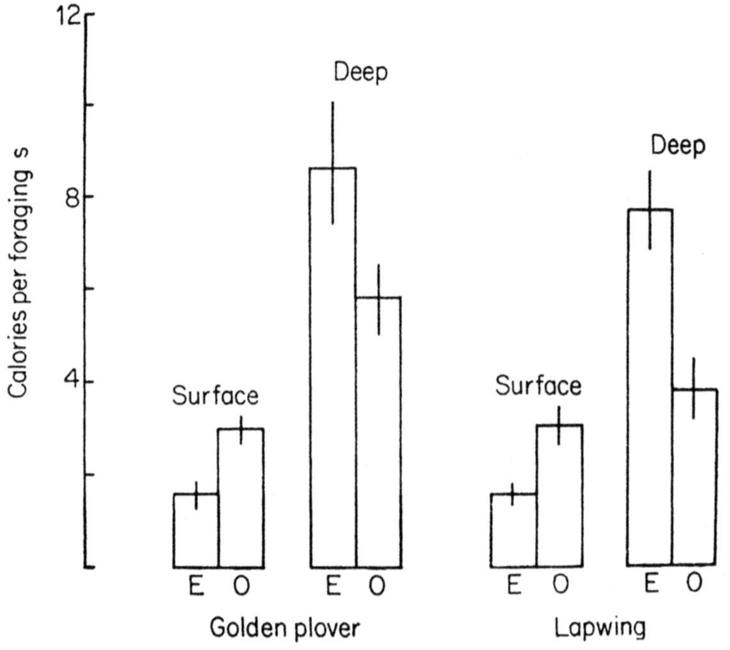

Figure 6.2: The rate of energy intake obtained by plovers from surface and deep layers of turf; E, intake expected from random sample of worms from the two layers, O, intake from mixture observed to be taken. Bars represent standard errors

Source: Thompson (1983)

We showed in Figures 5.2 and 5.3 that crouch duration increases with the size of worm taken. This is due to the positive correlations between worm size and depth and worm depth and crouch duration (Table 5.1). We do not assume that the relationship between crouching and worm size arises through the direct assessment of the latter by plovers. It may do, but a more likely explanation is that birds simply go for deep worms when they can be detected. 'Go for deep worms' is a useful rule of thumb (see Barnard 1984b) which usually procures more profitable prey. By this argument, the reduction in worm size taken when gulls are present is due to crouching increasing the vulnerability of plovers to attack and the time devoted to it consequently being reduced. The birds' ability to locate deep worms decreases and deep worms become less profitable.

The apparent use of depth as an index of prey size raises the question of how prey are detected. The answer is not clear, but there is a number of possibilities. Some authors (e.g. Perry 1945, Fallet 1962, Lange 1968, Vaughan 1980) have suggested auditory detection. Perry (1945) based his assumption on the stances adopted by birds. Fallet (1962), on the other hand, presented evidence that golden plovers could locate prey by sound; she suggested that this is accomplished in the pauses which usually precede a peck. Pienkowski (1983a), however, considers it unlikely that pauses reflect listening, partly because they occur very infrequently at night when acoustic detection is likely to be particularly important. Plovers also turn to take prey behind them. Lange (1968) and others consider these to be outside the birds' visual field and therefore detected auditorily. Again this is open to doubt

Table 6.5: Mean (± se) time spent in different activities and worm sizes taken by plovers with and without gulls

	Without gulls		With gulls		
	Lapwing	Golden plover	Lapwing	Golden plover	d.f.
Crouch duration (s)	1.01±0.11	1.61±0.09$$	0.91±0.08	1.30±0.12*	118, 153
Peck duration (s)	5.87±1.10	1.25±0.16$$$	2.48±0.22**	1.14±0.17	127, 278
Worm length taken (mm)	30.3±2.55	29.5±2.16	25.5±1.83	20.4±2.64**	102, 119
Handling time (secs/mm)	.087±.006	.069±.007$	.074±.003*	.054±.002*	127, 278
% pecks + crouch	46.0±5.3	89.5±1.4 $$$	27.8±12.2**	84.8±1.7*	114, 360

$ p < .05, $$ p < .01, $$$ p < .001; t-test comparing golden plovers and lapwings.
* p < .01, ** p < .01, *** p < .001; t-test comparing flocks with and without gulls.
d.f., degrees of freedom for each species respectively.

Source: Thompson (1983)

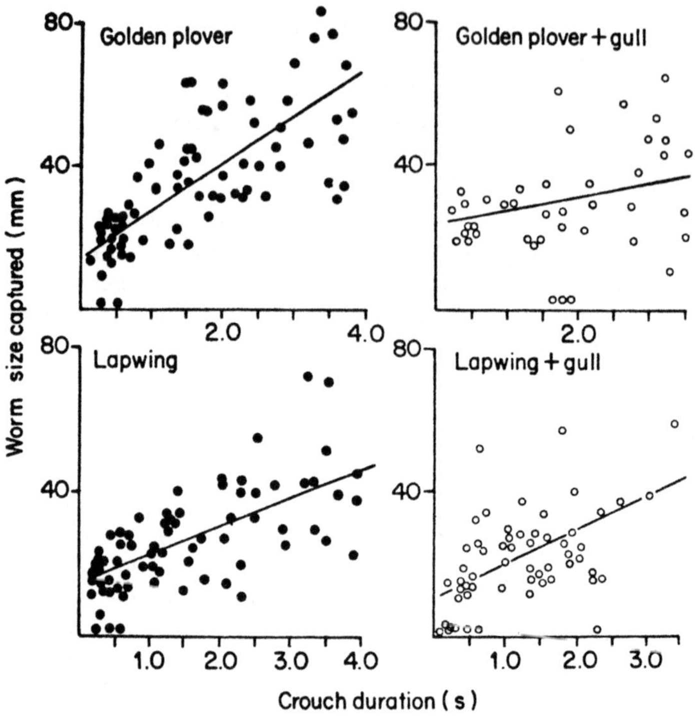

Figure 6.3: The relationship between crouch duration and worm size captured in lapwings and golden plovers. Closed circles, birds without gulls ($r = .48$, $p < .001$ for golden plovers, $r = .38$, $p < .001$ for lapwings), open circles, birds with gulls ($r = .06$, n.s. for golden plovers, $r = .13$, $p < .005$ for lapwings). Lines fitted by regression

Source: Thompson (1983)

(Pienkowski 1983a). Plovers have large eyes which give almost all-round vision because they are placed at the widest part of the head. Nevertheless, Lange (1968) provided experimental evidence that several species of plover could detect 'noisy' prey such as earthworms and mealworms by sound, though this does not mean they do so in the field (Pienkowski 1983a).

Heppner (1975) studied American robins (*Turdus migratorius*), which forage in a manner very similar to plovers, and concluded that earthworms were located visually and not auditorily or olfactorily. The capture rate of Heppner's robins was not affected by playing high-intensity white noise and the general background noise was more than sufficient to mask the low-intensity sound generated by burrowing earthworms. Perhaps the most telling circumstantial evidence from plovers is the reduced capture rate during night feeding (Pienkowski 1983a). Earthworms, particularly the larger species like *Lumbricus terrestris*, are more active on the surface at night. It is possible, however, that worms moving on the surface create less noise than those burrowing through the soil.

Crouch duration and feeding efficiency. Although plovers can improve their pecking success by crouching, it need not necessarily be the case that longer crouches always mean greater feeding efficiency. Crouching has costs as well as benefits. It is a time and energy cost which has to be offset against food intake and it is one of the cues used by gulls to select targets (see Chapter 8). The longer birds crouch, the less time they have for feeding and the more vulnerable they become to attack. It seems likely, therefore, that the probability of different crouch durations will change with flock composition. To test this, we divided recorded crouches into 0.5s classes and calculated expected profitability as:

$$E_c = \frac{E_i - E_l}{c_c + h_i} \tag{6.5}$$

where E_c is the expected rate of energy intake after a crouch of duration c, E_i is the gross energy content of the worm size i taken after the crouch, E_l the energy lost through theft by gulls, c_c the duration of the crouch, and h_i the time spent handling the worm. Figure 6.4 shows that, in both lapwings and golden plovers, there is a significant positive correlation ($p < .05$ for lapwings and golden plovers) between crouch duration and profitability when there are no gulls, but a significant negative correlation ($p < .01$ and $.05$ respectively) when gulls are present. The reversal occurs because gulls steal mainly the larger worms (with longer associated crouches) taken by plovers and their profitability is therefore reduced.

Because small to intermediate-sized worms are more abundant than large worms, we should not necessarily expect plovers to show only long crouches in the absence of gulls. We should expect them to crouch for the shortest times when gulls are present, but for only slightly longer when they are absent. The solid histograms in Figure 6.4 bear this out for both species in flocks with gulls, but only (weakly) for golden plovers in flocks without. Crouch duration is still short in lapwings (see Table 6.5 for means). The difference between the two species may be due to the greater speed and agility of golden plovers in the air which renders them less vulnerable to aerial predators and allows them to spend longer in non-vigilant postures (see also Chapter 8). Weak evidence for a vulnerability effect comes from lapwings feeding in small fields. In fields smaller than 3ha, lapwings are constrained to feed nearer to perimeter hedges than they would in larger fields. This is likely to make them more vulnerable to such predators as hawks and falcons which rely for success on a surprise approach (see Chapter 9). Birds foraging in these small fields tend to make shorter crouches (mean crouch duration in small fields = 0.74 ± 0.11s, in larger fields = 1.01 ± 0.11s), though the difference is not significant ($p < .08$).

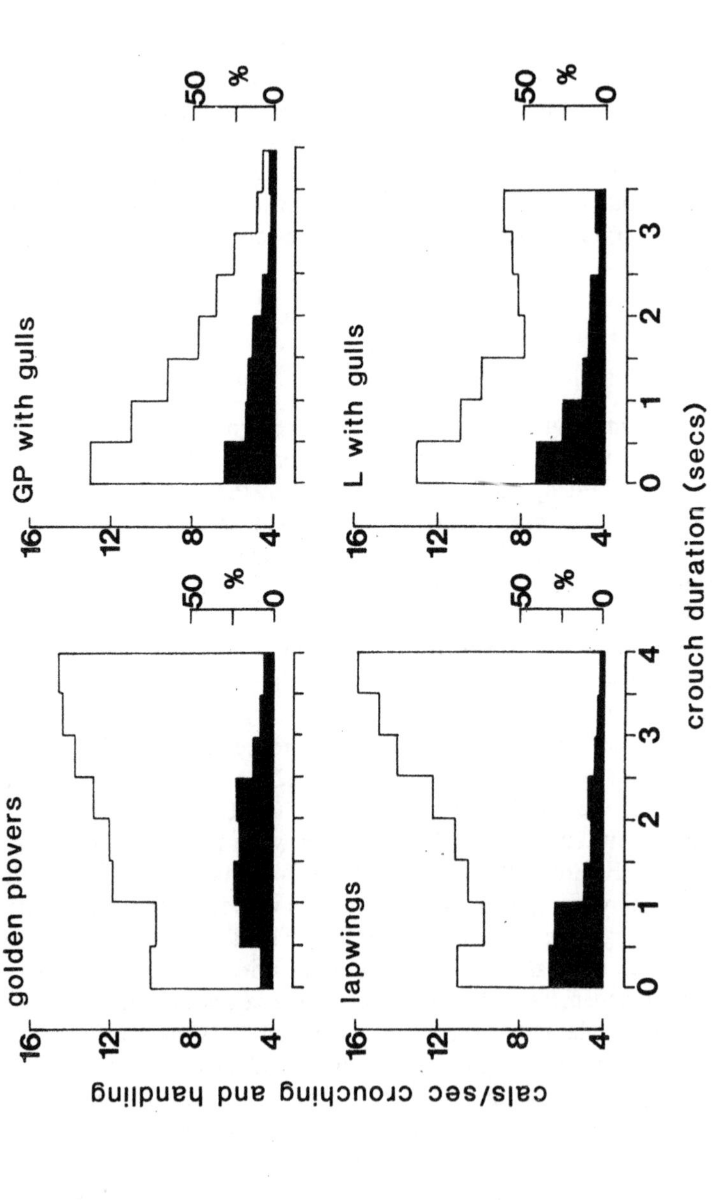

Figure 6.4: The relationship between crouch duration and prey profitability (open histograms) and the frequency distribution of crouch duration (shaded histograms) with and without gulls. Data for 153 and 118 crouches by golden plovers and lapwings respectively

Nevertheless, worms taken by lapwings in small fields are smaller than those taken in larger fields ($t = 3.52$, $p<.001$, d.f. 172). This is not due to any differences in worm density or size availability.

Food Availability and Feeding Efficiency

Our analysis so far has separated the effects of flock composition and environmental factors such as temperature and daylength on feeding efficiency in plovers. However, there is still one important environmental factor we have not taken into account. A small number of studies (e.g. Barnard 1980a, Waite 1981, Curtis and Thompson 1985) have shown that apparently direct effects of flock size on food intake may in fact be indirect results of correlations between flock size and food availability. We did not cater for food availability in the earlier analyses because, for reasons of time, access to fields and disturbance to birds, prey samples could be collected only for a subset of recorded sequences. Nevertheless, we know that variation in earthworm density affects flock formation and distribution in gulls and plovers and that it correlates positively with plover flock and subflock size. We now see how it affects feeding efficiency.

Effects of worm density. We used partial regression analysis to compare the effects of variables used in the earlier analyses and the density and distribution of worms in the turf. We also included the proportion of time spent crouching because of its importance in pecking success and because it correlates positively with worm density ($r = .47, p<.01$ for lapwings; $r = .62$, $p<.001$ for golden plovers).

Table 6.6 shows the result for lapwings. Although lapwing subflock size correlates positively with feeding efficiency (E_f), the relationship with worm density turns out to be stronger. Capture rate, E_f and E_b increase with worm density. As expected, the proportion of time spent crouching correlates with feeding efficiency independently of worm density because of its effect on prey size. Lapwings also take larger worms independently of density when the lapwing: gull ratio is high and individual risk of attack low. Once worm density is taken into account, the effects of temperature and daylength on feeding efficiency (see Tables 6.1, 6.2) disappear. In part, therefore, these seem to be a consequence of environmental effects on worm availability.

Table 6.7 shows a similar analysis for golden plovers. Again, worm density correlates with feeding efficiency more strongly than any measure of flock composition, although efficiency increases with the number of golden plovers per gull. Birds also take fewer and smaller worms as the number of gulls increases. In contrast to lapwings, daylength still has a positive effect on capture rate and feeding efficiency and there is only a weak effect of the proportion of time spent crouching.

We also measured the availability of worms in the parts of fields where plovers were feeding (see Chapter 3) and compared it in flocks with and without gulls. Table 6.8 shows that birds tend to feed on higher worm densities

Table 6.6: Beta values for the relationship between feeding efficiency and flock composition, environmental variables, worm density and % time spent crouching, for lapwings in mixed flocks

Dependent variables	Worm density	No. of lapwings	Lapwing: gull ratio	% time crouching
	Independent variables			
E_f	.29**	.26*	ns	ns
E_b	.33**	ns	ns	.27*
Mean worm length taken	ns	ns	.36*	.31*
Capture rate	.41**	ns	ns	ns

Data for 41 flocks. For other details see Table 5.1a.

Table 6.7: As Table 6.6 but for golden plovers in mixed flocks

Dependent variables	Worm density	GP + lapwings	GP:gull ratio	No. of gulls	Daylength	% time crouching
	Independent variables					
E_f	.19*	ns	.11*	ns	.22**	ns
E_b	.18*	ns	.10*	ns	.25**	ns
Mean worm length taken	ns	.39**	ns	−.21*	ns	ns
Capture rate	.22**	ns	ns	−.14*	.34***	.11*

Data for 168 flocks.

when gulls are absent, with lapwings on significantly higher densities than golden plovers. Reduced feeding efficiency when gulls are present may therefore be due partly to birds feeding on lower worm densities. While worm density tends to be lower in the presence of gulls, patchiness is greater (Table 6.8) and may help to offset the effects of reduced availability.

As with the studies of Barnard (1980a) and Waite (1981), these analyses underline the importance of caution in implying direct effects of flock size and composition on individual feeding efficiency. Intercorrelations between flock characteristics and environmental factors tend to confound apparent effects of flocking.

Feeding efficiency, pasture age and functional responses. Worm density has a significant effect on feeding efficiency. It is also the major factor distinguishing old from young pastures (see Chapter 3) in terms of the quality of feeding site. Do plovers therefore feed more efficiently on old pasture?

Table 6.8: The density (mean ± se no. worms/m²) and index of dispersion (variance:mean ratio) of worms in areas occupied by lapwings and golden plovers in flocks without and with gulls

	Worm density (and variance:mean ratio) in areas occupied by		
	Lapwings	**Golden plovers**	**t-test**
Flocks without gulls	212 ± 13.3 (2.49)$$ (n = 47)	174 ± 11.7 (1.71)$ (n = 35)	2.15*
Flocks with gulls	148 ± 16.5 (6.55)$$$ (n = 61)	139 ± 12.6 (3.33)$$$ (n = 51)	ns
t-test	3.01**	ns	

* $p < .05$, ** $p < .01$; t-test comparing mean worm densities.
$ $p < .05$, $$ $p < .01$, $$$ $p < .001$; X^2 testing for non-randomness (in this case increased patchiness) in the variance:mean ratio (see Southwood 1978).
Comparisons between flocks with and without gulls were made in the same fields.

We compared E_b for lapwings and golden plovers in old and young pastures. As expected, both species achieve a greater E_b in old pasture, although the difference is significant only in golden plovers (mean E_b for lapwings on old pasture = 3.21 ± 0.18 cals/s, on young = 2.53 ± 0.29 cals/s, t = 1.89, n.s.; for golden plovers on old pasture = 2.49 ± 0.19 cals/s, on young = 1.65 ± 0.22 cals/s, t = 2.88, $p < .01$). Interestingly, the relationship between worm density and E_b in old and young pastures is very different in the two species. In lapwings, the relationship is positive in both types of pasture. In golden plovers, it is positive in young pasture but negative in old (Figure 6.5a). Thus, although golden plovers do better on old pasture, they do not respond to changes in worm density in the same way as lapwings. One explanation for this difference might be that, on old pasture, lapwings tend to exclude golden plovers from areas with higher worm densities. Two pieces of evidence support this. Firstly, whereas the ratio of lapwings to golden plovers increases with worm density on both types of pasture, it increases more sharply on old pasture (Figure 6.5b), resulting in a heavily disproportionate number of lapwings on the highest densities. Secondly, the rate at which lapwings attack golden plovers is significantly higher on old pasture (Figure 6.5c) and, as we shall see later, attacks tend to result in victims leaving the vicinity of the attacker.

An alternative explanation is that golden plovers do feed on high worm densities but, for some reason, their capture rate does not increase. To test this, we examined the relationship between worm density and capture rate (the so-called functional response: Solomon 1949, Holling 1959) in the two plovers. Three different types of functional response are usually recognised (see Begon and Mortimer 1981 for a good up-to-date treatment) depending on the nature of prey availability. When only one sort of prey is sought, the

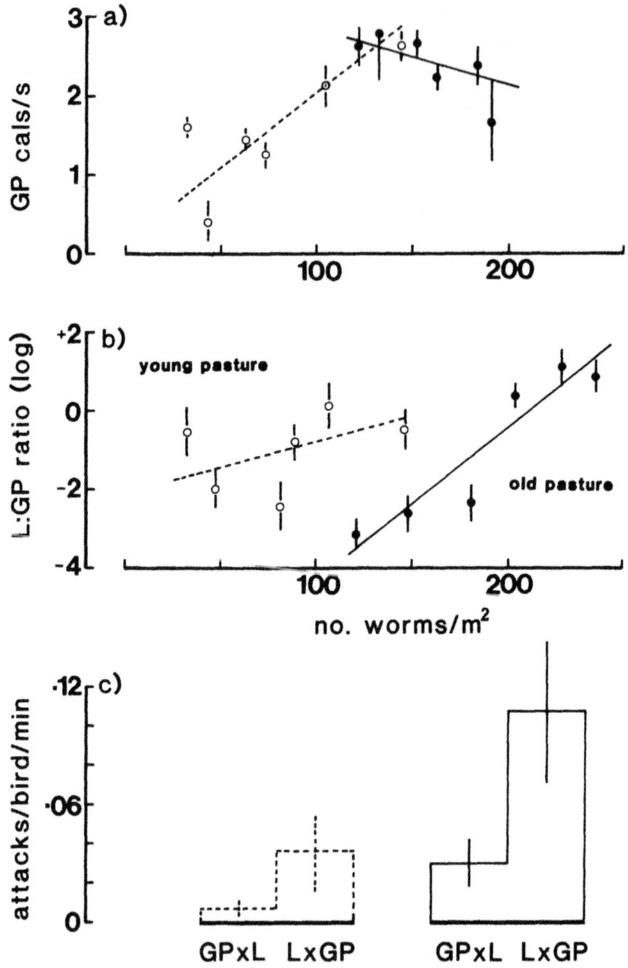

Figure 6.5: Plover feeding efficiency and species ratios on young and old pasture. (a) The relationship between E_b *and worm density for golden plovers,* $r = .30, p < .02$ *on young pasture,* $r = -.17$*, n.s. on old (for lapwings,* $r = .34, p < .05$ *on young pasture,* $r = .41,$ $p < .02$ *on old). Data for 39 lapwings and 130 golden plovers. (b) The relationship between lapwing:golden plover ratio and worm density (*$r = 13$*, n.s., and* $r = .37, p < .01$ *on young and old pasture respectively). Data for 66 and 130 observations on young and old pasture respectively. (c) Rates of interspecific aggression between plovers. There are more attacks overall on old pasture (*$t = 2.29, p < .05$*) and disproportionately more initiated by lapwings (*$t = 2.57, p < .02$*). Sample sizes from left to right, 26, 47, 29, 78 attacks*

response is most commonly a decelerating rectangular hyperbola known as a 'Type 2' response. Type 2 responses have been found in a number of bird species (e.g. Murton 1968, 1971b, Goss-Custard 1977b, 1980, Barnard 1980a, Sutherland 1982). Figure 6.6 shows that the relationship between worm density and capture rate resembles a Type 2 response in both species on young pasture (Figure 6.6a,b), but only in lapwings on old pasture (Figure 6.6a,c). While there is no positive relationship for golden plovers on old pasture, their capture rate tends to be as high as the maximum on young pasture (Figure 6.6b,c).

Aggression in Mixed Flocks

In addition to the activities discussed in the previous section, plovers are also occasionally aggressive towards one another. Aggression occurs both within and between species and takes one of two forms (see also Cramp and Simmons 1983): chases, which usually result in the displacement of one bird by another and appear to be the result of the aggressor defending a feeding area within the flock (see later), or kleptoparasitism, the stealing of prey from a successful forager. In some wader species, for example oystercatchers, kleptoparasitism is a major form of aggressive interaction (e.g. Vines 1980, Goss-Custard 1980, Goss-Custard *et al.* 1982a,b, Sutherland and Koene 1982). In lapwings and golden plovers feeding in the study area, however, it is relatively rare. In lapwings, only 28% of recorded interspecific and 12% of intraspecific attacks were apparently kleptoparasitic. The corresponding figures for golden plovers are 11% and 9% (see Figure 6.7 for sample sizes).

Because aggression between plovers is rare, we did not record it in the same way as the other activities. Instead, we observed flocks for 5-15 minutes and recorded the number of each type of attack and the species of the attacker and victim. Background data for flock composition and environmental conditions were recorded for each count as for previous activities.

Effects of Flock Composition and Environmental Conditions

In plovers, flock composition and environmental factors influence individual feeding efficiency, and aggression appears to reflect attempts to secure food. It is therefore reasonable to expect aggression to change with those characteristics of flocks and the physical environment which affect feeding efficiency. Table 6.9 suggests that this is so. Following cold nights or during days when ambient temperature is low, rates of interspecific aggression (both chases and kleptoparasitism) tend to be high. Aggression between lapwings is also higher after a low minimum overnight temperature. As in some other species (Patterson 1975, Barnard 1980c, but see Caraco 1979a), therefore, aggression is commoner when it is cold and feeding priority likely to be high.

Rates of aggression are also influenced by flock composition, but only

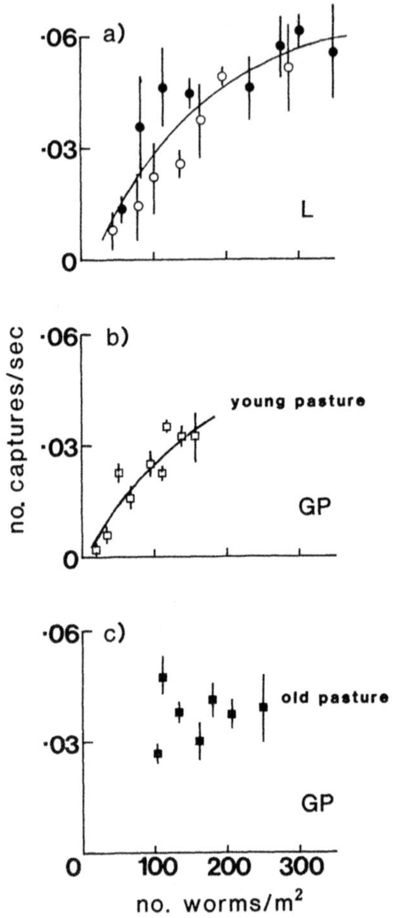

Figure 6.6: The relationship between mean worm-capture rate and worm density. (a) Lapwings in old (closed circles) and young (open circles) pasture, n = 39 and 31 observations respectively. (b) Golden plovers in young pasture, n = 66 observations. (c) Golden plovers in old pasture, n = 130 observations. All lines fitted by regression. Bars represent standard errors

between species. Lapwings are more likely to attack golden plovers when the density of plovers is high and when there are lot of gulls in the flock. Golden plovers are more likely to attack lapwings when their own subflock density is high and there are lots of gulls relative to lapwings. From what we know of the effects of flock composition on feeding efficiency, it appears that plovers are more aggressive when their feeding efficiency is likely to be low. Thus, lapwings tend to attack golden plovers when the latter are abundant and when lapwings are under pressure from gulls. Similarly, golden plovers attack lapwings more often when intraspecific competition is likely to be high (see later)

Table 6.9: Beta values for the relationships between flock composition, environmental variables and rates of intraspecific and interspecific aggression (attacks/plovers/s) in mixed flocks

	Independent variables					
Dependent variables	**GP/ha**	**(L+GP)/ ha**	**No. gulls**	**Gull:L ratio**	**Min. °C**	**Ambient °C**
L attacks GP	ns	.19*	.17*	ns	ns	−.32**
GP attacks L	.26**	ns	ns	.20*	−.16*	ns
L attacks L	ns	ns	ns	ns	−.24**	ns
GP attacks GP	ns	ns	ns	ns	ns	ns

* p < .05, ** p < .01, see Table 6.1a for details of analysis.
Data for 55 and 42 time-budget observations of lapwings and golden plovers respectively.

and when they are more likely to be attacked by gulls (because high gull:lapwing ratios reduce the success rate of gulls stealing from lapwings). Figure 6.7 illustrates the effect of gulls on interspecific aggression by comparing rates of aggression between plovers before and after the arrival of one or more gulls. The rate increases significantly in both species but, as might be expected, the increase is more pronounced in lapwings, which are preferentially attacked by gulls.

Apart from gulls, flock density is also an important factor determining rates of aggression between plovers. Why does aggression increase with bird density? One possibility is that increased density reduces individual feeding efficiency, perhaps through interference or prey depletion. Figure 6.8 shows that both lapwings and golden plovers tend to be clumped within flocks (i.e. most inter-neighbour distances (equation (6.2)) are short). Clumping is, however, more pronounced after the arrival of additional lapwings or golden plovers (Figure 6.8b). The figure also shows that the rate of aggression between plovers is negatively correlated with inter-neighbour distance and increases for any given distance after the arrival of more birds. The most pronounced increase is for birds less than 2m apart.

We recorded time budgets for focal birds and kept track of the number of con- and heterospecifics within ten lapwing-lengths of them by counting every 20s (see earlier). Among subgroups where the mean inter-neighbour distance is less than 1m, there is a significant decline in E_b compared with subgroups where neighbours are between 1m and 2m apart (lapwings $t = 2.57$, $p < .02$, d.f. 18; golden plovers $t = 2.12$, $p < .05$, d.f. 41). Thus increased density appears to reduce individual feeding efficiency. This does not necessarily conflict with the earlier finding that birds tend to aggregate where prey availability and feeding efficiency are highest. Birds may suffer reduced feeding efficiency as local density increases, but still be doing better than their companions elsewhere in the flock.

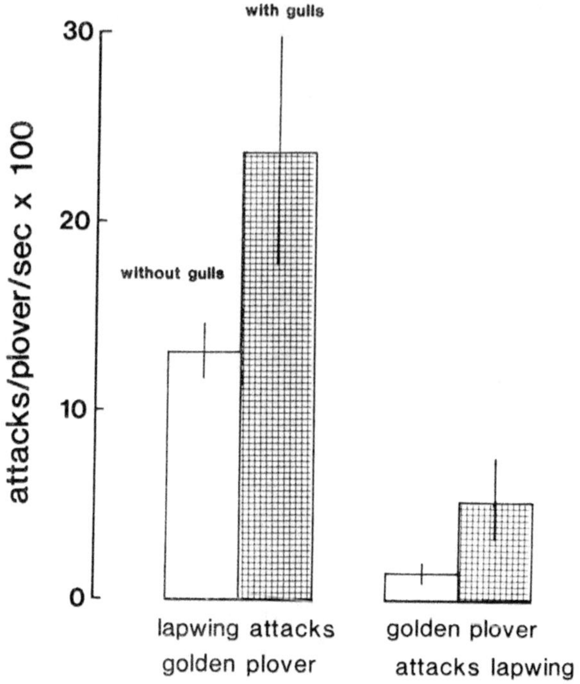

Figure 6.7: Rates of interspecific aggression between plovers before and after the arrival of gulls. Data for 97 time budgets. Sample sizes, 24, 31 for lapwings with and without gulls; 23, 19 for golden plovers with and without gulls. Bars represent standard errors

The effects of flock density and the presence of gulls (Table 6.9) may be interrelated. Inter-neighbour distance between plovers tends to decrease when gulls arrive. The frequency distribution of plover inter-neighbour distance therefore becomes more skewed towards short distances after the arrival of gulls (skewness statistic S before arrival = −0.85, after = 2.58; values above 0 indicating skewing to the left of the mean, see Snedecor 1956, Nie *et al.* 1975).

We have shown that the density of plovers within flocks correlates negatively with individual feeding efficiency (when local fluctuations are taken into account, cf. Tables 6.1 and 6.2) and positively with the rate of interspecific aggression. Is there a direct relationship between aggression and feeding efficiency? Such a relationship might arise in at least three ways: firstly, aggression may be a response to increased prey depletion at high feeding densities (Curtis and Thompson 1985); secondly, it may be a response to a decrease in prey availability near the soil surface as a result of disturbance at high densities (e.g. Goss-Custard 1970a, Waite 1983, Curtis and Thompson 1985);

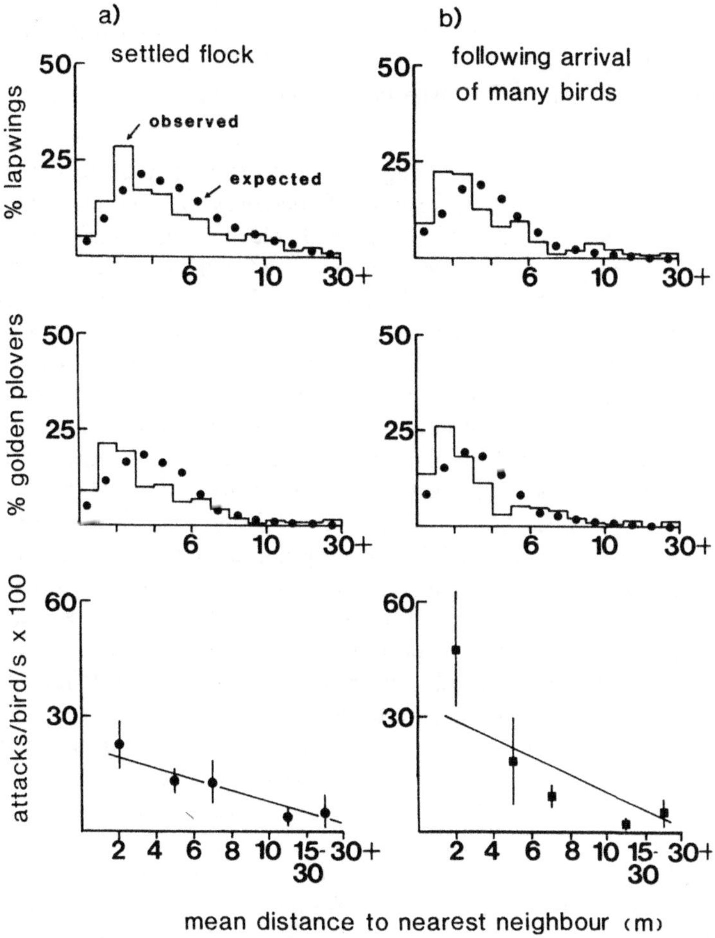

Figure 6.8: Frequency distribution of inter-neighbour distances between lapwings and golden plovers and the mean rate of interspecific aggression in relation to nearest neighbour distance. (a) Data for settled flocks (> 10 minutes since alighting, $r = -.37$, $p < .05$, d.f. 32), closed circles = expected (Poisson) distribution. X^2 for comparison between observed and expected distributions = 17.6, $p < .001$ for lapwings, and 23.8, $p < .001$ for golden plovers. Data for 192 lapwings and 159 golden plovers. (b) Data collected within 10 minutes of the arrival of lapwings and/or golden plovers, correlation between rate of aggression and mean nearest-neighbour distance significant ($r = -.31$, $p < .05$, d.f. 41), X^2 for comparison between observed and expected distributions = 20.4, $p < .001$ for lapwings, and 28.2, $p < .001$ for golden plovers. Data for 232 lapwings and 201 golden plovers

and, thirdly, it may be a response to reduced searching efficiency through interference (e.g. Goss-Custard 1976, Smith 1977, Curtis and Thompson 1985). Clearly, these effects are likely to be interrelated and difficult to distinguish (e.g. Goss-Custard 1980, Sutherland and Koene 1982). Nevertheless, we carried out some simple observations and experiments to try to separate them.

Prey depletion. To see whether aggressive interactions were related to prey depletion, we picked mixed flocks in which there was considerable variation in local bird density and observed lapwings and golden plovers in areas where (a) at least five birds (of either plover species) were foraging within 2.5m of their nearest neighbour (*high-density subgroups*) and (b) five or more birds foraged more than 3m from their nearest neighbour (*low-density subgroups*). We recorded the capture rate of focal birds within the two types of subgroup and then hand-sorted turf samples to record worm availability. We also took control turf samples from regions 5m from observed subgroups where no birds had been feeding during the observation period. Using scale maps of foraging flocks (see Chapter 4), we were able to estimate the number of worms removed per unit time and area in each type of subgroup. If increased flock density results in greater depletion of prey and aggression is a response to depletion, we should expect (a) the ratio of the number of worms removed to the number available per unit area to be greater in high-density subgroups, and (b) aggressive interactions to be more frequent in high-density subgroups.

Table 6.10 shows the results lumped for lapwings and golden plovers (because the density of feeding birds tends to influence the rate of interspecific aggression similarly in the two species). The decline in worm density attributable to birds is greater where high-density subgroups have been feeding (15.3%) than where low-density subgroups have been feeding (5.8%). An increase in local flock density therefore does appear to correlate with increased prey depletion. This complements the findings in relation to flock distribution within fields in Chapter 4. An important point, however, is that the number of worms removed by birds exceeds the decline in worm density in the regions of both high- and low-density subgroups (although the ratio of the number of worms taken to the control worm density is still greater in high-density subgroups (0.01 against 0.004). As expected from the analysis in Table 6.9, rates of interspecific aggression are higher in high-density subgroups ($p < .001$).

Depression of prey availability. Since our estimates of prey depletion by birds calculated from turf samples are smaller than those calculated from observed intake, it seems unlikely that increased local flock density depresses surface worm availability (i.e. drives worms down as opposed to depleting them by predation). Also, as we pointed out in Chapter 3, soil and weather conditions made it difficult for worms to move quickly through the soil. Nevertheless, there

Table 6.10: Changes in worm density, capture rate and interspecific aggression where the local density of plovers is high and low (see text)

	Density of plovers: (i) high	(ii) low	Notes
Worm density (no./m^2)	193 ± 10.4 (228 ± 09.8)	96 ± 13.8 (102 ± 11.3)*	1, 2, 3
Average decline in no. of worms/minute/m^2	1.16	0.20	
Worms captured/plover/min	1.93 ± .17	1.32 ± .08**	4, 5
Total no. of worms removed by plovers/min/m^2	2.31	0.42	
Rate of interspecific aggression (attacks/plovers/min)	.248 ± .012	0.006 ± .004***	1, 2, 5

Notes:
1. Worm density in brackets is that measured in the control area (see text).
2. All means are given as mean ± standard error.
3. Data for 21 and 23 experimental, and 26 and 24 control 0.25m-sq. turf samples.
4. Data for 63 time-budget observations: 41 'high', 22 'low'.
5. * $p < .05$, ** $p < .01$, *** $p < .001$; t-test.

is good evidence from other species that foraging birds disturb surface-dwelling invertebrate prey and drive them down. Goss-Custard (1970a), for example, found that the littoral crustacean *Corophium volutator* tends to burrow out of reach of foraging redshank. Goss-Custard (1976) and Curtis and Thompson (1985) suggest this as a reason for the negative relationship between capture rate of *C. volutator* by redshank and gulls and the density of foraging birds. We performed a simple experiment to see whether earthworms burrowed away from the turf layer in response to the activity of plovers.

It was not possible to sample an area of turf before and after plovers fed on it, so, instead, we simulated the presence of foraging birds by tapping a 4mm-diameter steel rod on the surface of the soil. We simulated a high-density subgroup by tapping at a rate of 120 taps/min and a low-density subgroup by tapping at 60 taps/min. Each period of tapping lasted one minute and was performed within a 0.25m x 0.25m area. Immediately after tapping, we took 3cm-deep turf samples within the trial area and from a control area 1m away.

Table 6.11 shows a non-significant tendency for worm density to be greater in trial than control areas after tapping. Furthermore, the difference was marginally greater after high-density subgroup simulations. Although caution is necessary in comparisons between different areas, the apparent increase in worm density after tapping helps to explain the discrepancy between the intake of worms by birds and the apparent decrease in worm density in the soil (Table 6.10). Why, however, should worm density increase as a result of tapping the ground? One explanation is that tapping resembles rainfall and worms move upward in the expectation of encountering moist soil (e.g. Edwards and Lofty 1972). This may be why gulls and plovers often 'foot-tremble' on pastureland (Plate 5, Simmons 1961, Lange 1968, Burton 1974, Cramp and Simmons 1983, see also Pienkowski 1983b for similar behaviour in intertidal habitats).

Interference, area-copying and aggression. Feeding efficiency may also be reduced in high-density subgroups through physical interference between birds and a reduction in searching efficiency. In boat-tailed grackles, for instance, Smith (1977) found that area-restricted searching (the tendency to search in the immediate vicinity of a capture — see Tinbergen 1960, Smith 1974a,b) was reduced when birds fed in high-density flocks. Area-restricted searching is a simple means of increasing feeding efficiency where food is patchily distributed, and the degree to which predators area-restrict their searching appears to correlate with the degree of clumping of their food supply (Smith 1974a,b, Zach and Falls 1977, 1979, Barnard 1978). Since the plovers in our study area feed on patchily-distributed prey, the reduction in feeding efficiency in high-density subgroups may result from interference with their search paths.

One way of measuring area-restricted searching is to calculate the so-called meander ratio (Williamson and Gray 1975) of a predator's search path. This is simply the ratio of the actual distance covered between any two points on the path and the straight-line distance between the points. The higher the ratio, the greater the tortuousness of the search path. We recorded the amount and direction of turning before and after captures in focal lapwings and golden plovers, but there is no tendency for the ratio to increase after a capture or as worm patchiness increases and no reduction when local flock density increases. There is thus no evidence that reduced feeding efficiency correlates with constrained searching ability.

Another possibility is that increased local flock density reflects area-copying (Barnard and Sibly 1981). Birds may use the success of others to indicate locally-rich food supplies and move in to search in the same area. Furthermore, area-copying may occur to a greater extent when birds capture the most profitable small to intermediate-sized worms (see Chapter 5). This might result in increased depletion of these worms and help to account for the discrepancy between observed intake and measured depletion (birds deplete worms near the surface, but their activity brings up others (larger) from deeper down). To test this, we compared changes in local flock density around focal birds within 30s of the latter capturing prey. Figure 6.9 shows the change in the number of heterospecifics within 2-3m of focal birds in relation to the size of worm captured by the focal bird. There is a tendency for birds to accumulate around successful heterospecifics, but only when worms between 18mm and 48mm (classes 2 and 3) have been caught. These sizes correspond closely with those which are most profitable to lapwings and golden plovers.

Plate 5. Lapwing 'foot-trembling' on pasture. Photograph courtesy of Professor Richard Vaughan

Table 6.11: The results of an experiment to simulate the potential effects of high and low densities of birds on the density of worms in the upper 3cm of soil

	Mean (± se) density of worms/m²		
	Trial area	Control area	Matched pairs t-test
High-intensity vibrations (120 taps/min)	120 ± 13.9	99.2 ± 15.5	1.63, ns
Low-intensity vibrations (60 taps/min)	108 ± 9.7	86.7 ± 13.6	1.38, ns

We also compared changes in local density between flocks with and without gulls (Figure 6.10). Interestingly, the increase in density following a capture is much more pronounced when gulls are present. One reason for this might be that area-copying for profitable worms is a way of offsetting the reduction in feeding efficiency when gulls are present. Increased clumping in flocks with gulls may therefore be a consequence of area-copying instead of, or as well as, the 'selfish herd' effect suggested earlier and in Chapter 8.

Curiously, although similar analyses were performed, there seems to be no tendency for conspecific density to increase following a capture. It appears, therefore, that intraspecific aggression in relation to bird density may be a response to prey depletion, but that interspecific aggression may be a response to both this and area-copying.

Feeding Efficiency and Daily Energy Balance

We have shown that several factors influence feeding efficiency in lapwings and golden plovers. While we have so far considered only short-term changes in feeding efficiency (resulting from changes in flock composition and environmental conditions), it is not difficult to see that these may affect a bird's chances of meeting its daily energy demand. The presence of gulls for much of the day, for instance, might impose quite a serious constraint on energy intake. To what extent do fluctuations in feeding efficiency affect the birds' daily energy balance?

Daily Energy Requirements

Two basic approaches can be used to see whether birds meet their daily energy requirement (see Walsberg 1983 for a recent review). One is to calculate total gross energy intake and compare it with estimated daily *existence energy requirement* (EER) (see below). The second is to compare total daily

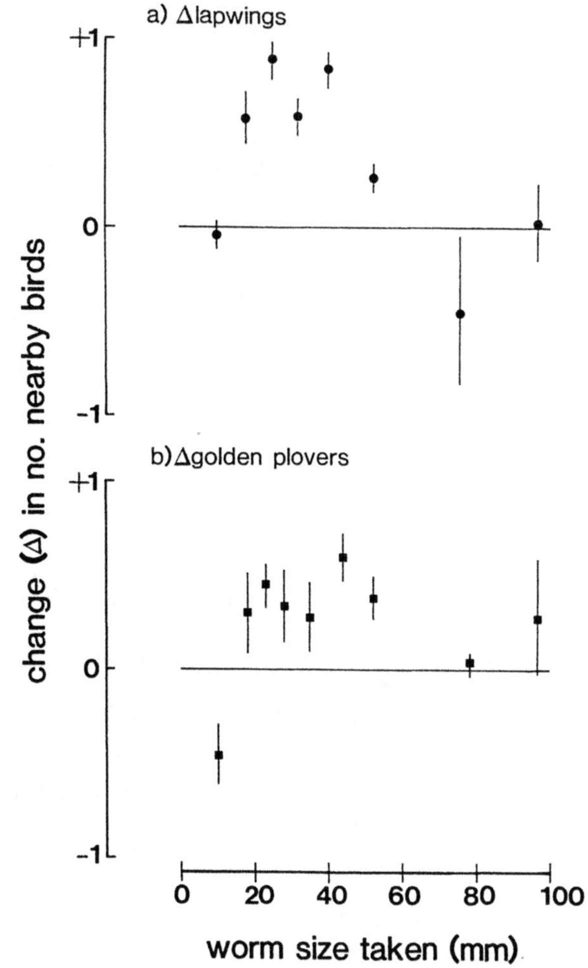

Figure 6.9: Changes in the number of birds within 2-3m of a successful focal plover during the 20s period after a capture. (a) Changes in the number of lapwings near golden plovers (n = 89 observations), (b) changes in the number of golden plovers near lapwings (n = 47 observations). Bars represent standard errors

Source: Barnard et al. *(1982)*

gross energy intake with estimated expenditure in activities performed during the day. In this case, activities are costed as multiples of *basal metabolic rate* (BMR). We used both approaches.

Existence energy requirement. Kendeigh (1970) defined EER in birds as the energy required per day for '... standard metabolism, specific dynamic

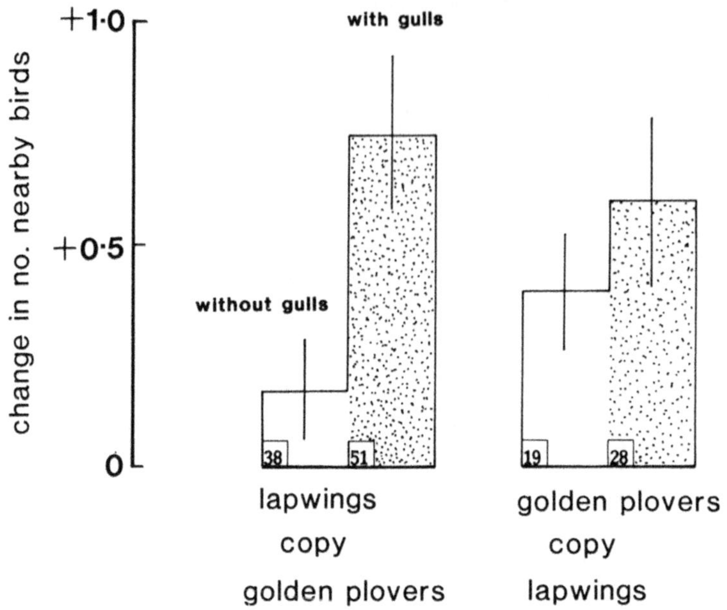

Figure 6.10: Changes in neighbour density as in Figure 6.9 but before and after the arrival of gulls. The difference between means for lapwings copying golden plovers is significant ($t = 2.97$, $p < .01$, d.f. 87). Bars represent standard errors

action and locomotory activity': in other words, the energy needed to carry out essential physiological processes and behaviour. Existence metabolism includes basal metabolism, suprabasal tissue metabolism for body-temperature regulation, the heat increment of feeding and the energy requirement for locomotor activity (Kendeigh and Blem 1974). The EER depends to a large extent on temperature, so that, for instance, it is calculated as:

$$\text{EER} = 4.337 \times Wg^{0.53}\text{Kcals} \tag{6.6a}$$

at 0°C and as:

$$\text{EER} = 0.54 \times Wg^{1.7545}\text{Kcals} \tag{6.6b}$$

at 30°C, where W_g is the wet weight of the bird in grams (see also King and Farner 1974). Good data are available from passerines. Kendeigh (1949) measured EER in house sparrows as the amount of food metabolised by caged birds maintaining constant weight. As long as any additional weight is metabolically inactive or of low activity (like fat and water), EER should be proportional to the lean weight of the bird. EER declines following pre-winter moult as a result of increased insulation and energy conservation. In house sparrows, the post-moult EER may be over 2 Kcal/bird/day lower than that preceding the moult. The insulation/conservation effect then dissipates during the following months so that, for house sparrows, EER rises to only 0.7Kcal below the pre-moult level by December (Kendeigh 1949).

Basal metabolic rate. Basal metabolism can be defined as standard metabolism (the rate of energy use by animals at rest and in a fasting condition) at an ambient temperature at which little or no extra heat is required for the maintenance of body temperature (Kendeigh 1970). Factors affecting basal metabolism are those correlating with changes in ambient temperature: time of day, season, geographical location and so on. Basal metabolism tends, therefore, to be higher through autumn and winter than through spring and summer (e.g. Miller 1939).

Gross energy intake by waders has been measured in several studies and expressed as a multiple of BMR (Lasiewski and Dawson 1967, Aschoff and Pohl 1970) where BMR is calculated as:

$$\mathrm{BMR} = 78.3 \times W^{0.723} \qquad (6.7)$$

where W is the weight of the bird in kilograms and BMR is measured in Kcals/24 hours. Most studies of waders have shown that daily energy intake is equivalent to between 3 and 6.8 x BMR (e.g. Hulscher 1974, 1982, Smith 1975, Goss-Custard 1977b,c, Pienkowski 1982), and that birds achieve this mainly during daylight hours (Goss-Custard 1969, Evans 1976, Baker 1981). There is, however, some debate about how much energy birds must assimilate in order to meet their requirement.

Ebbinge *et al.* (1975) suggested that birds needed an intake equivalent to between 2 and 4 x BMR per day to maintain body temperature, move and forage. Evans *et al.* (1979) revised this estimate to incorporate the potential costs of foraging and an assimilation efficiency of 80-90%, and suggested that 4.5 x BMR was more realistic. Actual energy intake during the day can deviate substantially from these estimates, particularly during very cold conditions when food is scarce. Pienkowski (1982) recorded winter daylight intake values as low as 2 x BMR in ringed plovers and almost 1 x BMR in grey plovers, despite the birds in his sample feeding throughout the day. Pienkowski suggested, however, that the latter multiple was near to the minimum for survival even if birds fed through the night.

To estimate intake relative to requirement in lapwings and golden plovers, we used already-published BMR costings for different activities. These are necessarily crude estimates, not least because they are taken across a range of different species and the categories of activity do not always exactly coincide, but they have been widely used in conjunction with time-budget data to calculate daily feeding efficiency in birds. The following costings are taken from Aschoff and Pohl (1970), King and Farner (1974), Maxson and Oring (1980) and Puttick (1982): night resting (= BMR), foraging (3 x BMR), preening (2.5 x BMR), day resting (1.25 x BMR), flight (12 x BMR), stepping (2.8 x BMR), ground chase (5 x BMR), aerial chase (15.2 x BMR).

The results of calculations based on these estimates are shown in Table 6.12. All measurements used in analyses were made between mid-December and mid-January and mean recorded daylight temperature was used to calculate the EER. The body weights in equations (6.6) and (6.7) were 260g for lapwings and 215g for golden plovers (after Goodyer 1976, Davidson 1981, Cramp and Simmons 1983, and Wader Study Group ringing returns for the Midlands). The energetic cost of each activity was calculated by multiplying the amount of time (hours) spent in each activity by the appropriate multiple of BMR (see above).

Table 6.12: Estimated daily energy requirement and intake (Kcals/24 hrs) for lapwings and golden plovers with and without gulls during the December-January period

	LAPWINGS		GOLDEN PLOVERS	
	with gulls	without gulls	with gulls	without gulls
ENVIRONMENTAL VARIABLES				
Ambient temperature (°C)	2.8 ± .26	3.1 ± .29	2.8 ± .26	3.2 ± .28
Daylength (hours)	9.25	9.30	9.27	9.31
BASIC ENERGY REQUIREMENTS				
BMR (Kcals/24 hrs)	29.56	29.56	25.77	25.77
EER (Kcals/24 hrs)	77.5	75.6	70.1	68.6
24-HOUR FORAGING COSTS				
Night roost	18.41	17.99	15.71	15.56
Loafing	1.27	2.00	1.71	2.17
Flight to feeding site	3.10	3.10	2.58	2.54
Active foraging	26.64	27.20	22.67	22.90
Incidental flights	11.82	7.38	10.82	7.21
Chases by gulls	—	.24	—	.10
TOTAL COSTS	61.24	57.91	53.49	50.48
ENERGY BALANCE				
Kcals assimilated/hour	9.22 ± 1.64	6.34 ± .85	6.30 ± .64	6.47 ± .72
Kcals assimilated/daylight	66.48	46.66	44.35	45.64
xBMR	x2.24	x1.58	x1.72	x1.77
Energy gain—total costs	+5.24	−11.0	−9.14	−4.34
Energy gain—EER	−11.0	−28.9	−25.8	−23.0

All means presented with standard errors.

In line with evidence from other species of wader during winter (e.g. McLennan 1979, Puttick 1980), lapwings and golden plovers appear to undershoot their daily EER during daylight feeding. Interestingly, however, the effect of gulls is different in the two species. In lapwings, the daily energy deficit becomes greater; in golden plovers it is reduced. Lapwings come closest to meeting their daytime requirement when gulls are absent for the entire day. Under these conditions their estimated daily assimilation is greater than twice BMR and greater than the total cost of all activities performed during the day. This difference in the effect of gulls appears to be due mainly to the reduction in crouching and hence selectivity in feeding in lapwings (see earlier). Although lapwings lose more time through attack than golden plovers, the average time cost per day is vanishingly small (mean time lost by lapwings = 48.5 ± 6.46s/day; mean lost by golden plovers = 20.9 ± 14.1s/day, t = 1.78, n.s.).

Since temperature is an important factor determining EER (Kendeigh 1949, 1970, Pienkowski 1982), we looked at the ability of plovers to meet their daily requirement as mean daylight temperature varied. Figure 6.11 shows the effect of temperature as total daily energy intake minus EER. As might be expected, plovers incur a greater deficit as mean daylight temperature decreases. Furthermore, their ability to meet their requirement at any given temperature is depressed by the presence of gulls. Again, gulls have a more serious impact on lapwings which do worse regardless of temperature when feeding with gulls. Golden plovers, on the other hand, do worse only on mild days (with a mean temperature of >5°C), which is also the time most gulls are present in flocks.

Compensating for an Energy Deficit

For obvious reasons, the incursion of an energy deficit can only be temporary and we should expect birds, where possible, to compensate in some way. One simple solution is to spend a greater proportion of the day foraging.

Spending more time feeding during the day. Feeding is only one of the important maintenance behaviours birds must perform during the day. Nevertheless, if energy (and/or nutrient) demand is high, a greater proportion of the day may have to be given over to it at the expense of other activities. Several studies of overwintering waders have shown that birds spend more time feeding and/or achieve higher rates of food intake when temperatures are low and windspeed (and hence 'chill factor') high (e.g. Goss-Custard 1969, 1983, Heppleston 1971a, Puttick 1980, Townshend 1981, Milsom 1984). This is supported by our own observations of lapwings and golden plovers. Taking short-term commitment first, there is a strong negative relationship between ambient temperature and both the proportion of birds seen feeding during an instantaneous scan of a foraging flock and the proportion of the day birds spend feeding (Figure 6.12). A comparison

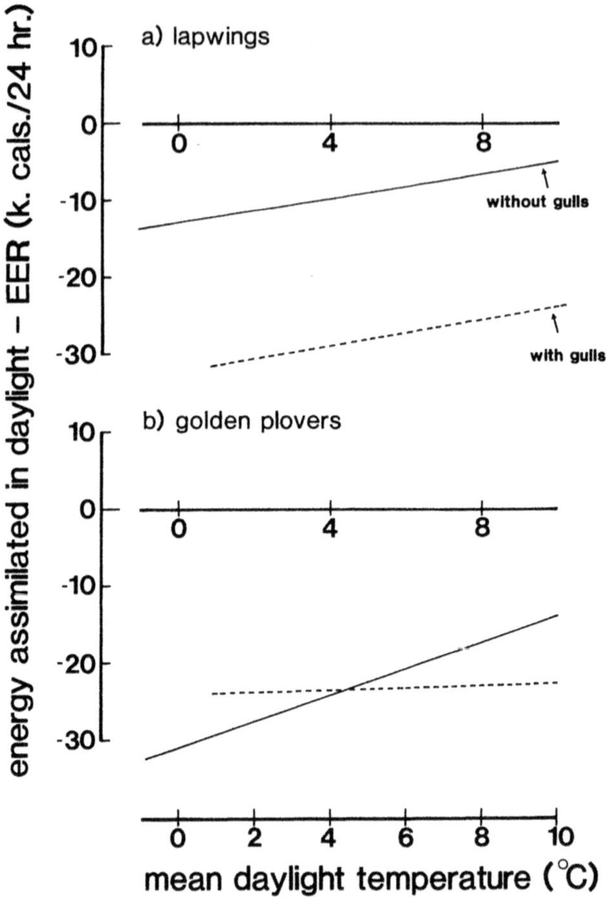

Figure 6.11: The effect of average daylight temperature and the presence of gulls on the ability of lapwings and golden plovers to meet their estimated energy requirement (total daily intake — EER). See text and Table 6.12 for details

between flocks with and flocks without gulls shows a significant increase in the proportion of lapwings feeding when gulls are present ($t = 2.59$, $p < .02$, d.f. 53), but no change in golden plovers. Lapwings thus appear to spend more time feeding to offset the negative effect of gulls on their feeding efficiency. Figure 6.1 and Tables 6.2 and 6.3 showed that an increase in the number of lapwings has a negative effect on the feeding efficiency of golden plovers in the absence of gulls, but that an increase in the number of golden plovers has a negative effect on lapwing feeding efficiency in the presence of gulls. The arrival of large numbers (>15) of heterospecifics, however, results in an increase in the proportion of birds feeding within a flock only in golden plovers in the absence of gulls ($t = 2.12$, $p < .05$, d.f. 26).

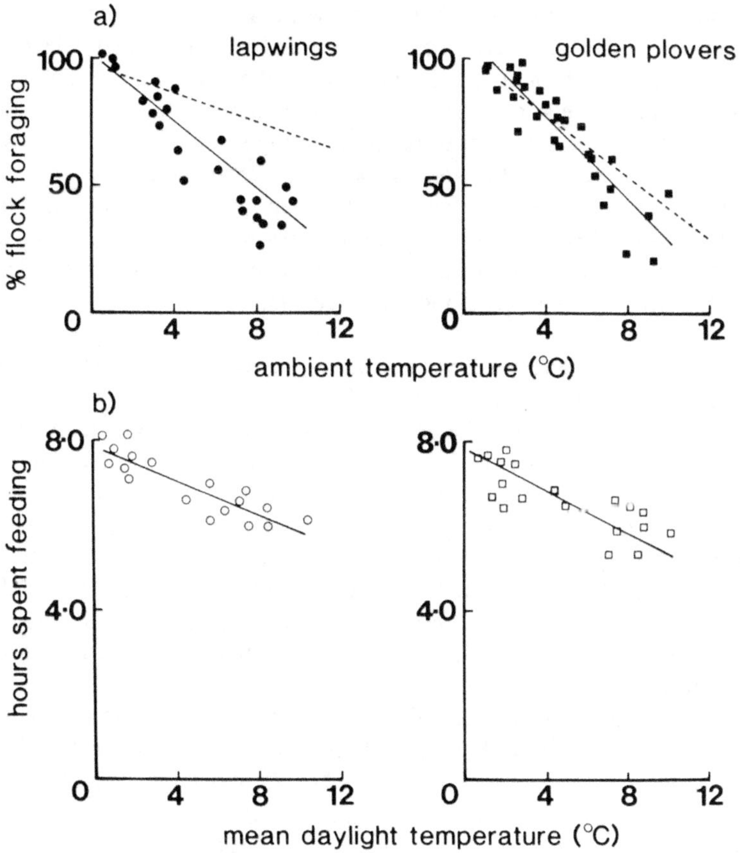

Figure 6.12: (a) The relationship between average daylight temperature and the percentage of birds within flocks feeding during the day, —— without gulls ($r = -.92$, $p < .001$ for lapwings, $r = -.88$, $p < .001$ for golden plovers), ---- with gulls ($r = -.74$, $p < .001$ for lapwings, $r = -.75$, $p < .001$ for golden plovers). Lines fitted by regression. Data for 29 flocks without gulls and 30 with. (b) The number of hours spent feeding in daylight in relation to average daylight temperature (see text), $r = -.96$, $p < .001$ for lapwings, $r = -.92$, $p < .001$ for golden plovers

The difference in the proportion of birds feeding when gulls are present is greatest on mild days when there are more gulls in the flock ($r = .21$, $p < .02$, d.f. 132). Also, we showed earlier (Tables 6.1 and 6.3) that the positive effect of temperature on feeding efficiency (E_f) is lost in golden plovers but not in lapwings when gulls are present. One explanation for this is that, on mild days, lapwings are able to compensate for the interference by and loss of prey to gulls by feeding more rapidly and/or spending more time feeding (increasing the numerator relative to the denominator in equation 6.1).

Temperature variation and the presence or absence of gulls clearly affect commitment to feeding in the short term. What about the long term, the proportion of the day spent feeding? We observed a number of flocks over 18 days and recorded the amount of time individual birds spent feeding. We then partialled out the relationships between the number of hours a day plovers spent feeding and both temperature and the proportion of time that gulls were present. There were no foraging flocks in which gulls were absent for an entire day. Table 6.13 shows that birds spend more of the day feeding when it is cold (see also Milsom 1984) and gulls are present for longer periods. Care is needed in interpreting this last point, however, because gulls may be attracted to fields where lapwings spend more time feeding.

Night feeding. Although birds appear to compensate for reduced feeding efficiency by spending more of the day feeding, it may sometimes be that this is not enough. What then? The obvious possibility is feeding at night. Surface worm availability tends to be higher at night (see Chapter 3) and McLennan (1979) found that lapwings achieved just over twice their diurnal rate of food intake by feeding at night during mid-winter. Several pieces of evidence suggest that lapwings and golden plovers sometimes feed at night in our study area (Chapter 3) and that low temperatures and the presence of gulls during the day are factors which encourage night feeding. Do birds therefore continue to feed into the night when they are failing to meet their energy requirement during the day?

A comparison of the number of foraging flocks which continued feeding on their pasture after dark showed that, following cold days (mean daylight temperature <4°C), 67% of lapwing and 27% of golden plover subflocks (data for 18 flocks) continued feeding. On milder days, only 28% of lapwing subflocks and no golden plovers continued to feed (data for 14 flocks). Under both conditions, therefore, lapwings are more likely to feed into the night than

Table 6.13: Correlation coefficients (r) and beta values (β) for relationships between the number of hours spent feeding by plovers, the length of time gulls are present and environmental conditions

		Independent variables				
Dependent variables		Average °t	Max °t	Min °t	Daylength	Time spent at feeding site by gulls
No. hours lapwings spend feeding	r	−.956***	−.944***	−.886***	.068	.898***
	β	−.95***	ns	ns	ns	.34*
No. hours golden plovers spend feeding	r	−.918***	−.893***	−.868***	.016	.839***
	β	−.92***	ns	ns	ns	ns

* $p < .05$, *** $p < .001$. See Table 6.1a for details of partial regression analysis. Data for 18 days (see text).

golden plovers. There is also a significant tendency for lapwings to feed into the night when at least one gull is present in the flock during the hour before nightfall (% lapwing subflocks feeding into the night without gulls = 25%; with gulls = 76%, $X^2 = 5.37, p < .05$). Although there is a similar trend in golden plovers, the difference is not significant (% without gulls = 10%; % with gulls = 28%, $X^2 = 1.69$ n.s.). On days when the mean temperature is below 4°C, almost all flocks containing gulls feed into the night.

Drawing on energy reserves. It is possible, of course, that plovers incur a steadily increasing energy deficit through the winter and survive by drawing on fat and protein reserves laid down in the autumn. Several studies suggest that waders deposit fat and protein as an insurance againt energy deficits in harsh conditions (Evans and Smith 1975, Evans 1976, 1979, Pienkowski *et al.* 1979, Pienkowski 1982, Davidson 1981). Pienkowski (1982) suggested that the peak mid-winter weights of plovers should enable birds to survive for up to a week without feeding. 'Topping-up' by inadequate feeding could allow survival over much longer periods. We might therefore expect to see plovers achieving a strong, positive energy balance and putting on weight in autumn/early winter (to build up reserves) and possibly also in late winter (to make up mid-winter loss).

Dugan *et al.* (1981) found that grey and golden plovers do tend to put on weight in the early winter, and Cramp and Simmons (1983) report maximum weights for golden plovers (ringed in the Netherlands) in December. December weights were 6% higher than those for November and 18% higher than those for February. Similar data for lapwings revealed highest weights in the September-December period and lowest weights in February-March, by which time body weight had fallen by 26%. Davidson (1981) found that the lipid index in several coastal wader species was highest (i.e. little fat had been metabolised) in the mid-winter period. He then compared lipid and muscle indices (muscle protein is metabolised only when food intake is virtually zero) and lean weight in birds before and after a six-day cold spell in northeast England. The results showed that, unlike the other wader species which foraged on the coast, golden plovers drew extensively on their reserves and showed a 9.3% decline in lean weight over six days. Some birds even moved down to the south and southeast coasts (see also Chapter 2). Both golden plovers and lapwings, however, appear to tolerate substantial decreases in body weight. Davidson (1981), for example, recorded decreases of 31-40% in lapwings and 33-34% in golden plovers before starvation.

We estimated the mean daily energy balance of the plovers in our study area during three periods: mid-November to mid-December (early winter), mid-December to mid-January (mid-winter) and mid-January to mid-February (late winter). Calculations showed that both species incur a deficit in early winter and that the deficit increases into mid-winter (Figure 6.13). By late winter, however, the deficit decreases and golden plovers achieve positive balance. This is suggestive of the trends predicted above, but implies that reserves are laid down earlier in the year. However, it is important to remember that daytime deficits may be offset by night feeding. Except during the mid-winter period, estimated assimilation by plovers is greater than 2.3 x BMR. Another important consideration is the changing membership of the lapwing and golden plover populations. Immigration and emigration occur periodically through the winter, especially during and after cold spells (see Chapter 3) and towards the end of winter (February/March). In February/March substantial numbers of new birds, some in breeding plumage, pass through the study area. Some of the transient golden plovers achieve in excess of 6 x BMR at this time of year, presumably to deposit fat prior to migration (see Chapter 2) and to cater for increased (5-30%) metabolic rate during the moult (Payne 1972, Cramp and Simmons 1983).

Figure 6.13: The mean ability of lapwings and golden plovers to meet their estimated energy requirement by daylight foraging at different periods through the winter. See text for details. Weights used in calculations differed between the three periods. For lapwings they were 235, 260 and 240g (see Goodyer 1976), and for golden plovers 202, 215 and 176g (see Cramp and Simmons 1983). Bars represent standard errors for hourly intakes

Individual Variation in Foraging Behaviour

An important point arising from recent studies of flocking and feeding in birds is that different individuals do not necessarily behave in the same way. For instance, birds may vary in their competitive ability or their tendency to exploit the success of others (e.g. Caraco 1979a, Rohwer and Ewald 1981, Barnard and Sibly 1981, Goss-Custard *et al.* 1982a, 1984, Ens and Goss-Custard 1984, see Chapter 1). Variation in foraging behaviour may also be due to age or sex (e.g. Norton-Griffiths 1967, Heppleston 1971b, Evans and Smith 1975, Goss-Custard and Durell 1983, Ens and Goss-Custard 1984, Gochfeld and Burger 1984). As in several other species (e.g. Beer 1961, Bateson *et al.* 1980), lapwings can be aged, sexed and recognised individually on the basis of plumage variation (Glutz *et al.* 1975, Cramp and Simmons 1983, K.R. Futter in prep.).

Birds of different age and sex differ particularly in their facial and dorsal plumage and wing-coverts. Juveniles have smaller black forehead patches and their crests and facial stripes are shorter. The dorsal feathers and wing-coverts of juveniles are tipped with buff in a barred pattern. Barring does not occur in adults, although first-year adults may have buff edges to some of their feathers. The dorsal plumage of juveniles is greyer than that of adults and lacks the sheen. Breast markings are also greyer and may have a streaked appearance. Juveniles also have shorter wings and tail. Differences between the sexes in adults occur mainly in facial markings, though the wings and tail are shorter in females. Females have a shorter crest and, typically, a buff-coloured face. As the breeding season approaches, the black breast markings of the male expand until the neck and the greater part of the head are black. These changes do not occur in females, although the black facial stripes may become more prominent. We used these differences to look for individual variation in foraging behaviour.

Age Effects

From Futter's observations of foraging behaviour in adult and juvenile lapwings, it appears that juveniles are, on the whole, less successful. Within any given flock, juveniles tend to achieve lower capture rates than adults of either sex, but are affected by flock composition in ways that are more similar to adult females. Partial regression analyses show that time budgeting and feeding efficiency in juveniles and adult females are influenced by both flock composition and environmental factors, whereas adult males are affected mainly by variations in temperature and daylength. Interestingly, juveniles are affected to a greater extent than adults by the presence of gulls. As before, time spent crouching decreases and capture rate increases. It may be that juveniles are forced onto poorer areas by adults and hence are more seriously affected by the activities of gulls. However, it may also be that they are more vulnerable. In common with juveniles of other species (e.g. Norton-Griffiths

1967, Buckley and Buckley 1974, Davies and Green 1976), there is a suggestion that young lapwings are less adept at capturing worms. Their handling times for given-sized worms tend to be slightly longer (at least over the range mainly taken by birds) than those of adults. This may provide gulls with a greater opportunity for attack (see Chapter 8).

Sex Effects

The sexes differ in their capture rate on any given worm density. We compared foraging sequences in males and females feeding simultaneously in the same field and sampled the worm community where each bird had been feeding. Females achieve a significantly higher capture rate than males (0.071 ± 0.014 against 0.047 ± 0.004 captures/s, $U = 36$, $p < .05$), even though they do not feed on higher worm densities ($t = 0.47$, n.s., d.f. 25). Although females also achieve greater feeding efficiency than males, the difference is not significant (mean E_b for females $= 4.49 \pm 1.56$ cals/s, mean for males $= 2.81 \pm 0.76$ cals/s, $U = 13$, $0.5 < p < 0.1$).

The sexes also differ in aggressive behaviour. One of the reasons why fewer lapwings feed in small (<3ha) pastures is that newly-arriving birds tend to be attacked. Examination of the initiators of attacks and the new birds that are prevented from landing suggests that both are usually male. Of observed initiators, 68% were male; and, of the birds attempting to land in one particular small pasture, 63% of females (15 out of 24) but only 18% of males (3 out of 17) were successful. Recent observations by K.R. Futter suggest that male lapwings defend small feeding territories within flocks more often than females. Lapwings defend feeding territories sporadically both within and between days (see also Brown 1926, Lind 1947). Defended areas vary in size from about $2m^2$ (under conditions of partial snow cover) to areas 10m in diameter. The majority (79%) of these areas are defended by males. Futter's observations suggest that disputes over territories are not always settled by overt aggression. Instead, an elaborate series of assessment displays appears to be used.

Adults, but especially males, perform a characteristic 'back-up' display which appears to function as a keep-out signal to approaching birds. The display involves a lowering of the anterior part of the body, with the head held erect, and a raising upwards of the back and tail. The tail is usually slightly fanned and the wings held apart to reveal the dorsal plumage (see also Glutz *et al.* 1975). If displays escalate, the wings may be held out and slightly raised. During back-up displays, the crest is erected and the black forehead patch emphasised. The resident may move towards the intruder, especially if the intruder is juvenile or has previously lost a dispute with the resident.

When birds have been feeding close together in a flock for some time, the back-up display is sometimes omitted. In these cases, interactions usually begin with contestants standing and facing in opposite directions. At this stage, the crest is not erected and there is no aggressive posturing, although there may be some redirected pecking in which clumps of turf are pulled up and thrown overhead or to the side (a behaviour characteristic of plovers). From here, a 'parallel walk' (Plate 6a, see also Spencer 1953) may be performed, in which the contestants walk in the same direction, maintaining a more or less constant distance (2-3 body widths) between them. The posture is now erect with the wings drooping slightly. The crest is also erect. If the interaction does not end there, parallel walking may be followed by a 'facing' display in which birds adopt an extreme upright posture and face towards each other (Plate 6b). The crest is now fully erected and the wings droop almost to the ground. Redirected pecking may again occur. If the interaction still continues, birds may fight using both bill and wings as weapons. Although fighting is rare, it may occur even without the prelude of conventional display described above. It occurs most frequently, however, between apparently closely-matched (similar in size and plumage characteristics) males. Juveniles rarely fight or perform any of the prelude displays, while females have been observed performing only the back-up display, which is directed towards other females or juveniles.

Intrasexual aggression may be due to competition for locally defendable food supplies. This is supported by an analysis of the distribution of worms in relation to territorial defence. Although there is no difference in worm density between defended and undefended areas (mean density in defended areas = 182 ± 20.3 worms/m, mean in undefended areas = 202 ± 18.4 worms/m, $t = 0.72$, n.s., d.f. 39), the distribution of worms in defended areas is significantly more patchy (variance:mean ratio in defended areas = 3.12, $p<0.05$ (X^2 test), in undefended areas = 1.88, $0.1>p>0.5$ (X^2 test)). Furthermore, territories are defended mainly (86%) in old pasture where worms are more patchily distributed (see Chapter 3).

There may also be differences in behaviour between lapwings which are not attributable to age or sex. Within one field over a 22-day period, for instance, five out of seven individually recognisable lapwings defended territories within the same $20m^2$ area of the field, despite other birds moving around the field as the horizontal distribution of worms changed (Chapter 4). These same birds also tended to remain behind when the flock was disturbed. Although birds are known to remain behind in flocks of other species (e.g. Barnard 1980b), evidence suggests that these might be individuals of low competitive ability. The limited data so far available for lapwings, however, suggest that remainers may be birds at both extremes of the competitive ability spectrum. Two of the five birds initiated significantly more ($p<.01$ and $p<.05$) and two significantly fewer ($p<.001$, $p<.05$) attacks against con- and heterospecifics than the average for the flock.

(*a*)

Plate 6. Aggressive displays between male lapwings: (a) 'parallel walking'; (b) 'facing' (see text). Photographs courtesy of Professor Richard Vaughan

(a)

Time Budgeting and Feeding Efficiency: Résumé

The way lapwings and golden plovers allocate time to different activities and the consequences of this for feeding efficiency are clearly influenced by flock size and composition. The relationship, however, is not simple. Lapwings tend to do better when conspecific density is high, but this appears to be a consequence of the positive relationship between lapwing density and worm availability. Feeding efficiency in golden plovers, however, is more closely related to the overall density of both plover species, presumably because golden plovers use lapwing density as a guide to good feeding areas.

In both species, feeding efficiency is influenced by the way time is allocated between activities. It depends particularly on the amount of time birds spend assessing prey prior to pecking. The fact that most earthworms are concealed beneath ground means that they are difficult targets and take time to locate. In golden plovers, there seems to be a direct trade-off between prey assessment and vigilance for predators. In situations where commitment to scanning can be reduced, therefore, the amount of time devoted to the crouching assessment posture tends to increase. In lapwings, the relationship between scanning and crouching is less clear-cut and birds tend to crouch more when they are on higher worm densities and can afford the extra time for assessment.

The effect of gulls on plover feeding efficiency is due largely to the reduction in the amount of time plovers spend crouching. Crouching increases a bird's chances of attack, mainly because it indicates the imminent procurement of prey and attracts the gulls' attention, but also because it is likely to reduce the bird's ability to detect an approaching attacker (see Chapter 8). When crouching is inhibited, plovers are less able to locate concealed worms and end up taking smaller items, mainly from the soil surface. The effect of gulls is more pronounced on lapwings because they respond less quickly and are less agile than golden plovers. Both species fail to meet their apparent energy requirement by daytime feeding alone and, in lapwings, the deficit is greater when gulls are present. Birds compensate for this by spending a greater proportion of the day feeding and by feeding into the night when daytime feeding efficiency is low.

Aggression (mainly chases and kleptoparasitism) occurs both within and between plover species. Interspecific aggression is commoner and tends to occur when feeding priority is likely to be high and the composition of the flock unfavourable to the attacker. Increases in the local density of birds within the flock result in a reduction in individual feeding efficiency, even though there is a positive correlation between local bird and worm densities and birds in high-density areas do better than those elsewhere in the flock. Aggression appears to be related to prey depletion as the local density of birds increases. Successful foragers attract neighbours and copying is most pronounced after birds have found the most profitable worm sizes. There is no evidence that the activity of birds depresses worm availability; indeed, there is a suggestion that availability may increase near feeding birds.

As in other species, there are behavioural differences between individual lapwings, in this case in terms of feeding efficiency, aggressive behaviour and the tendency to defend feeding territories within flocks. Differences appear to hinge on age, sex and variation in competitive ability.

Summary

1. Feeding efficiency in lapwings and golden plovers is influenced by flock composition and physical environmental conditions.

2. Variation in feeding efficiency is closely correlated with changes in the allocation of time to different behaviours, especially postures associated with the assessment of prey size.

3. The presence of gulls has a more serious impact on the time budgeting and feeding efficiency of lapwings than on those of golden plovers.

4. Interspecific aggression appears to be related to feeding priority and increased prey depletion as local flock density increases.

5. Plovers fail to meet their energy requirement by daytime feeding alone. Temperature, daylength and the presence of gulls all affect the proportion of the day plovers spend feeding and the tendency to feed at night.

6. Age, sex and other factors result in differences in foraging behaviour between individual lapwings.

Chapter 7

Flock Dynamics: Patterns of Arrival and Departure

Being in a group affects the way individuals behave. In our case, single- and mixed-species flocking affect the way birds allocate time to behaviours associated with feeding and looking for predators, and this directly affects their feeding efficiency (Chapters 5 and 6). Since the effect on individual feeding efficiency depends on the size and species composition of the flock, we might expect birds to be selective about the flocks they join or remain in to feed. Indeed, from a knowledge of the relationship between flock size and composition on the one hand and individual time budgeting and feeding efficiency on the other, we should be able to predict the probability that a bird will join, remain in or leave any given flock. This should then allow us to predict the way particular flocks will build up. In other words, the dynamics of a feeding flock should be predictable from its effects on individual behaviour.

In Chapter 1, we discussed some models which integrated the effects of single- and mixed-species flocking on individual behaviour and made predictions about flock dynamics. We have also seen that predictable relationships between time budgeting and flock dynamics have been demonstrated in some species (e.g. Caraco 1979b, Caraco *et al.* 1980a, Barnard 1980b), although these studies were concerned only with single-species flocks. In this chapter, we investigate the relationship between time budgeting, feeding efficiency and flock dynamics in mixed flocks.

In order to relate individual behaviour and flock dynamics within the same flock, we observed flocks in arbitrarily-chosen pastures from the moment the first bird arrived from the roost or pre-feeding flock in the morning until the last bird departed at dusk, or it became too dark to record the birds' activities. During observation, we recorded all arrivals and departures at the flock, as well as the number and species of birds flying over it without landing. We also counted the birds of each species in the flock at hourly intervals to check the correspondence between actual flock size and composition and that estimated from arrival/departure data. In addition, we periodically recorded time budgets for a number of arbitrarily-chosen focal birds as described in Chapters 5 and 6. The earthworm community in the top 3cm of soil was sampled as before with 0.25m-sq quadrats both in the vicinity of some focal birds and elsewhere in the flock, and environmental factors affecting choice of feeding site and individual feeding efficiency (see Chapter 6) were recorded for each flock and focal observation.

Types of Foraging Flock

A number of studies have attempted to classify bird flocks on the basis of their dynamics. Ashmole (1971) and Hoffman *et al.* (1981), for instance, distinguished flocks on the basis of size, longevity (persistence) and the type of food they exploited. Newton (1972), Rubenstein *et al.* (1977), Macdonald and Henderson (1977), Caraco (1979a) and Caraco *et al.* (1980a,b) discriminated passerine flocks on the basis of mobility and size, while Goss-Custard (1970a, 1976, 1980, 1983) classified wader (charadriiform) flocks in terms of density, mobility and mode of foraging. Most feeding associations between ducks and geese (Anatidae) have also been distinguished by density and mobility (Zwarts 1976, Ogilvie 1975, Bryant and Leng 1975, Drent 1980, Thompson 1981, Prater 1982). We distinguish flocks and subflocks of plovers and gulls on the basis of their long-term variation in size. There are three categories.

Equilibrium flocks/subflocks (EFs). Equilibrium flocks/subflocks occur when more birds are attempting to feed at a site than can be accommodated at any one time (Barnard 1980c, Chapter 1). Flock/subflock size increases from zero because the rate of arrival exceeds the rate of departure; most arrivals can presumably find somewhere profitable to feed. As flock size increases, competition for available food is likely to become more severe until a point is reached when additional arrivals will decrease the mean feeding efficiency of birds in the flock and it will pay some individuals to leave (or not to join) and search for an alternative site. Since it may take time to assess whether or not a site is profitable, birds will continue to arrive and depart and flock/subflock size will oscillate about a mean. EFs therefore reach a dynamically stable size which we expect to be determined by the range of social and physical environmental factors affecting individual feeding efficiency. An example from a lapwing subflock is shown in Figure 7.1c.

Static flocks/subflocks (SFs). In some cases, fewer birds may be attempting to feed at a site than can be accommodated. Under these conditions, flocks/subflocks build up until the last bird has arrived and then remain at a

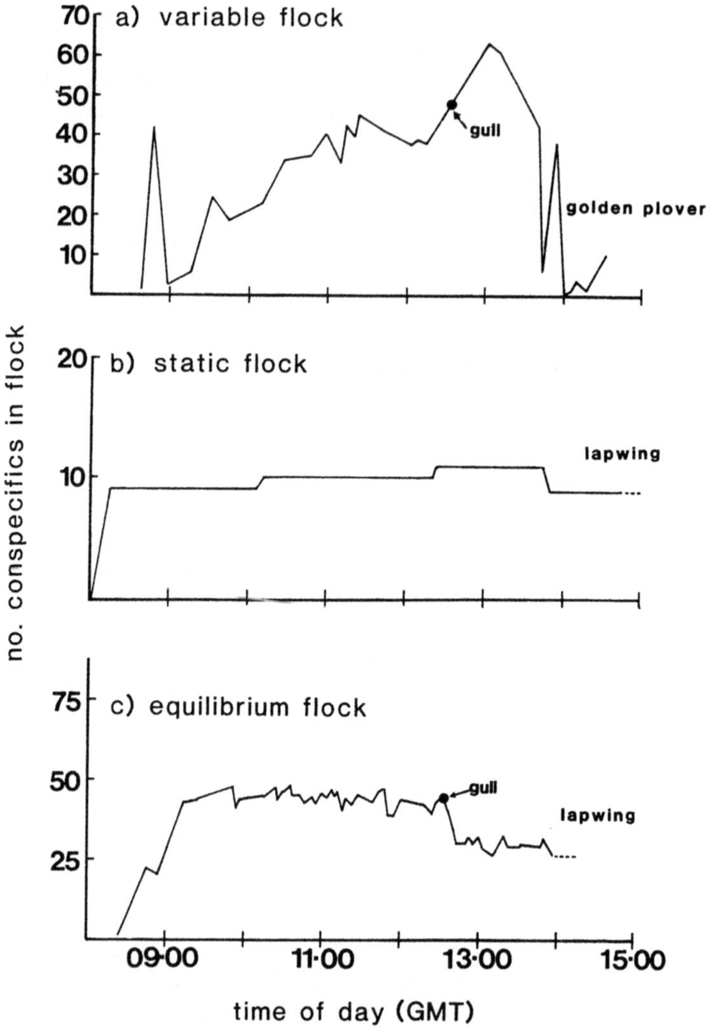

Figure 7.1: Examples of different types of flock/subflock: (a) variable (VF), (b) static (SF), (c) equilibrium (EF), including change after the arrival of a gull (solid circle). See text for details

more or less constant size until disturbed or departing to roost (Figure 7.1b, Chapter 1). Departure rates are low because there is little competition (as judged from the number of aggressive interactions, Chapter 6), presumably because the number of birds is below the carrying capacity of the site.

Variable flocks/subflocks (VFs). Variable flocks/subflocks are simply those which do not fall into the previous categories. They are characterised by large and unpredictable changes in size over relatively short periods. Most plover subflocks are of this type (Figure 7.1a, Table 7.1).

The Occurrence of Different Types of Flocks

Table 7.1 compares the occurrence of the three types of flocks in lapwings and golden plovers. Data for old and young pastures (see Chapter 3) are presented separately. There are considerable differences between the two species. The most striking is in the tendency to form EFs. Over 26 % of lapwing flocks/subflocks become EFs, but in golden plovers the figure is only just over 2 %. Furthermore, golden plover EFs last a much shorter time than those of lapwings (mean duration for lapwings = 78 ± 19.4 mins; for golden plovers = 12.7 ± 3.91 mins; t = 3.69, $p<.005$, d.f. = 25). At least in old pastures, golden plovers are also less likely to form SFs (Table 7.1). There are differences in the types of flock formed on old and young pasture. In lapwings, the proportion of each type of flock/subflock varies significantly with pasture type, with EFs occurring significantly less often than expected in young pasture. Lapwings, and to a much lesser extent golden plovers, tend to form SFs more frequently on old pasture than on young (for lapwings $X^2 = 4.45$, $p<.05$; for golden plovers $X^2 = 0.52$, n.s.). The apparently much larger effect of pasture age on lapwing SF formation is partly due to bias, with a large proportion (38%) of lapwing flocks on fields smaller than 3ha being SFs. Golden plovers rarely feed in these small fields.

If, as we suggest, EFs are the result of more birds attempting to feed at a site than can be accommodated at the time, we should expect *most* EFs to occur in small pastures which contain high worm densities. These should attract a large number of birds but accommodate only a few. We should, however, expect the *largest* EFs in large pastures with high worm densities,

Table 7.1: The flock dynamics of lapwing and golden plover flocks/subflocks in young and old pasture

		% flocks/subflocks of each type					
		Lapwings			Golden plovers		
		EF	SF	VF	EF	SF	VF
Young pasture n = 94, 72	%	8.5**	4.3	87.2	2.3	6.1	91.6
Old pasture n = 125, 63	%	32.3	17.1	50.6	2.7	8.9	88.4
TOTAL	%	26.1	13.1	60.8	2.5	7.9	89.6

EF = equilibrium flock, SF = static flock, VF = variable flock. See text.
Observations were made over periods of at least 20 mins.
n, number of lapwing and golden plover flocks/subflocks respectively.
Data for the December 1980/January 1981 period.
** $p < .01$; X^2 test comparing observed and expected occurrence of EFs on young and old pasture.

because these should both attract and accommodate a large number of birds. We tested these predictions using lapwing flocks/subflocks feeding in preferred old pastures.

Comparing the proportion of flocks/subflocks which reached an equilibrium size in three small (<3ha, range = 1.4-2.9ha) and three large (>3ha, range = 3.1-6.9ha) pastures, it turns out that EFs are just over twice as likely to occur in small fields as in large fields (proportion of flocks becoming an EF in small fields = 0.468; proportion in large fields = 0.202; n = 136 flocks/subflocks. $X_2 = 4.82$, $p < .05$). To see whether field size and worm density influence EF size in the predicted manner, we carried out stepwise partial regression analysis and included temperature measurements, time of day and daylength as independent variables (the first affects the density of worms in the top 3cm of soil, and the second two affect feeding priority in birds (Chapter 6)). The analysis was carried out only for flocks without gulls, because gulls affect plover feeding efficiency (Chapters 5 and 6) and tend to avoid fields smaller than 3ha. EF size was calculated as the mean number of lapwings present in the period over which arrival and departure rates were approximately equal (see Barnard 1980b).

Table 7.2 shows that field size and worm density are indeed the best predictors of EF size in lapwings. EFs tend to be bigger in bigger fields and when more worms are available. Figure 7.2, however, suggests that EF size may tend to level off when worm density exceeds about 200 worms/m^2. As expected, daylength correlates negatively with lapwing EF size because feeding priority is likely to be greater on short days (Chapter 6). Surprisingly, however, temperature does not appear to influence EF size (mean), though the variance of EF size is lower on cooler days. For comparison, Table 7.2 also includes analyses of golden plover subflocks (all SFs or VFs). Here, there is no effect of field size or worm density. Instead, as expected from Chapters 3 and 4, subflock size increases with the size of the lapwing EF although there tends to be fewer golden plovers later in the day and when days are long.

Size variation in VFs. Variable flocks (VFs) fluctuate in size unpredictably over short periods. One golden plover subflock, for instance, sampled seven times through a 12-minute period contained 3, 68, 295, 293, 9, 3 and 46 birds at each successive sampling. This, however, is an extreme and few flocks/subflocks vary to that extent. Most short-term large fluctuations of this sort are due to various forms of disturbance, the source of which is often unidentifiable. More commonly, VFs fluctuate less widely and over longer periods. Here it is possible that changes may be due to measurable environmental factors. In juncos, for instance, Caraco (1979a) found that ambient temperature was negatively correlated with flock-size variance. In house sparrows, on the other hand, Barnard (1980b) found that flock-size variance correlated positively with daylength and negatively with time of day.

As we might expect, the effects of environmental factors on VF size in

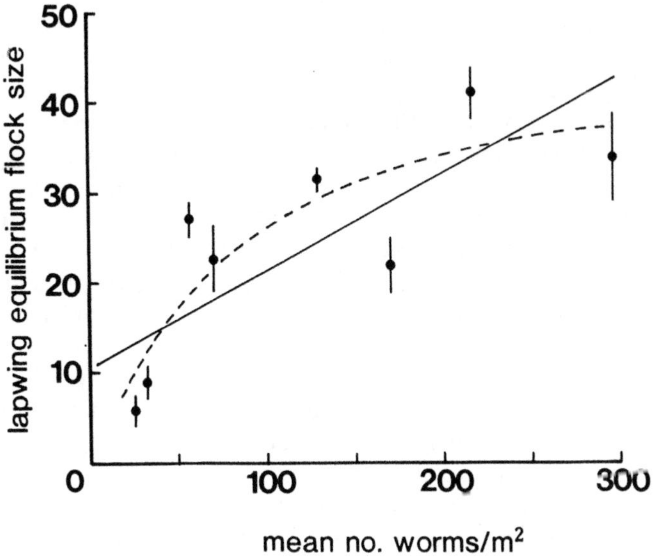

Figure 7.2: The relationship between EF size in lapwings and worm density. $r = .73$, $p < .05$. Data for 8 flocks/subflocks. Solid line fitted by linear regression, broken curve by second-order regression. Bars represent standard errors

Table 7.2: Beta values from stepwise partial regression analysis of the relationship between environmental factors and equilibrium flock/subflock (EF) size in lapwings and mean subflock size in golden plovers

		Independent variables					
Dependent variables		Worm density	Field size	Max °C	Daylength	EF size	Time of lapwing Lapwing EF formation
Lapwing EF size	(i)	.87**	.59**	ns	−.50*	—	—
	(ii)	—	.31*	ns	ns	—	—
Golden plover subflock size	(i)	ns	ns	ns	−.42*	ns	−.83**
	(ii)	—	ns	ns	ns	.39*	ns
Variance in lapwing EF size	(ii)	—	ns	−.39*	ns	ns	.29*

* $p < .05$, ** $p < .01$; significance levels for F-ratio associated with beta value, ns = not significant, — not included.

Data for 37 EFs: (i) worm samples also collected, (ii) worm samples not collected.

Variables originally included in the analysis: minimum overnight temperature, maximum daylight temperature, ambient temperature (°C), daylight, time of day, field size, worm density.

mixed plover flocks are not simple. Figure 7.3 shows that variance in lapwing VF size correlates negatively with minimum overnight temperature, but there is no similar correlation in golden plovers ($r = -.195$, n.s., $n = 61$ days of observation). Neither daylength nor daytime temperature (maximum or ambient) appear to influence VF size variance in the two species.

Caraco (1979a) explained the negative effect of temperature on junco flock-size variance in terms of increased aggression when it was warm and energy demand low. Increased fighting led to a more even distribution of high- and low-status birds between flocks. In Chapter 6, however, we showed that temperature correlates *negatively* with rates of aggression in plovers (see also Barnard 1980b). Caraco's explanation is therefore unlikely to apply, although, as Barnard (1980b) points out for house sparrows, the differences between juncos and plovers in the effects of temperature may reflect the more extreme low temperature range in Caraco's study rather than a difference in the relationship between temperature, feeding priority and aggression. Nevertheless, a more likely explanation is that reduced worm availability and increased rates of depletion lead to more movement between feeding sites on cold days. In addition, we have seen (Chapter 3) that plovers tend to form flocks in a greater number of fields on cold days. It might be that increased interspecific aggression in low temperatures (Table 6.9) results in some lapwings being forced to form small flocks in less profitable fields.

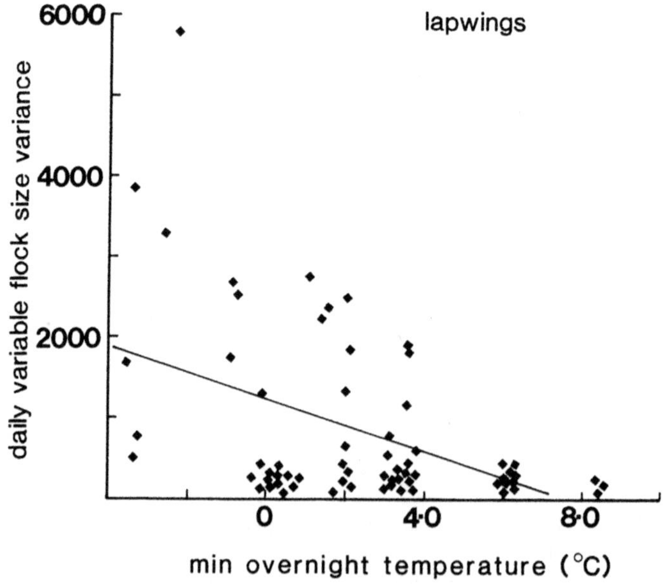

Figure 7.3: The relationship between the variance of VF size in lapwings and minimum overnight temperature. $r = -.38$, $p < .05$. Data for 61 days

At a broad level, therefore, the type of flock at a feeding site varies as we should predict. To understand why we need to examine the factors influencing the arrival and departure of birds at different feeding sites.

Factors Affecting Arrival and Departure

The dynamics of a feeding flock are a function of the rate at which birds arrive and depart. It seems reasonable to suppose that birds will decide to join or leave a flock on the basis of their expected feeding efficiency and security from predators within the flock. How might they assess these? From what we know already, birds may assess flocks in at least two ways. Firstly, they may use the presence of birds to indicate the availability of food between and within fields (Chapters 3 and 4). Secondly, they may use flock size and species composition as a guide to the levels of competition and feeding efficiency they can expect within the flock (Chapters 5 and 6). We examined the factors influencing arrival and departure in plover flocks by recording each lapwing, golden plover and gull joining or leaving given flocks (of all three types) over periods of a half to three hours. For each arrival and departure, we noted the number and species of birds concerned. If two or more birds landed or took off within five seconds of each other, we regarded them as a group. Within groups, the mean inter-arrival interval was 3.6 ± 1.7 secs and the mean inter-departure interval 1.4 ± 2.2 secs.

The Size and Composition of Arriving and Departing Groups

The first lapwings usually arrive at a field in a group. On average, these initial groups comprise about 15-20 birds and range in size from two to over 100 birds. Subsequent lapwings tend to arrive singly or in small groups of two to four. Figure 7.4a compares the observed size distribution of arriving groups with that expected on a random basis, and shows a significant bias towards small groups. A similar comparison for golden plovers shows a tendency to arrive in larger groups than lapwings, but the distribution is still significantly skewed to the left. Both species tend to leave in small groups, and over 60 % of lapwings leave singly (Figure 7.4). Departures in large groups ($>$ 10 birds) are usually the result of disturbance.

Table 7.3 compares the composition of groups (therefore excluding solitary birds) in which lapwings, golden plovers and gulls arrive and depart. The most striking point is that all three species predominantly arrive and depart in single-species groups. When only mixed plover groups are considered, however, the tendency is significantly greater for departures and is most pronounced in golden plovers. Although most black-headed gulls arrive and depart with conspecifics, gulls are more likely than either of the plover species to be in mixed groups, and they are significantly more likely to arrive ($p < .01$, X^2 test) and depart ($p < .05$, X^2 test) with golden plovers than with lapwings.

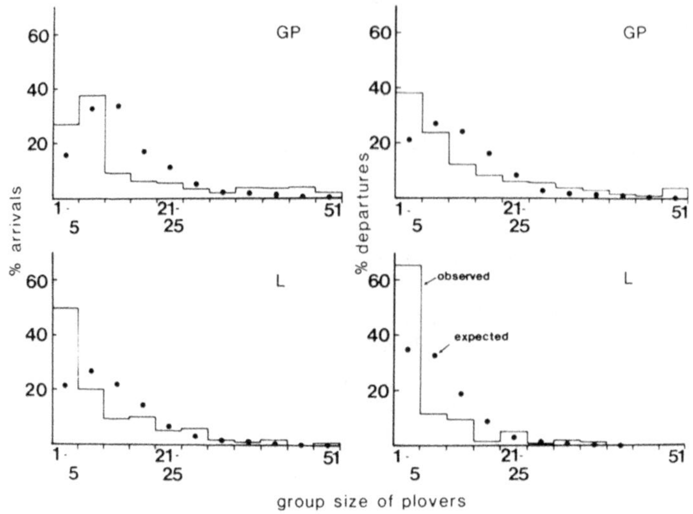

Figure 7.4: The size of arriving and departing groups of plovers in pasture fields: histograms, observed frequency distribution; circles, expected (Poisson) distribution. Top, arriving (left) and departing (right) groups of golden plovers, X^2 comparing observed and expected distributions = 59.4 and 41.0, $p < .001$, respectively; bottom, the same for lapwings, $X^2 = 39.2$ and 22.6, $p < .001$. Data for 525 golden plovers and 754 lapwings

Comparisons of arriving and departing groups show that gulls tend to arrive with larger than average groups of both plover species (for golden plovers, t = 2.77, $p<.01$, d.f. = 226; for lapwings t = 2.02, $p<.05$, d.f. = 2.02), but depart with larger than average groups only of golden plovers (t = 2.84, $p<.01$, d.f. = 135). We shall discuss the reason why gulls might selectively join golden plovers later.

Variation in the Arrival and Departure Rates of Plovers

If birds decide to join or leave a flock on the basis of expected benefit, we should be able to predict arrival and departure rates from the factors influencing individual time budgeting and feeding efficiency. Table 7.4 shows an analysis relating these to the rates at which lapwings and golden plovers join, depart from and fly over particular flocks. As we might expect, the arrival rate of golden plovers is correlated positively with lapwing density and field size. Golden plovers use local concentrations of lapwings to indicate areas of high worm density (Chapter 3) and prefer to feed in larger fields (Chapter 2). Surprisingly, there is no clear predictor of lapwing arrival rate. Departure rate in lapwings, however, increases with time of day and the number of gulls in the flock. Lapwings are therefore more likely to leave later in the day as they

Table 7.3: Species composition of groups arriving and departing at feeding flocks

Species composition	% Arriving groups	% Departing groups
LAPWING		
L	99.0	96.2
L + GP	0.5	3.5**
L + B-HG	0.5	0.3
n = 413, 231		
GOLDEN PLOVER		
GP	96.2	91.1
GP + L	0.5	6.7***
GP + B-HG	3.3	2.2
n = 347, 178		
BLACK-HEADED GULL		
B-HG	84.3	83.9
B-HG + L	2.4	3.2
B-HG + GP	13.3$	12.9
n = 83, 31		

L = lapwing, GP = golden plover, B-HG = black-headed gull.
**p <.01, ***p < .001; X^2 test comparing % of lapwings and golden plovers.
$ p < .05, X^2 test comparing % gulls arriving with lapwings and golden plovers (for departing groups X^2 = 1.99, n.s.).
n = number of observations of arriving and departing groups respectively.
All departures were from 3-species flocks.

satiate and depart to roost, and when gulls are likely to have a serious effect on their feeding efficiency, forcing them to move on.

As expected, there is no effect of gulls on golden plover departure rate. Golden plovers also depart towards the end of the day, but there is an additional positive effect of field size. While curious at first sight, the effect of field size makes sense if large fields hold bigger flocks from which a lot of birds can depart at any given time. There is indeed a positive relationship between field size and flock/subflock size in both species (for lapwings r = .564, p<.001, d.f. = 175, for golden plovers r = .621, p<.001, d.f. = 97; see also Crooks and Moxey 1966). Although there is no correlation between flock/subflock size and arrival rate, fewer lapwings and golden plovers fly over (rather than land in) large flocks. Fewer birds also fly over pasture late in the day, but this is more likely to reflect the approach of roosting when birds are moving to roost or pre-roosting sites. More birds might be expected to fly over pasture early in the day when they are hungry and looking for somewhere to feed. There is also a positive correlation between daylength and the rate of flying over in golden plovers (and gulls). Birds may be more selective about where they feed when a lot of time is available, and so move about more between feeding sites.

Table 7.4: Beta values for factors influencing the rate at which plovers and gulls join, depart from, or fly over given fields

	Independent variables							
Rates of:	**No. B-HG**	**No. GP**	**GP/ B-HG**	**L + GP**	**L/ha**	**Field size**	**Daylength**	**Time of day**
ARRIVAL								
L	ns	ns	ns	ns	ns	ns	ns	ns
GP	ns	ns	ns	ns	.88**	.22*	ns	ns
B-HG	ns	ns	ns	ns	ns	ns	ns	ns
DEPARTURE								
L	.27**	ns	ns	ns	ns	.14	ns	.21*
GP	ns	ns	ns	ns	ns	.19*	ns	.17*
B-HG	ns	1.16***	−.76**	ns	ns	ns	ns	ns
FLYING OVER								
L	ns	ns	ns	−.15*	ns	ns	ns	−.11
GP	ns	ns	ns	−.29***	ns	ns	.25*	−.18*
B-HG	ns	ns	ns	ns	ns	ns	−.51**	ns

Data for 196 flocks. See Table 7.2 for other details.

Edge effects on arrival and departure The advantages and disadvantages of feeding in a flock are likely to depend on where you are within the flock. Birds on the edge, for instance, may be more vulnerable to predators than those in the centre, simply because they provide an easier target. If food is patchily distributed and birds compete for access to the best patches, edge birds are also likely to end up on the poorest food supplies. There is some evidence for changes in individual behaviour in birds as a function of their position in a flock. In starlings (*Sturnus vulgaris*), Jennings and Evans (1980) found that the amount of time birds spent scanning increased with distance from the centre of the flock. Inglis and Lazarus (1981) suggest that in some cases, the effect of flock size on time budgeting may be caused by size-related changes in the proportion of birds which are at the edge of the flock. Since there is good reason to suppose that the risk of predation to plovers is greater at the edge of a flock, especially in small fields (see Chapter 9), we might expect flock size to be regulated by departures from the edge of the flock but arrivals at the centre. We define 'edge' birds as those with no companions between themselves and the border of the field.

Analysis of arrivals and departures at lapwing EFs (we analysed data for EFs rather than all flock types because of their more or less continuous arrivals and departures) show that 68 % of departures are from the edge, but 56 % of arrivals are also at the edge (data for 168 arrivals and departures). There is therefore a slight bias towards departures from the edge ($X^2 = 2.90$, $0.5 < p > 0.1$), but no tendency for birds to prefer alighting in the centre. Why should this be so? One possibility is that worm availability relative to position in the flock varies in a way which tends to counteract increased risk at the edge. In fact, contrary to expectation (Chapters 3 and 4), worm density tends to *increase* towards the edge of lapwing EFs (mean density at edges = 218 ± 14.3 worms/m^2, mean in centre = 179 ± 9.3 worms/m^2; t = 2.29, $p < .001$, d.f. = 61, n = 29 and 33 0.25m-square turf samples respectively) (cf. Waite's 1984b study of corvids). In addition, the variance:mean ratio (and thus patchiness) of worm density increases towards the edge (ratio at edges = 15.4 ± 4.6, range = 9.1-41.1; in centre = 3.6 ± 1.45, range = 0.16-13.0; t = 3.93, $p < .01$, n = 15 EFs sampled). At first sight, this seems to conflict with the trends in arrival discussed in Chapter 4. There we showed that golden plovers tend to alight within fields where the local density of lapwings, and therefore worms, is greatest. The discussion in Chapter 4 does not, however, take into account the position of locally-high bird and worm densities within flocks. As we have seen (Figure 4.1a), lapwings tend to end up on the periphery of mixed flocks. This means that newly-arriving golden plovers landing where lapwing density is greatest will tend to land near the edge.

Increased worm density near the flock periphery may account for half the birds alighting there, but why is there a bias towards departures from the periphery? One answer may be that the increase in worm patchiness towards the periphery makes locally-rich food supplies more defendable and that the increase in departures reflects displacement by defending lapwings (we have already seen in Chapter 6 that aggressive defence is more likely when worms are clumped). To test this, we observed arbitrarily-chosen lapwings in 11 EFs on old pasture when ambient temperature ranged between 3°C and 6°C (to control for the confounding effect of temperature on aggressive interaction (Chapter 6)). We noted whether birds were central or peripheral and recorded their behaviour for ten minutes. Table 7.5 shows a tendency for rates of aggression within and between plover species to increase towards the flock periphery. Lapwings at the edge are more likely to attack other plovers (lumping species, t = 2.14, $p < .05$). In particular, they are more likely to attack golden plovers than conspecifics (Table 7.5). Almost 90 % of intraspecific and 70 % of interspecific attacks in these cases are lunges or chases which do not appear to be attempts to steal food (see Chapter 6 and Cramp and Simmons 1983). Between 12 % and 18 % of interactions, however, result in the recipient taking off. Increased aggression by peripheral lapwings may be the reason why golden plovers tend to wind up concentrated in the centre of mixed flocks (Figure 4.1).

Arrival and Departure in Gulls

The factors affecting arrival and departure in plovers are those which determine the relative availability of prey within and between feeding sites. In this case, of course, birds are searching for their own prey in the ground. The situation with black-headed gulls is different. Here, birds are depending on

Table 7.5: The mean rate (± se) at which lapwings attacked (x) conspecifics or golden plovers in the centre or at the edge of equilibrium flocks

	Central lapwings		Edge lapwings	
	x L	x GP	x L	x GP
Mean rate of attack/min	.079 ± .017	.146 ± .029	.108 ± .022	.182 ± .021
no. observations	24	20	23	36
t-test	—	1.99	—	2.43*

* $p < .05$; t-test comparing mean rates of intra- and interspecific aggression.

other species to locate and provide prey which can then be stolen. While kleptoparasitism reduces some of the costs of obtaining prey, gulls are still faced with the cost of locating suitable flocks. One way to reduce this might be to associate with plovers which are moving between sites and are therefore likely to locate rich feeding areas. Although we have seen in Table 7.3 that most gulls arrive at flocks on their own (or in single-species groups), when they do arrive with plovers there is a significant (>5 fold) tendency to arrive with golden plovers rather than lapwings. Similarly, gulls are significantly more likely to leave with golden plovers. At first sight, it seems odd that gulls should prefer to associate with golden plovers, since their success rate is much higher against lapwings (Chapters 6 and 8). Golden plovers, however, depend for their feeding efficiency on locating flocks of lapwings. Since they appear to be good at choosing flocks where food availability and feeding efficiency are high, it may pay gulls to use them as a quick means of locating profitable flocks. We saw in Chapter 3 that gulls commonly associate with pre-foraging, though not post-foraging, plover flocks. One reason for this might be that gulls use pre-foraging aggregations as a means of joining up with birds which are likely to depart for good feeding sites. In other words, pre-foraging flocks might act as a form of information centre (Ward and Zahavi 1973, Chapter 1).

Gulls, information centres and assembly points. We have outlined the rationale and some of the evidence for the information centre hypothesis in Chapter 1. As dietary opportunists entering into a wide range of inter- and intraspecific exploitative relationships (see e.g. Brockmann and Barnard 1979, Burger and Gochfeld 1979, 1981), gulls are good candidates for testing the idea. Several studies have already done so and it is worth reviewing some of their conclusions.

Evans (1982a,b) studied the formation of feeding flocks in black-billed gulls (*Larus bulleri*). Black-billed gulls nest in dense inland colonies in New Zealand. They typically forage in flocks and exploit unpredictable but temporarily-rich patches of food, including earthworms and other small invertebrates, that are exposed on open pasture after rain or turned up in fields by ploughing (Dawson 1958). They are thus very similar in aspects of their feeding ecology to the European black-headed gull.

Evans (1982a) found that, when flocks of black-billed gulls leave a colony to feed, the first bird out (the 'leader') often advertises its departure by emitting loud calls that attract and apparently recruit other birds ('followers') into the flock. This suggests that leaders derive some benefit from having others around them when they leave the colony. It would seem very easy for them to forego calling if recruitment was costly or not advantageous. As Evans points out, however, the precise nature of the advantage is not clear. Birds may benefit through increased protection from predators while in the air or foraging, or they may benefit from increased feeding efficiency at the feeding site. Indeed, in a second study, Evans (1982b) showed that gulls locate good feeding sites by alighting where they see other conspecifics already foraging and capitalise on local increases in food availability by copying neighbours. Evans therefore sees colonial roosts in the gulls not so much as information centres as assembly points where foraging birds reunite from dispersed feeding sites and create local concentrations which facilitate the formation of flocks at the beginning of the next day. The assembly point hypothesis assumes that neither leaders nor followers are necessarily determined by previous foraging success (as expected if roosts were information centres).

Andersson *et al.* (1981) made a similar study of black-headed gulls in Sweden. They set up an artificial feeding station (a white, baited raft) on a lake near a gull colony and recorded the visits of different gulls to the station. If the colony acted as an information centre, gulls returning from this unusually rich food source should be followed out by other colony members when they next leave. Contrary to the information centre hypothesis, however, returning gulls were not followed by others, even though they arrived at the colony with conspicuously-expanded crops and fed their young copiously with high-quality food. There was also no evidence for an alternative possibility that birds followed only groups rather than singletons out from the colony, despite there being a significant tendency towards temporal clumping in departing birds. However, it may have been that food was generally easy to find and that no birds were doing badly enough to follow successful foragers. An experiment with model gulls nevertheless suggested that, as in the black-billed gulls, birds were attracted to groups of foraging conspecifics. Aggregation on departure may be a means by which birds ensure they have at least some neighbours to copy at a feeding site. The evidence from Andersson *et al.*'s study was thus against the information centre hypothesis, but consistent with Evans's suggestion of colonies as assembly points.

Pre-foraging Flocks as Information Centres/Assembly Points

Although the initial concept and most tests of the information centre hypothesis have focused on communal roosts, the pre-foraging (or post-roosting (Ydenberg and Prins 1983)) flocks formed by several species of socially-feeding predators may play a similar role. There is evidence that the equivalent of post-roost gatherings in hamadryas baboons (*Papio hamadryas*) operate as centres where animals 'vote' on where to forage (Sigg and Stolba 1981). Some support for the idea in birds is provided by Ydenberg and Prins's (1983) investigation of post-roosting flocks of barnacle geese (*Branta leucopsis*).

Ydenberg and Prins studied post-roost gatherings on the Dutch island of Schiermonnikoog and found that they acted as centres for dispersal to and from feeding grounds on the island and elsewhere. Barnacle geese are specialist grazers exploiting the new growth of the food plants (here the grasses *Lolium perenne* and *Poa pratensis*). Ydenberg and Prins see the gatherings as allowing birds to decide whether or not to stick with the current feeding site or choose a different one. Information could come from several sources, including the number of geese in the assemblage (and therefore the likelihood of gaining a good foraging position at the feeding site), temperature and the rate of arrival of geese from other areas (a high rate might indicate generally poor feeding conditions). In loose support of the information centre idea, it turns out that the pattern of dispersion from the island changes with temperature, so that geese are more likely to leave when it is cold. Geese also spend longer in post-roost gatherings if they have departed from feeding areas late the previous day. This is explicable if the information on which decisions to persist or change are based becomes more valuable, and thus worth taking more time over, as the probability of changing increases. The probability is likely to increase after late departures, which usually signify reduced food availability.

Ydenberg and Prins's study focuses on single-species pre-foraging aggregations. There is no reason to suppose, however, that mixed flocks cannot act in the same way. As we have suggested, kleptoparasites like black-headed gulls may benefit from joining mixed pre-foraging host flocks by reducing the time spent searching for rich feeding sites.

If gulls do capitalise on pre-foraging flocks in this way, we can make two predictions:

1. Gulls should leave with (groups of) birds which are likely to find the most profitable feeding sites (for gulls).
2. Gulls should do better in foraging flocks when they arrive with these birds than when they arrive on their own.

Unfortunately, it was not possible to test whether gulls associated with the *individual* birds which had fared best the previous day.

We observed birds in 18 pre-foraging flocks which built up in a field of winter wheat. The arrival and departure of all birds was recorded from the moment the first bird arrived from the roost until the last bird departed to feed. For each departing gull, we recorded the size and species composition of the groups with which it left and the order in which groups departed (it may be, for instance, that gulls always depart with the first group because this is likely to contain the hungriest birds or those which are departing for previously-located rich feeding sites (de Groot 1981)). We also recorded maximum, minimum and ambient temperature and daylength because any of these may have influenced the gulls' food requirement and therefore probability of departure. For those gulls which landed in pastures adjacent to the pre-foraging site, we recorded the amount of time spent in the foraging flock and prey-capture rate (number of worms stolen per minute). In the same fields, we also recorded the density of small (<32 mm) and large (>32 mm) worms in the immediate area in which groups of plovers and gulls landed.

Table 7.6 shows that a greater number of golden plover than lapwing groups are likely to leave pre-foraging flocks during any given period (the difference in Table 7.6 also reflects a difference per head in the flock in the same direction). Furthermore, gulls are about four times more likely to associate with departing golden plovers than with departing lapwings. Gulls were never seen to leave pre-foraging flocks on their own, or, interestingly, with mixed groups of plovers (even though these made up 7.6 % of departing groups). An important point, however, emerges from the analysis of the sequence of group departures. Table 7.6 shows that gulls tend to join the *first* departing plover group on only about one in every two days. They are, however, significantly (over three times) more likely to join the first group of golden plovers than the first group of lapwings. An obvious explanation is that the first lapwings to leave the pre-foraging flock stand a strong chance of

Table 7.6: The tendency for gulls to leave pre-foraging flocks with groups of lapwings or golden plovers. Data for 18 days

	Species departing		
	Lapwing	**Golden plover**	X^2
No. departing groups available to gulls	36	44	—
No. departing groups with gulls (%)	3 (8)	15 (34)	5.62*
No. days on which gulls joined the first group of plovers departing (%)	2 (11)	7 (39)	7.68**
Mean no. birds in pre-foraging flock	43.5 ± 7.15	61.0 ± 11.0	—

* $p < .05$, ** $p < .01$; X^2 test comparing results for lapwings and golden plovers, d.f. = 1. Expected frequencies based on the proportion of departing groups containing lapwings or golden plovers (*), and on the number of days on which lapwings or golden plovers were first to depart (**).

being the first birds onto the pasture. They are likely to have to locate food by sampling rather than using the presence of other foragers. Gulls associating with them may thus waste time locating profitable feeding sites. Golden plovers, on the other hand, rarely feed on their own or in single-species flocks. When they depart from pre-foraging flocks they are likely to seek out lapwings which have already begun to forage. By associating with golden plovers, gulls may reduce their searching time in locating profitable lapwing flocks. A prediction arising from this is that gulls should become less fussy about the species with which they depart from pre-foraging flocks as time goes on. Lapwings departing later are likely to encounter birds already foraging and spend less time searching for feeding sites. Although pre-foraging flocks last only a relatively short time (average of 0.54 hours), preference in gulls does appear to change. The significant difference between the tendencies to join lapwings and golden plovers early on disappears towards the end of the pre-foraging period ($X^2 = 3.65$, n.s.).

The results of partial regression analysis suggest that group size is an important factor determining the gulls' choice of departing group (Table 7.7) and that the effect of group size is much stronger in golden plovers. So far, therefore, it does seem as if, at least initially, gulls selectively associate with golden plovers rather than with lapwings, and that the bigger the group the more attractive it is. While there are good reasons for expecting association with golden plovers (see above), is there any evidence that gulls locate better feeding sites as a result of the association? Table 7.7 shows a significant

Table 7.7: Beta values for relationships between the departure of gulls from pre-foraging flocks, the composition of departing groups and the gulls' subsequent foraging success

Dependent variables		Independent variables: No. golden plovers in departing group	No. lapwings in departing group	Mean worm density
Tendency for gulls to leave	(i)	.66**	.25*	—
with plovers	(ii)	ns	ns	.93*
Average amount of time spent by gulls in feeding flock	(ii)	ns	.94*	ns
Gull's average capture rate (worms/min)	(ii)	ns	.75	ns

Data for 18 pre-foraging flocks: (i) for 71 departing groups; (ii) for 70 turf samples.
Additional independent variables included in the analysis: total no. departing plovers, golden plover:lapwing ratio, order of departures, temperature (min. overnight, max. daylight and ambient), daylength, variance in worm density, density of small (< 32 mm) and large (> 32 mm) worms.
* $p < .05$, ** $p < .01$.

positive relationship between the tendency for gulls to join a departing group and the density of worms in the immediate area in which it eventually lands. The most preferred groups therefore appear to end up on the best feeding areas. However, it is not the case that preferred groups end up where the density of *large* worms is high. This is not surprising, of course, since the choice of feeding site by plovers is likely to be geared to the availability of small to intermediate-sized worms (Chapter 5). Since all gulls appear to leave pre-foraging flocks with plovers, it follows that most joining foraging flocks early in the day do so in association with plovers. Later in the day, however, when gulls move between foraging flocks, they tend to leave and arrive on their own, hence the strong bias towards single-species arrivals/departures in Table 7.3 (probability of a gull arriving or departing alone at the first foraging flock of the day = .103; probability of arriving alone between 11.00 and 15.00 hours GMT = .873; $X^2 = 45.5$, $p < .001$, d.f. = 1 for 105 gulls arriving at flocks). Prediction 1 above is therefore borne out to the extent that gulls locate rich feeding sites as a consequence of their preferred association with golden plovers at the beginning of the day, but worm availability is not biased towards the sizes preferred by gulls. Do gulls therefore do better in foraging flocks when they join with plovers?

We examined the relationship between the size and species of departing group joined by gulls and both the subsequent amount of time gulls spent in the foraging flock and their capture rate in the flock. Table 7.7 shows that gulls spend longer in a foraging flock when they arrive with large groups of plovers, but that time spent in the flock correlates with the number of lapwings rather than the number of golden plovers. The reason may be that larger groups of lapwings also provide gulls with more preferred hosts at the feeding site, and their greater stay time may reflect increased feeding opportunity (see the capture rate in Table 7.7). There is no reason to suppose that gulls would benefit from increased numbers of golden plovers in the flock. Indeed, as we shall see in the next chapter, feeding efficiency in gulls tends to be compromised by the presence of golden plovers. Interestingly, Table 7.4 shows that, while the composition of a foraging flock has no effect on the arrival rate of gulls, gulls are more likely to depart when the flock contains a large number of golden plovers. That gulls clearly benefit from arriving with plovers (prediction 2 above) is shown by comparing the stay times and capture rates of birds arriving in plover groups with those of gulls arriving singly or with other gulls. The latter stay for significantly shorter times and achieve less than half the capture rate of plover-group birds (mean stay time for gulls with plovers = 47.8 ± 13.45 mins, mean time for gulls without = 8.56 ± 1.96 mins, $t = 2.89$, $p < .01$, d.f. = 19; mean capture rate for gulls with plovers = 0.37 ± 0.03 worms/min, mean rate for gulls without = 0.15 ± 0.02, $t = 5.65$, $p < .001$, d.f. = 12).

Flock Composition and Equilibrium Flock Size

In the introduction and Chapter 1, we suggested that flock size, composition and dynamics should be predictable from a knowledge of the effects of flocking on individual time budgeting and feeding efficiency. If this is the case, we should be able to predict attractiveness of different flocks and changes in species flock/subflock size as a consequence of changing flock composition. Predicting changes in flock/subflock sizes which are the result of adaptive individual decisions, however, presents a problem if the number of birds varies widely for other reasons (such as high rates of disturbance or a tendency for birds to arrive in large groups). For this reason, we chose to examine changes in the size of equilibrium flocks/subflocks, where arrival and departure rates remain more or less constant over a long period and are therefore likely to reflect the utility of the feeding site rather than chance perturbations. If this is so in plovers, then we can predict changes in EF size in response to particular changes in flock composition. Because of differences in the frequency of occurrence of different flock/subflock types in the two plover species (Table 7.1), however, we confine our analysis of changes in EF size to lapwings.

Flock Composition and EF Size in Lapwings

As we have seen in Chapter 6, time budgeting and feeding efficiency in lapwings are influenced by the presence of gulls and golden plovers in ways which are dependent on flock composition. Thus gulls decrease lapwing feeding efficiency in all flocks, but golden plovers cause a decrease only in flocks containing gulls. The predicted consequences of the arrival of gulls and golden plovers on lapwing EF size are therefore:

(a) The arrival of one or more gulls should result in a decrease in EF size.
(b) The arrival of golden plovers should result in a decrease in EF size only when gulls are present.

As might be expected, the number of occasions on which flocks/subflocks remain undisturbed long enough to reach an equilibrium size before and after the arrival of given birds is small. Nevertheless, we were able to record 15 cases in which the number of lapwings reached an equilibrium before and after the arrival of gulls. As Figure 7.5a shows, there was, in all cases except one, a decrease in lapwing EF size following the arrival of one or more gulls. An example is shown in Figure 7.1c. Interestingly, the effect of gulls appears to be all-or-nothing; there is no significant correlation between the number of gulls arriving and the magnitude of the subsequent decrease in lapwing EF size ($r = -.384$, n.s., $n = 15$ samples).

To test the second prediction (b), we analysed data for flocks in which golden plover subflock size changed by different numbers of birds but the

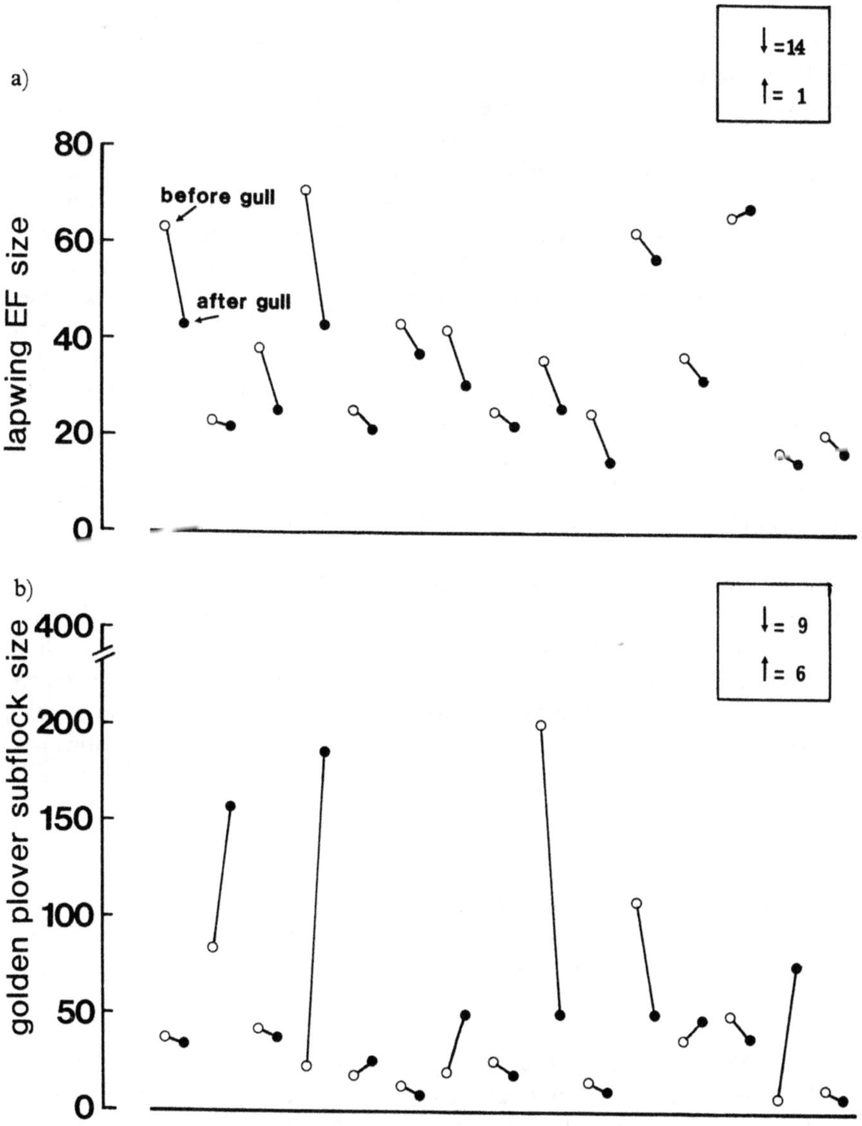

Figure 7.5: Changes in plover flock/subflock size after the arrival of one or more gulls. (a) Changes in lapwing EF size; open circles, EF size before arrival, closed circles, EF size after. (b) Changes in golden plover VF size; symbols as above

Source: modified after Barnard et al. *(1982)*

number of gulls remained constant (to control for any confounding effect of gulls arriving or departing). Figure 7.6 shows that, as expected, lapwing EF size decreases with increasing numbers of newly-arriving golden plovers when gulls are present and golden plovers have a negative effect on lapwing feeding efficiency. The relationship, however, is not a simple one. When only a small number of golden plovers arrive, lapwing EF size tends to *increase.* The most likely reason is that the addition of only a few golden plovers results in little interference with lapwing feeding efficiency, but reduces the number of kleptoparasitic attacks by gulls against lapwings (see Chapter 8) and creates a bigger flock to attract more birds. When a lot of golden plovers arrive, however, increased interspecific aggression initiated by golden plovers (as a consequence of the increased golden plover:lapwing ratio — see Table 6.9) may cause some lapwings to leave. When gulls are absent, the number of golden plovers has no effect on lapwing EF size (Figure 7.6). Overall, however, the number of lapwings almost always decreases. This may be because, without the benefit of reduced kleptoparasitism, the disturbance effect of alighting birds (e.g. Figure 4.1) is sufficient to cause some lapwings to leave, even though there is no evidence of reduced feeding efficiency.

Flock Composition and Subflock Size in Golden Plovers

Although golden plovers are less likely to form EFs, there should still be a tendency for changes in subflock size to reflect expected individual feeding efficiency. As with lapwings, we can make some predictions about the effect of new arrivals on golden plover subflock size. In this case, the arrival of gulls does not affect individual feeding efficiency (Table 6.4a). Additional lapwings, however, reduce golden plover feeding efficiency in flocks without gulls, though not if gulls are present (Tables 6.3, 6.4). Our predictions are therefore:

(c) The arrival of gulls should have no effect on golden plover subflock size.
(d) The arrival of lapwings should result in a decrease in golden plover subflock size when there are no gulls present, but not when flocks contain gulls.

Figure 7.5b shows that, indeed, there is no consistent change in golden plover subflock size after one or more gulls arrive. Just under half the observed subflocks increased in size and just over half decreased. There is a significant negative correlation between the number of lapwings joining the flock and the change in golden plover subflock size when there are no gulls (Figure 7.7a), but no correlation when gulls are present (7.7b). As with lapwing EFs however, the relationships are not simple. Figure 7.7a shows an *increase* in the number of golden plovers when only a few lapwings arrive. This may be because a few additional lapwings do little to change the lapwing:golden plover ratio and thus the frequency of aggressive and other interactions initiated by lapwings against golden

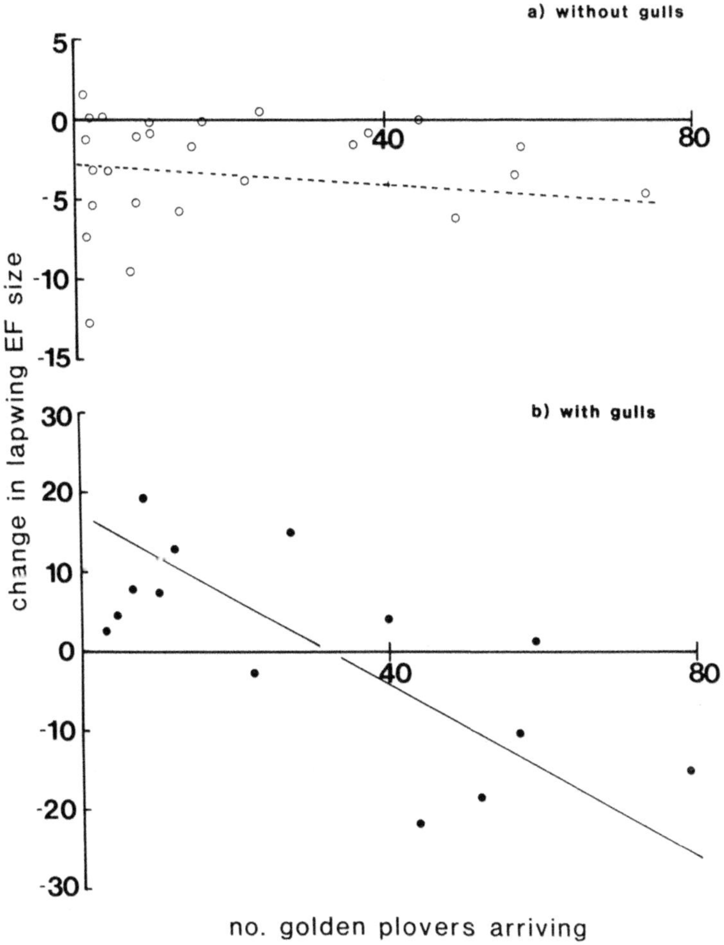

Figure 7.6: Changes in lapwing EF size after the arrival of golden plovers. (a) Without gulls, $r = -.19$, n.s., $n = 26$ flocks; (b) with gulls, $r = -.54$, $p < .05$, $n = 14$ flocks. Lines fitted by regression

Source: as Figure 7.5

plovers. Small increases in lapwing number, however, may allow golden plovers to crouch more and scan less (Tables 6.3b, 6.4b) and hence do better than they would by leaving. It may be, however, that golden plovers simply do not respond to small changes in the number of lapwings when there are already a lot of lapwings present (J. Lazarus pers. comm.).

In both lapwings and golden plovers, therefore, subflock sizes show a tendency to change with flock composition as we should expect from the effects of flock composition on individual feeding efficiency.

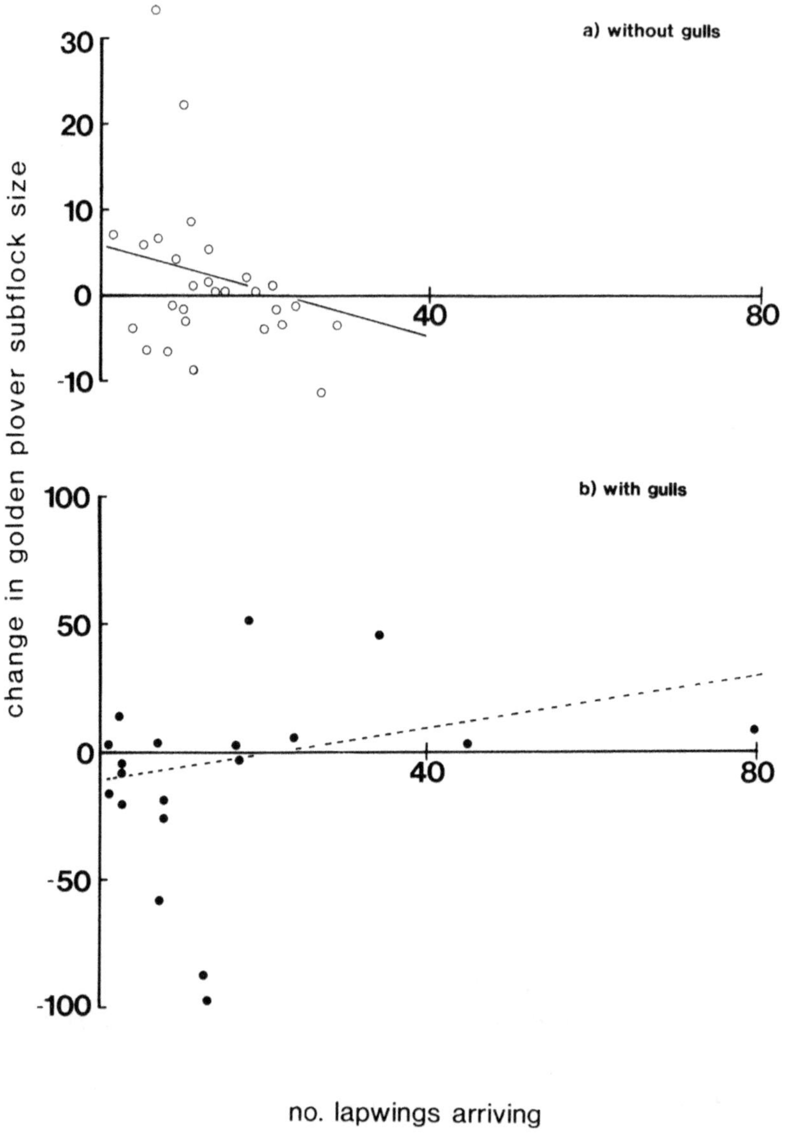

Figure 7.7: Changes in golden plover VF size after the arrival of lapwings. (a) Without gulls, $r = -.39$, $p < .05$, $n = 28$ subflocks; (b) with gulls, $r = .29$, n.s., $n = 19$ subflocks. Lines fitted by regression

Source: as Figure 7.5

Time Budgeting, Feeding Efficiency and Flock Dynamics: Résumé

Our assumptions about the predictability of foraging flock dynamics in gulls and plovers appear to be borne out. As in earlier studies of other species (e.g. Caraco 1979b, Barnard 1980b), the attractiveness of a flock depends on the expected effects of joining or remaining in it.

Arrival and departure rates at mixed plover and gull flocks are not random. All three species are affected by aspects of flock composition and the physical environment, which are among those determining individual time budgeting and feeding efficiency. Golden plovers are more likely to land in large fields where the density of lapwings is high, while gulls tend to arrive with golden plovers but leave if the latter are too numerous in the flock. These observations fit with what we know about the feeding strategies of the three species and the apparent risks of predation in the study area. That the pattern of lapwing arrival does not seem to depend on flock composition also fits. Lapwing feeding efficiency is influenced to a much smaller degree by flock composition (except for the presence of gulls) than that of the other species. Since lapwing flocks often build up independently of golden plovers and gulls, we might expect criteria other than flock composition to determine the attractiveness of a feeding site. Departure rates in lapwings are, however, influenced by flock composition, and birds are more likely to leave when pressure from kleptoparasitic gulls is high. The same is not true for golden plovers, which are under much less pressure from gulls. While both plover species tend to arrive and depart in single-species groups, they are significantly more likely to depart in mixed groups. If they are about to leave, it may pay birds to join departing groups of other species, simply through safety in numbers (Barnard 1982, Evans 1982a,b). Since their ultimate feeding requirements will be different, however, we might expect segregation when they decide on their next feeding site.

There is an interesting edge effect in plover arrivals and departures; both are more likely to occur at the flock periphery. While edge effects which have been demonstrated in other species seem to be based on increased vulnerability to predators at the flock periphery (e.g. Jennings and Evans 1980, Inglis and Lazarus 1981), in the present case the edge effect appears to be brought about by the distribution of birds relative to worm density. Worms increase both in density and patchiness towards the flock periphery. The greater defendability of food supplies at the periphery results in increased aggression and displacement of birds, particularly golden plovers. This has three effects. Firstly, both lapwings and golden plovers tend to be ousted from locally-rich feeding areas and leave; secondly, golden plovers that do not leave tend to end up in the centre of the flock; and, thirdly, both lapwings and golden plovers may land near the periphery because that is where lapwing density (and therefore worm density) is greatest.

While the two plover species tend to move between flocks in single-species groups, gulls move either singly or in association with groups of plovers. In the latter case, they show a pronounced preference for golden plovers. The most obvious reason is that golden plovers also rely (in a different way) on locating flocks of lapwings, and appear to be good at choosing flocks which are feeding on high worm densities. By tagging along with them, gulls may be able to reduce the amount of time spent searching for profitable feeding sites. Support for this argument comes from pre-foraging flocks, where gulls always leave with plovers and tend to delay their departure until a group of golden plovers leaves. By being selective in their choice of departing groups, gulls end up on better feeding sites. While we do not have the data to see whether gulls are following the individual plovers which fared best the day before, it seems in any case unlikely that pre-foraging flocks act as information centres *sensu* Ward and Zahavi (1973). A more plausible explanation is that they are assembly points (Evans 1982a) where birds can capitalise on the benefits of grouping both in transit to feeding sites and while feeding.

The difference in arrival and departure patterns in lapwings and golden plovers means that the dynamics of the two species' subflocks are also different. Because they tend to move between fields in a wider range of group sizes, golden plovers are much less likely to form equilibrium subflocks. Nevertheless, despite the differences in flock formation, variation in both lapwing and golden plover subflock size is explicable in terms of the effects of flock size and composition on individual feeding efficiency.

Summary

1. Plovers form three types of flock, distinguishable on the basis of their long-term variation in size. Lapwings are more likely to form equilibrium flocks/subflocks than golden plovers.

2. Arrival and departure rates in lapwings, golden plovers and gulls are influenced by aspects of flock composition which affect individual time budgeting, feeding efficiency and apparent predation risk.

3. While all three species tend to arrive and depart on their own or in single-species groups, gulls show a significant preference for associating with golden plovers. This may reduce their searching time.

4. Gulls appear to use pre-foraging flocks as a form of assembly point where they can select groups of plovers with which to depart for feeding.

5. Variation in species flock/subflock size is predictable from a knowledge of the effects of flock size and composition on individual behaviour.

Chapter 8

Kleptoparasitism: Host and Prey Selection by Gulls

Black-headed gulls steal food from plovers. As we saw in Chapter 2, they seem to be able to obtain in excess of their daily energy requirement solely by stealing earthworms from foraging lapwings. In the preceding chapters, it has also become clear that the arrival of gulls in a flock and their kleptoparasitic activity has a complex effect on time budgeting, feeding efficiency and flock dynamics in both lapwings and golden plovers.

Although gulls steal food from other birds rather than search for it themselves, they are, nevertheless, faced with problems similar to those of any other predator. The profitability of stealing is likely to depend on the sizes of prey procured by potential hosts and the rate at which prey is procured. The distance to target birds, the probability that they will escape with the prey and the time and energy costs of attack are additional factors which will influence the profitability of kleptoparasitism. In this chapter, we examine the costs and benefits of kleptoparasitism in black-headed gulls and see whether gulls choose the form and timing of attacks so as to maximise their feeding efficiency.

Attack Efficiency in Gulls

The behaviour of gulls in the study area was similar to that described by Källander (1977) (see Chapter 2). Throughout the winter season, gulls are present in varying numbers within plover flocks. The number of gulls per flock varies considerably (from one to 40-50) depending on the distribution of potential hosts between and within pasture fields and the rate at which they make appropriate-sized worms available. Gulls are usually scattered sparsely within flocks and often stand on local high points (mole hills, tussocks or ridges in the pasture) between attacks. Particular vantage points are used throughout the winter in some pastures. All plovers in a flock are potentially vulnerable to attack by gulls and in some cases, where few gulls are present and the flock is not partitioned into small gull territories, lapwings are attacked over distances of up to 65m. Even casual observations, however, suggest that gulls do not pick victims arbitrarily. Several factors appear to interact in determining the profitability of attack and the targets which are chosen.

The Energetic Cost of Attack

Clearly, attacks cost a gull time and energy and this has to be debited from any energetic or other returns from stolen prey. It is reasonable to suppose that the cost of an attack will be related in some way to its duration and the distance over which it is carried out.

Estimating cost. The energetic cost of flight in birds has been calculated by a number of authors (see e.g. Hart and Berger 1972, Tucker 1972, Greenwalt 1975, Butler 1979) using a variety of techniques, including direct measurement of oxygen consumption (Tucker 1968, Butler *et al.* 1977) and biotelemetry (Torre-Bueno and La Rochelle 1978, Butler and Woakes 1980). Tucker (1969, 1972) has made detailed estimates of the cost of flight in laughing gulls (*Larus atricilla*) using oxygen consumption as a measure of energy expenditure, and we have used these as a basis for estimating costs in the slightly smaller black-headed gull.

Working with free-flying barnacle geese, Butler and Woakes (1980) showed a strong positive correlation between wingbeat frequency (wingbeats/min) and the frequency of both breathing and heartbeat (indices of energy expenditure). Since there is also a good correlation between energy expenditure and wingbeat frequency in laughing gulls (calculated from data in Tucker 1972), we used wingbeat frequency as an index of energy expenditure in black-headed gulls.

While wingbeat frequency provides a convenient measure in the field, it does have limitations. One problem is that it does not reflect increased energy expenditure caused by intention movements before, and panting during, flight. Butler and Woakes (1980) recorded an average increase of 218% over resting state in the heartbeat of barnacle geese during the two seconds immediately before take-off. This increased to 280% two seconds after take-off, and increased further during subsequent flight. During flight, the mean rates of breathing and heartbeat were, respectively, 11.7x and 7.2x the resting rate. When birds landed, heartbeat rate returned to the resting level within three minutes, but breathing rate gradually increased as birds panted. Since we estimated energy expenditure for flights of only 0.3 to 84.7 seconds' duration (compared with an average of 14.4 minutes in the goose study), however, the effects of additional (particularly post-flight) expenditure are likely to have been reduced.

We filmed attacks using a portable colour video recorder and camera fitted with a zoom lens. Wingbeat frequency was then measured for each 0.5s interval throughout the attack flight. The energetic cost per wingbeat (obtained from Tucker's study) was multiplied by the number of beats per 0.5s to estimate the cost of attack flights. A few additional data were obtained directly in the field with a stopwatch. Assuming a correlation between wingbeat frequency and energy expenditure similar to that obtained by Tucker, Figure 8.1a shows a positive relationship between flight duration and estimated energy expenditure, levelling off at about 6 seconds. Since flight duration is strongly positively correlated with the total distance travelled during a flight (see Figure 8.3a), flights over long distances are more expensive. Furthermore, attack flights are significantly more expensive than relocation flights (where gulls move from one vantage point to another between attacks) of the same duration (Mann-Whitney $U = 124$, d.f. $= 19{,}20$, $p < .05$, see Figure 8.1a). The main reason for the difference is that attack flights begin with a burst of rapid acceleration (and hence a high wingbeat frequency) compared with the more gentle take-off and gliding flight associated with relocation. A greater proportion of short attack flights is taken up with acceleration, so two attacks of 4s cost more than a single attack of 8s. Paradoxically, there is also a tendency for long attack flights ($>$20m) to begin with less rapid acceleration than shorter flights, again reducing energy expenditure.

Figure 8.1 also shows the frequency distribution of attack-flight duration for both plover species. Since the peak of both distributions falls within the accelerating part of the cost curves, we can assume that, for most attacks, cost increases with duration and distance. Factors which tend to increase flight duration or distance, therefore, will increase the cost and decrease the profitability of attack. What are the circumstances in which attack flights become prolonged?

Detection of Attack

One immediate problem facing an attacking gull is detection by the target plover. A potentially short attack may end up as a protracted aerial chase if the plover sees the gull coming. Detection may therefore substantially increase the cost of attack. We recognised three categories of attack based on whether or not an approaching gull was detected by its target (Figure 8.2).

Undetected attacks. Approximately half the attacks made against lapwings and golden plovers go undetected until the target bird is reached (Figure 8.3b). As expected, undetected attacks tend to be shorter than detected attacks (Figure 8.3a). Undetected attacks are usually launched when the target is extracting or mandibulating a worm, although occasionally they are launched during the pre-extraction crouching posture (see later).

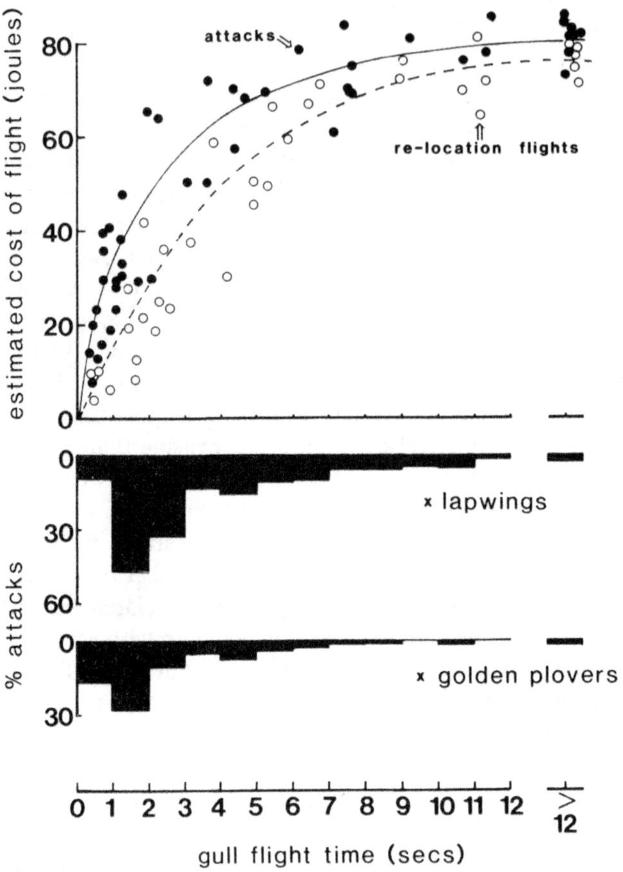

Figure 8.1: The estimated energetic cost of attack and relocation flights in relation to distance from the target plover (data for 47 attacks and 35 relocations) and the distribution of attack distances (histograms) against lapwings and golden plovers (data for 162 and 72 attacks respectively)

Type 1 detected attacks. Sometimes when an approaching gull is detected, the target flies off after first discarding its prey. The gull then alights where the target has been feeding and picks up the deposited worm. Fewer than 20 % of recorded attacks against each species fall into this category (Figure 8.3).

Type 2 detected attacks (aerial chases). More often, detection results in the target taking off with its prey and aerial (sometimes lengthy) chases by the gull. Aerial chases are more usual, and longer, when gulls are detected early in their approach. This happens either when attacks are launched over a long

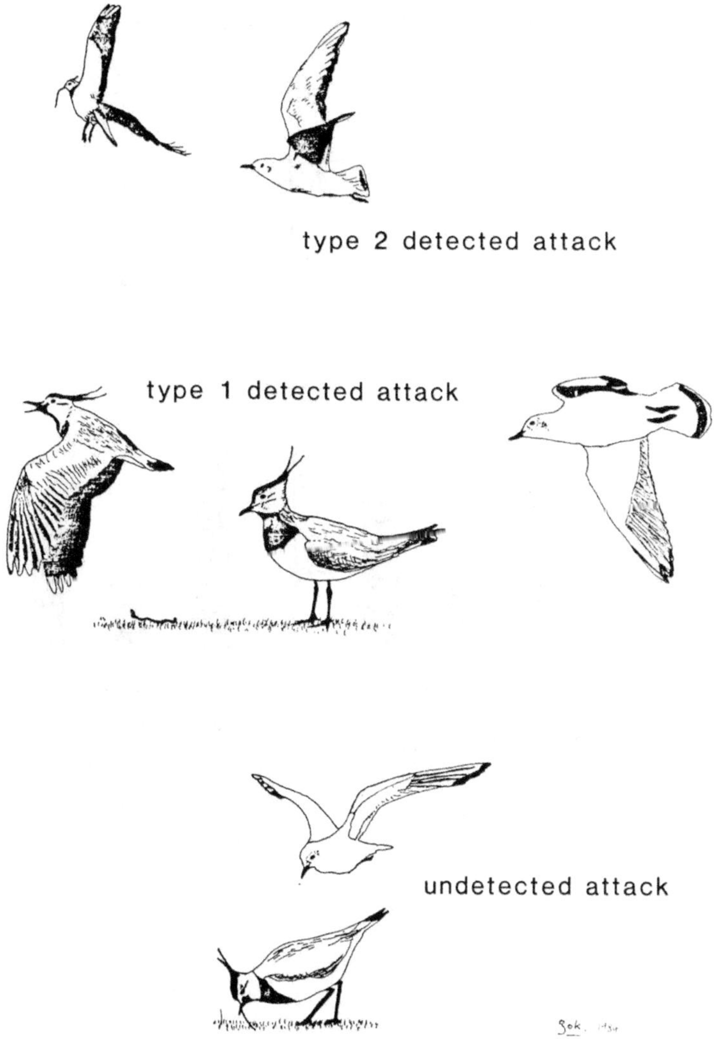

Figure 8.2: The three main forms of attack by gulls. Drawings for Type 2 detected attack and chapter vignette after Källander (1979)

distance or when target birds happen to scan in the gull's direction early in its flight. The relationship between target vigilance and attack success will be discussed later. Obviously, Type 2 attacks are usually much longer than the other two categories (Figure 8.3a). Sometimes the target escapes with its prey, but more often the prey is dropped during the chase and taken by the gull (either while still in the air or from the ground).

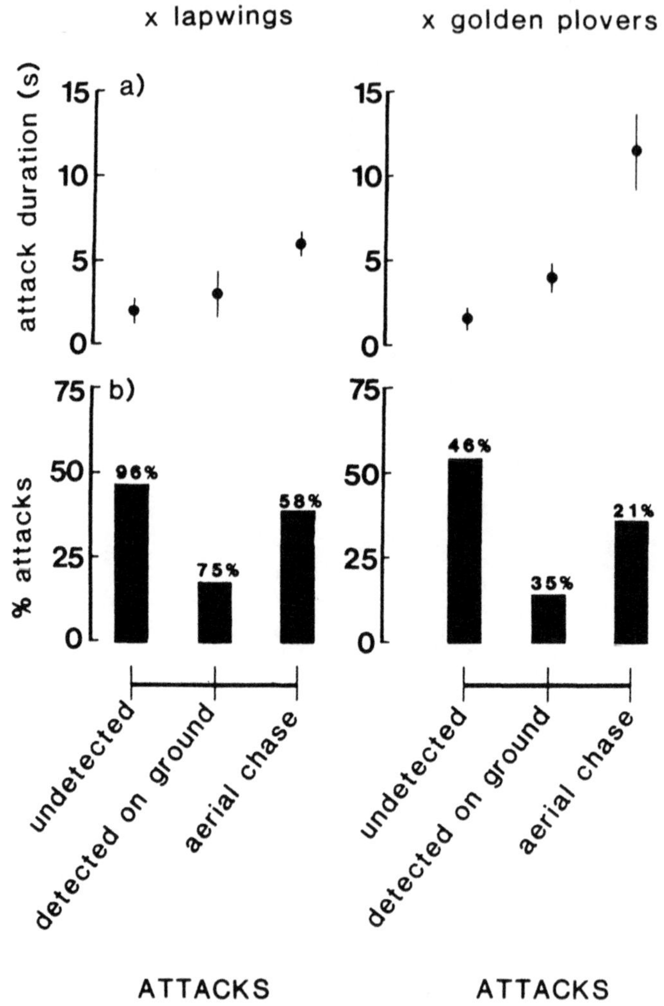

Figure 8.3: The duration (a) and frequency (b) of each type of attack against lapwings and golden plovers. Data for 194 attacks against lapwings and 101 against golden plovers. Percentage success for each type also shown. Bars represent standard errors

Clearly, then, detection during attack is likely to increase the gull's expenditure of time and energy. The extent to which it does is illustrated in Figure 8.4a, which compares the times spent flying for detected and undetected attacks launched at given distances from a target. For distances over 12s against golden plovers, for example, detection results, on average, in an almost two-fold increase in flight duration. Attacks over long distances are also more likely to be detected and result in an aerial chase. The relationship between

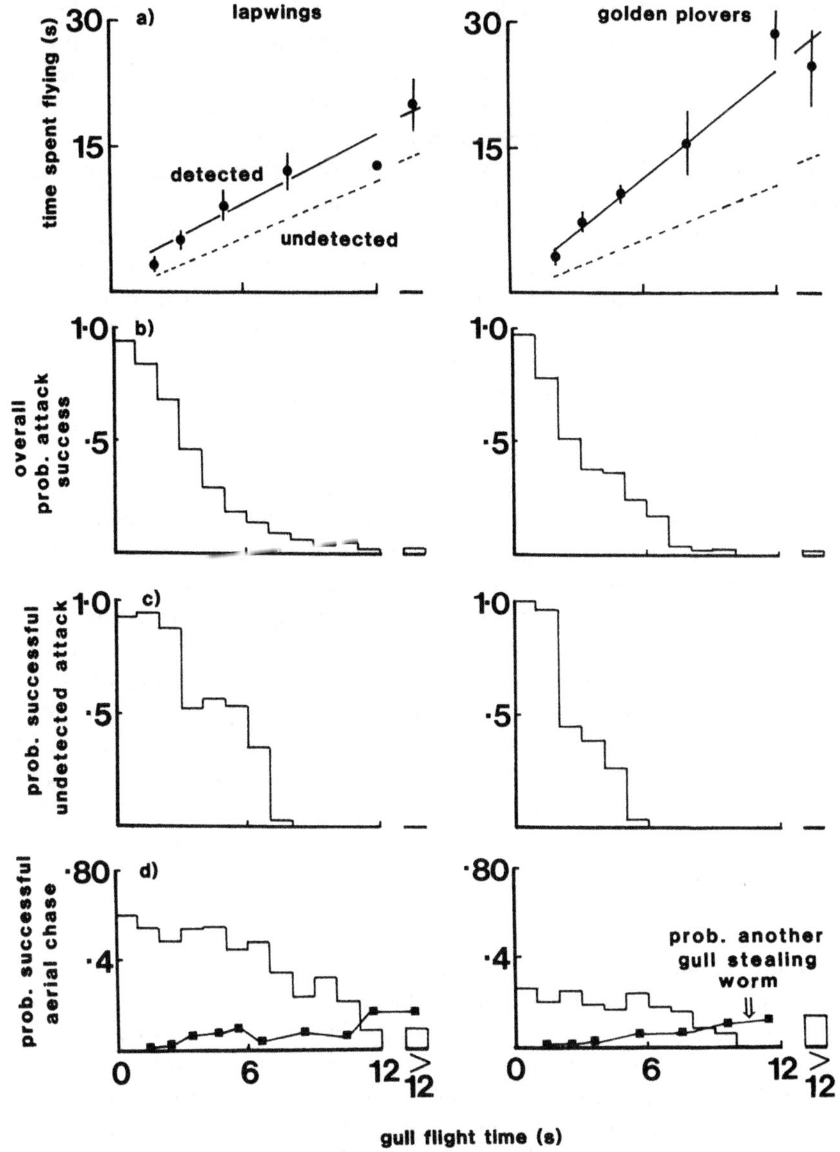

Figure 8.4: Factors affecting attack success in relation to distance from the target (measured as flight time). (a) The amount of time spent flying in detected and undetected attacks, data for 29 and 43 attacks respectively (only regression line shown for undetected attacks); (b) the probability of capturing a worm (all types of attack), sample sizes as in Figure 8.3; (c) the probability of capturing a worm in an undetected attack, data for 73 attacks against lapwings and 49 against golden plovers; (d) the probability of capturing a worm during an aerial chase, data for 83 attacks against lapwings and 49 against golden plovers

attack distance and the probability of a gull being detected is discussed later.

As well as increasing the cost of attack, detection also reduces the probability of a gull obtaining prey. Figure 8.4b shows that attack success declines rapidly with increasing distance between gull and target. That this is not due solely to target birds handling worms before the gull reaches them can be seen from Figure 8.4c,d. Over short distances ($<$ 6s flight time), gulls are more likely to obtain worms in undetected attacks than in attacks resulting in aerial chases. The probability of success over a given distance in both undetected attacks and aerial chases declines with distance, and is greater against lapwings than golden plovers. The only exceptions are extremely long chases (up to 200m from the flock), in which golden plovers are more likely to give up prey than lapwings (see also Källander 1977). However, energy expenditure by gulls for any given length of chase is marginally greater against golden plovers than against lapwings (see Figure 8.5a), apparently because the greater speed and manoeuvrability of golden plovers in the air demands faster and more erratic flight by gulls. We might thus expect gulls to be more selective when launching attacks at golden plovers. Figure 8.5b shows, however, that, while there is a significant positive relationship between attack time and the calorific value of worms procured by lapwings, there is no significant relationship for golden plovers (see Chapter 5 for calculation of calorific values). Finally, aerial chases reduce attack success in a second way. If chases are protracted, other gulls frequently join in and usurp the worm when it is dropped. The risk of losing a worm to another gull is small for short attack distances (Figure 8.4d) but rises to 19 % for attacks launched over 14m or more. Curiously, considering the energetic cost of prolonged chases, gulls often pursue targets for longer than they should in terms of their expected success rate (Figure 8.6, see also Plate 7). This might be expected, however, if gulls continuously reassess the probability of obtaining a worm within the next few seconds as the chase progresses.

Spacing between Gulls and Plovers

In Chapter 3 we showed that golden plovers and gulls prefer to feed in the pastures which are used most intensively by foraging lapwings. Newly-arriving golden plovers also tend to land within fields where the local density of lapwings is greatest, because this is a useful guide to the location of high worm densities (Chapter 4). The same is true for gulls. The mean inter-neighbour distance between lapwings within 10 m of an alighting gull is significantly higher than that taken across all lapwings within the flock (mean distance within 10m %= 2.24 ± .14m; mean across flock = 3.10 ± .26, $t = 2.91$, $p < .01$, d.f. = 62). Since the presence of gulls reduces lapwing feeding efficiency (Chapters 5 and 6), we might expect lapwings to move away from the vicinity of gulls in order to reduce the conspicuousness of their feeding activities.

Plate 7. Black-headed gull chasing a golden plover. Note the size of the worm held by the plover. Photograph courtesy of Professor Richard Vaughan

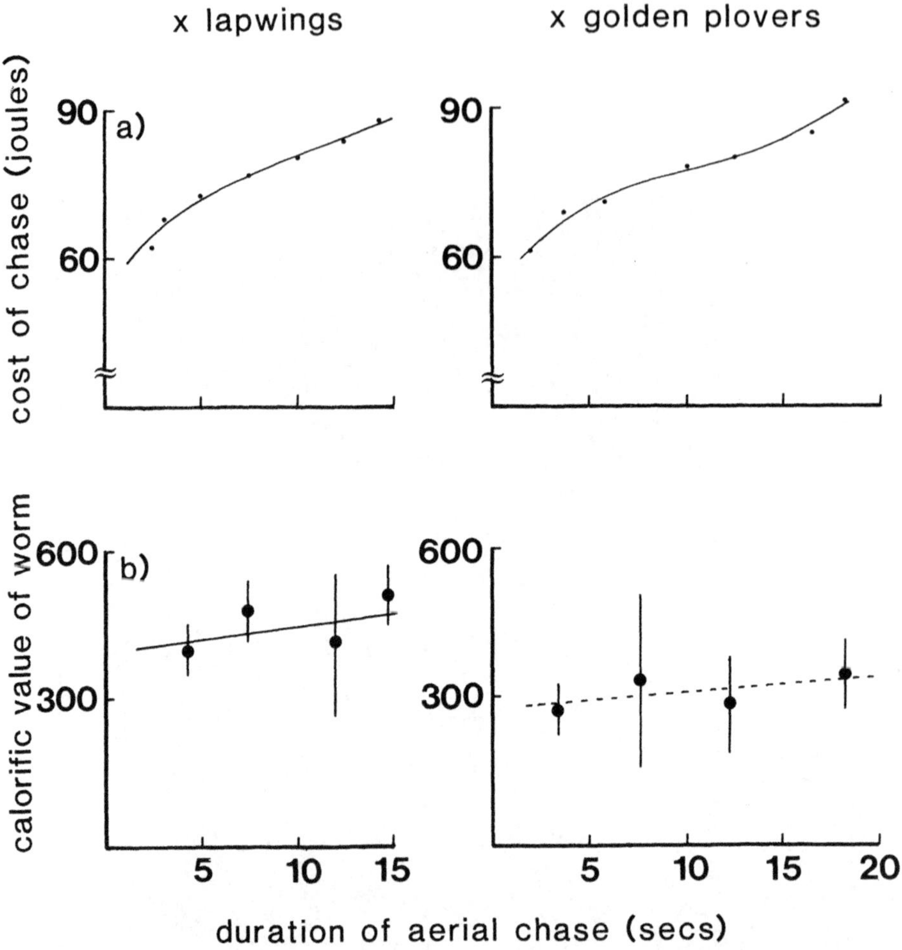

Figure 8.5: (a) The estimated energetic cost of an aerial chase in relation to its duration; sample sizes for attacks against each target species as in Figure 8.4d. (b) The length of chase in relation to the mean energy content of the target's worm; for lapwings r = .43, p < .05, d.f. = 21, for golden plovers r = .36, n.s., d.f. = 16. Data for successful and unsuccessful attacks. Lines fitted by regression. Bars represent standard errors

Moving away from gulls. To test this, we measured the distance from a gull to the nearest lapwing or golden plover at the time the gull alighted and again at different time intervals after alighting. Table 8.1 shows that, on average, lapwings have moved over 2m further away from the gull by the time 5-8 minutes have elapsed, though, as might be expected from Chapters 5 and 6, the effect is less pronounced in golden plovers. The number of plovers within negligible attack distance (< 1s gull flight time) therefore tends to

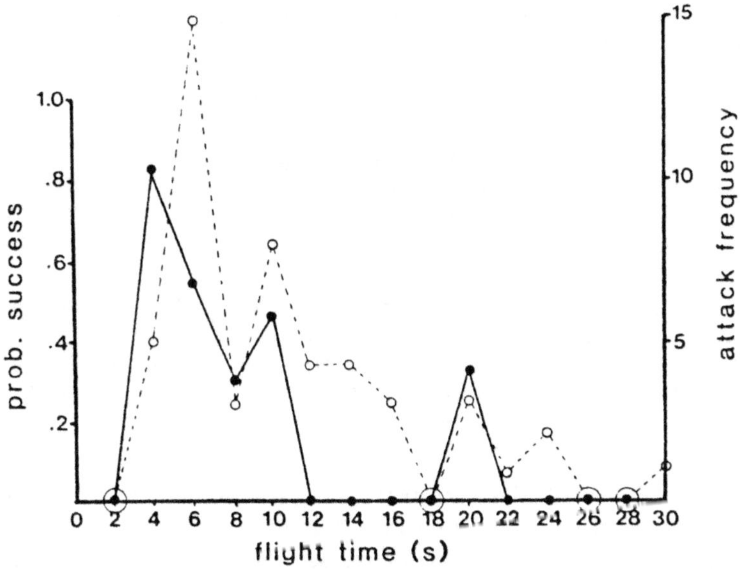

Figure 8.6: The probability of success in attacks (solid line) against golden plovers and the frequency distribution of attack duration (broken line) in relation to attack flight time. Flight times lumped in 2-second intervals

Source: Barnard (1984a)

decline with time. If the distance to the nearest plover increases beyond 4.5-5m, gulls often leave and move elsewhere in the flock (Table 8.1). Analyses show that, as with initial arrival, gulls usually alight in areas of locally-high lapwing density when making relocation flights within flocks (mean no. lapwings within 10m of gull at take-off = 1.77 ± 0.08; mean no. at landing = 4.64 ± 0.35). Relocation flights therefore appear to be a means of maintaining potential targets within a short attack distance. Further evidence that relocation flights are related to feeding efficiency is shown in Figure 8.7. Gulls wait longer before moving from a depleted area if their capture rate up to that point has been low. This is similar to the relationship between so-called giving-up times and food-patch quality shown in great tits by Krebs *et al.* (1974) and in pied flycatchers (*Ficedula hypoleuca*) by Bibby and Green (1980). Interestingly, gulls with a zero capture rate have highly variable waiting times. One reason may be that they have so little information on which to base their decision (J. Lazarus pers. comm.) If a locally-high density of plovers provides food at a low rate, the chances are that others in the field will too. It may therefore be worth staying longer in the current area.

Moving away from gulls is one way in which plovers may reduce their vulnerability to attack. Vulnerability also appears to be reduced by a variety of concealment tactics.

Table 8.1: Mean (± se) distances (m) between gull and the nearest lapwing or golden plover

		Mean distance between		
		Gull and nearest lapwing	Gull and nearest golden plover	Gull and nearest plover
(i)	When gull landed	2.14 ± .16	3.19 ± .12	2.11 ± .14
(ii)	300-500 s after gull landed	4.51 ± .31***	4.22 ± .46*	4.15 ± .27***
(iii)	When gull took off	5.64 ± .26***	4.34 ± .28**	4.25 ± .24***

* $p < .05$, ** $p < .01$, *** $p < .001$; t-test comparing (i) with (ii) and (iii).
Sample sizes for comparisons between lapwings and golden plovers respectively are: (i) 25, 18; (ii) 29, 22; (iii) 16, 11.

Reducing the visibility of prey. The spatial distribution of plovers changes in another important way when gulls arrive: it becomes more clumped (see Chapter 6). Although increased clumping may reduce the likelihood of successful plovers being spotted by a gull, gulls still occasionally (and successfully) attack birds in tight clumps. The visibility of procured prey, however, may also be reduced in a second way. Depending on the orientation of the long axis of the body relative to the position of a scanning gull, procured worms are either obscured from the gull's view or fully or partly visible. Figure 8.8a shows the range of orientations which, to the human observer, appears to obscure prey from a gull because the plover is facing away from it during extraction and mandibulation. To see whether there is any relationship between distance from a gull and the tendency to take up particular orientations relative to it, we recorded the orientation of all plovers in a number of sample flocks and simultaneously recorded their distance from the gull. Where more than one gull was present we assumed an equal number of plovers were vulnerable to each gull. Figure 8.8b shows that both lapwings and golden plovers are more likely to face away from gulls during handling when they are within 4-7m (2-4s flight time) of them (see also Plate 2a). The range of probabilities expected by chance is represented by the horizontal arrow (obtained from recording the position of foraging plovers with respect to a gull-sized rock). Interestingly, while birds face away from gulls when they are nearby, they face *towards* them more often than expected when they are far away. The change occurs at shorter distances in golden plovers (see later). We shall discuss a possible reason for facing gulls at greater distances in Chapter 9.

The cost of attack to gulls therefore depends on the distance to be covered, and a number of factors tend to increase the distance over which attacks have to be launched. Firstly, plovers are more likely to detect approaching gulls if the initial attack distance is long, and detection often results in long aerial chases.

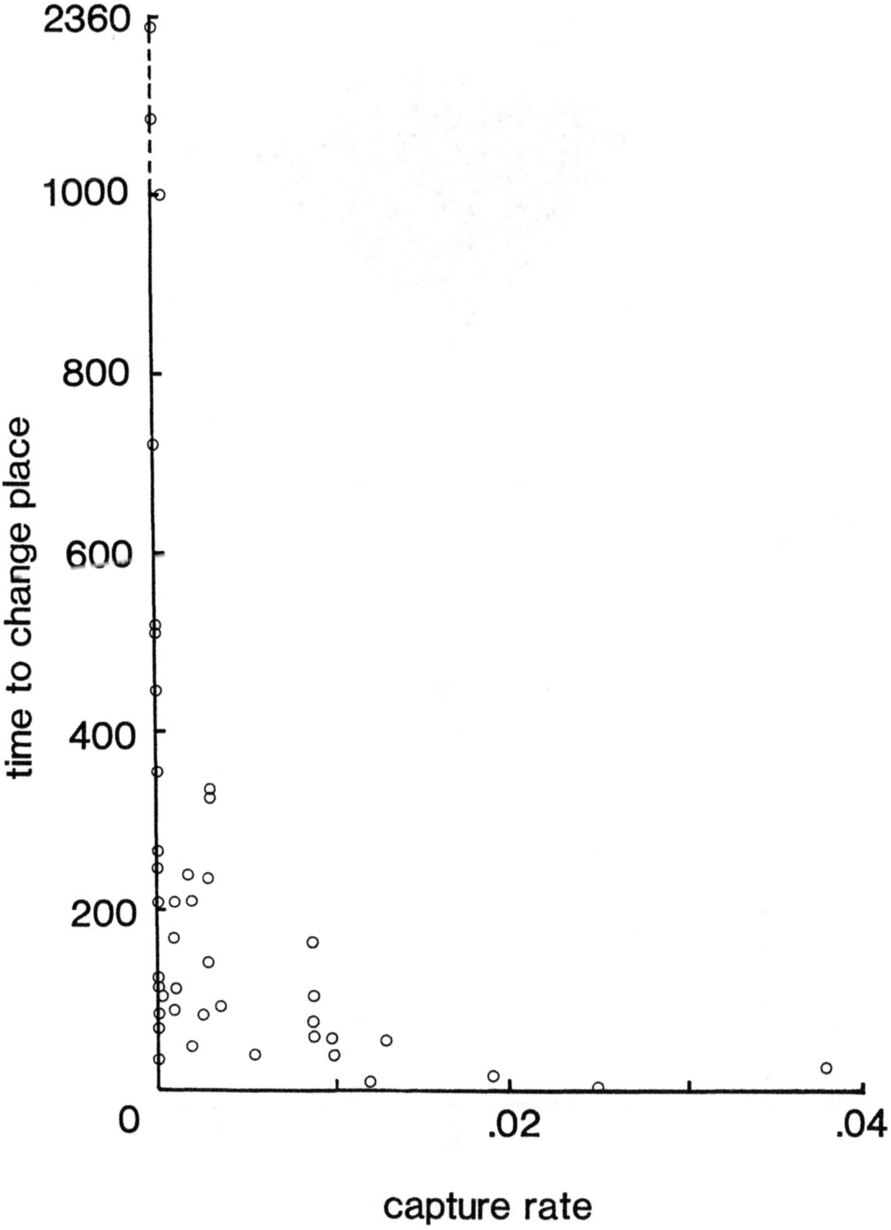

Figure 8.7: Stay time (s) before making a relocation flight as a function of preceding capture rate (worms per s)

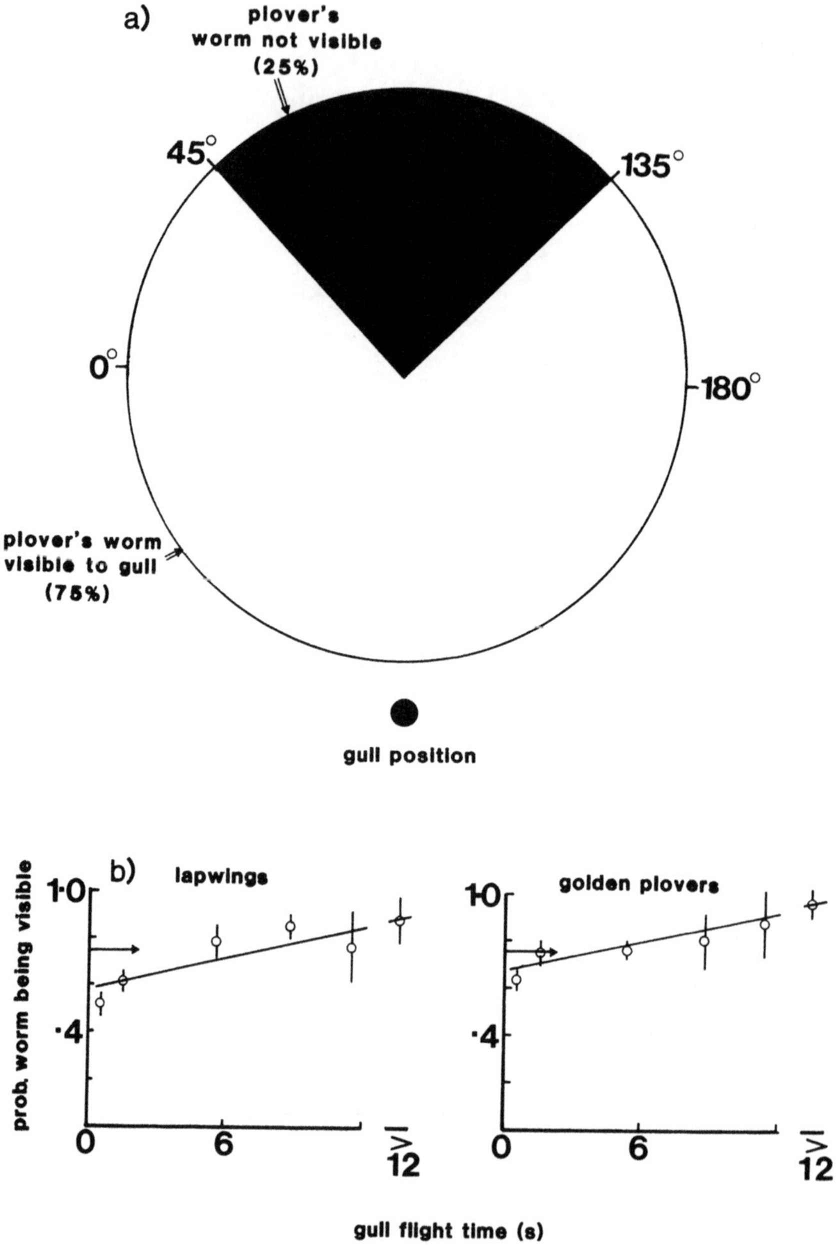

Figure 8.8: (a) Worms procured by plovers were assumed not to be visible to gulls (●) if plovers faced in any of the directions within the shaded sector; (b) the probability of a procured worm being visible to a gull as a function of the distance between plover and gull. Data for 49 observations of lapwings and 62 of golden plovers. Arrow indicates the probability if the orientation of plovers is independent of the position of a gull

This is exacerbated by the fact that gulls tend to pursue birds for longer than they should in terms of their expected success rate. Secondly, plovers increase attack distance by moving away from gulls when they arrive in a flock. Thirdly, effective attack distance is increased by birds close to gulls concealing their prey by facing away. While all these factors tend to increase the cost of attack, the importance of attack distance is likely to vary with the quantity and quality of prey being procured by potential targets.

The energy content of prey. In Chapters 5 and 6, we saw that lapwings and golden plovers preferentially take worms between 17 mm and 48 mm in length. A comparison with the size range taken by gulls (Figure 8.16c) shows that the latter steal mainly larger worms (>32 mm). The reason is obvious. Handling times in both lapwings and golden plovers increase sharply with worm length (Figures 5.2c and 5.3c). The average handling time for worms less than 32 mm in length is 2s or below. Unless gulls are within 4m of their target, therefore, prey-handling time is likely to be shorter than attack time and worms thus unobtainable. Even if handling time exceeds attack time, small worms present a smaller target and are likely to be less visible. Since the vulnerability of lapwings and golden plovers is likely to increase with the size of worm procured, we might expect birds close to gulls to avoid taking large worms. Figure 8.9 shows the mean calorific value of worms taken at different distances from gulls. In both species there is a sharp decrease in the value of worms taken within 5m (4s flight time) of a gull. Beyond 5m, there is little change with distance. These dietary shifts are another factor likely to force gulls into making long-distance attacks. We shall consider worm selection by gulls in more detail later.

So far, we have considered the effects of attack distance on the cost of

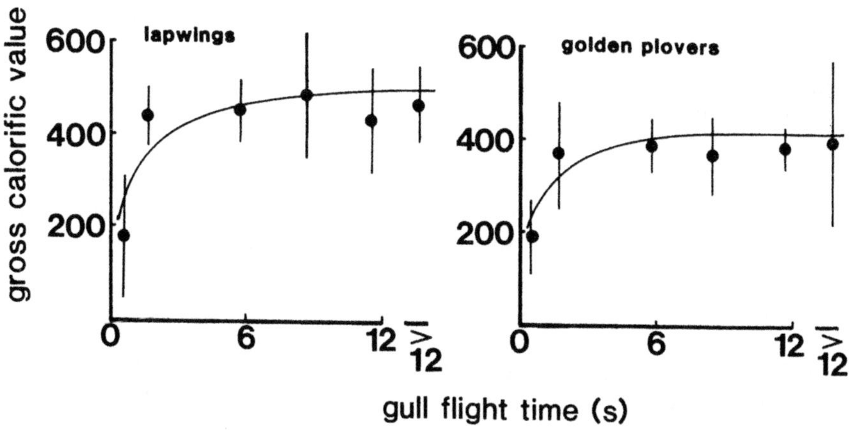

Figure 8.9: The mean gross energy content of worms procured by plovers at different distances from gulls. Sample sizes as in Figure 8.8b

attack and the way the responses of plovers tend to increase attack distance. We are now in a position to ask whether gulls take these effects into account when deciding to attack.

Choosing Attack Distances

The decision to attack any given plover clearly involves a trade-off between worm profitability (which increases with distance close to gulls) and the cost of attack (including energy expenditure, which increases with distance, and probability of success, which decreases with distance). We can therefore calculate the expected profitability of attacks over a given distance as:

$$E_t = (((E_i \cdot ps_u) - C_t)p_u/t + (((E_i \cdot ps_d) - C_t)\,1 - p_u/t' \qquad (8.1)$$

where E_t is the expected energy intake per unit time from an attack of flight time d seconds, E_i the corrected calorific value of a worm of size class i (see Chapter 5) procured by a target plover, ps_u the probability of success if the attack goes undetected, ps_d the probability of success if the attack is detected (see Figure 8.4), p_u the probability that the attack will go undetected, C the energetic cost of the attack, t the flight time if the attack is undetected and t' the flight time if the attack is detected.

The relationship between E_t and attack distance (gull flight time) between gull and target is shown in Figure 8.10. There is a pronounced peak for attacks against both species, with gulls doing best from attack flights lasting about 1.5-2.5s against lapwings and 1.5s against golden plovers. Gulls can expect to make a loss against golden plovers if flights last more than 5s. The decline in profitability is less sharp against lapwings, and gulls do not make a loss until flights are longer than 10s. If gulls maximise their attack efficiency, therefore, we should expect most attack flights to last for one or two seconds. Figure 8.10 shows that this is the case. The frequency distribution of attack-flight duration peaks at 1-3s against lapwings and 1-2s against golden plovers. There is, however, considerable spread in both cases. One reason might be that it pays gulls to attack over a range of distances rather than only over the single most profitable distance. Although short flight times are the most profitable, we have already seen that the responses of plovers tend to increase flight time. Often, therefore, the number of potential targets within the optimal distance may be small. In this case, we should seek the optimal *breadth* of attack-flight time in an analogous way to seeking optimal-diet breadth for plovers in Chapter 5.

Optimal breadth of attack time. As in the calculation of optimal-diet breadth in plovers, we calculated expected returns in a cumulative fashion, starting with flights of the most profitable duration only and successively incorporating flights of less profitable duration. The equation was therefore as follows:

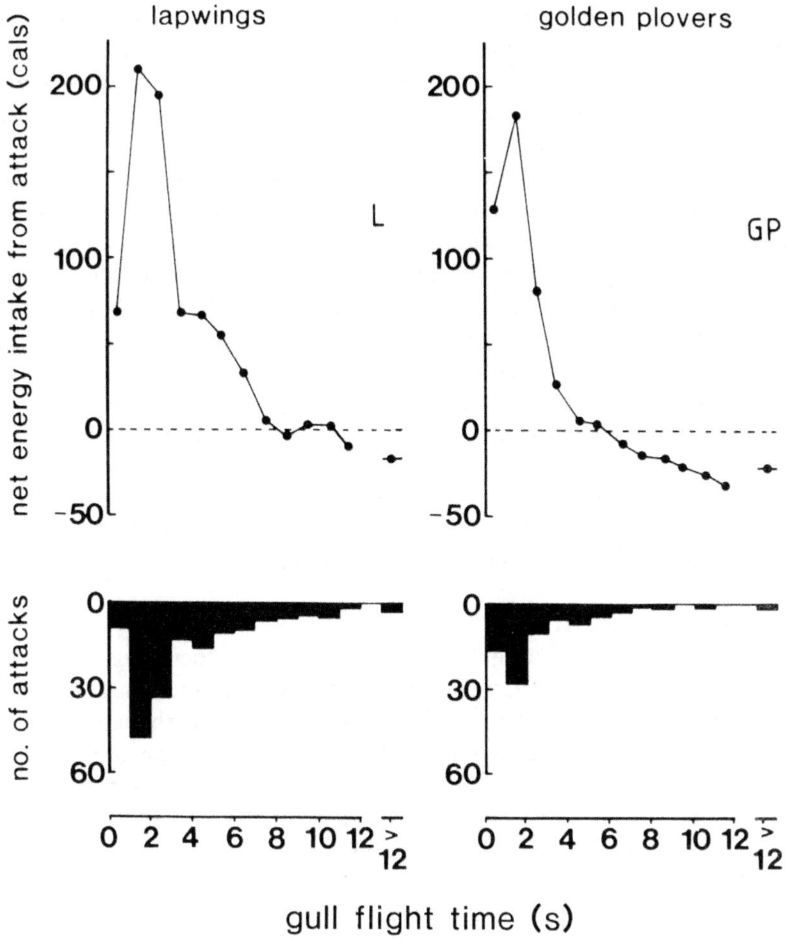

Figure 8.10: The rate of energy intake expected from attacks over different distances (solid lines), see equation (8.1) and the number of attacks over each distance (histograms) for each target species. Dotted line represents maintenance requirement. Data for 162 attacks against lapwings and 72 against golden plovers

$$E_{fr} = \sum(E_{t1} + E_{t2} + \ldots . E_{tn}) \qquad (8.2)$$

where E_{fr} is the energy intake per unit time expected from flight time range *fr*, and $E_{t1 \ldots n}$ is as in equation (8.1) for flights of the most profitable duration ($t1$) seconds and so on to the nth most profitable duration (t_n) seconds. Figure 8.11a shows that the gulls' expected rate of energy intake increases with increasing breadth of flight time to a maximum when flights of 8s are included for lapwings and flights of 6s for golden plovers. In other words, it pays gulls to attack lapwings so long as flights are shorter than 8s and to attack golden plovers so long as they are shorter than 6s. We should therefore expect most attack flights to be within these times. This is largely so (Figure 8.11b): 82.5 % of attacks against lapwings and 78.3 % against golden plovers fall within the predicted range of flight times.

There are several possible reasons why some attacks are longer than expected. Firstly, the probabilities used in equation (8.1) are averaged across gulls and may not be the best estimates for all individuals in all circumstances. Secondly, gulls may well have imperfect or biased information on which to base their decision to attack and may therefore make mistakes. Thirdly, depending on variation in the vigilance of plovers (for instance, in relation to flock composition — see Chapters 6 and 9) and the availability of different worm sizes, long-distance attacks may sometimes be necessary or, indeed, profitable. In the following sections, we shall discuss the influence of some of these factors on a gull's decision to attack. Although the correspondence between predicted and actual flight duration is not perfect, Table 8.2 shows that, by biasing their attacks towards those of shorter duration, gulls do significantly better than if they simply attack at random.

Choosing When to Attack

So far, we have shown that gulls take the distance of their target and their probability of success into account when deciding whether to attack. At any given distance, the gull's chances of success, however, clearly depend on the vigilance of the potential target. If the target is scanning, the gull's chances are likely to be smaller than if it is crouching or pulling a worm out of the ground. Gulls should therefore fine-tune their decision to attack by taking into account the target's posture, and thereby their likelihood of detection and the probability that they will reach the target before it has handled its prey.

In Chapters 5 and 6, we suggested that plovers are more likely to be attacked if they crouch before pecking, because crouching results in larger,

Table 8.2: Observed rate of energy intake by gulls compared with that expected if gulls attack plovers at random

	Intake per attack (cals)		
Attack made at	**Observed**	**Expected**	**t-test**
Lapwings	124.9 ± 9.7	85.6 ± 7.7	3.17**
Golden plovers	100.3 ± 6.8	52.9 ± 8.3	4.43***

** $p < .01$, *** $p < .001$; comparing the observed and expected means. In each case d.f. = 12, 12.

Data for 162 and 72 attacks involving lapwings and golden plovers respectively.

Expected data calculated from the frequency distribution of inter-neighbour distances between plovers (lapwings, n = 175, golden plovers, n = 121) and gulls.

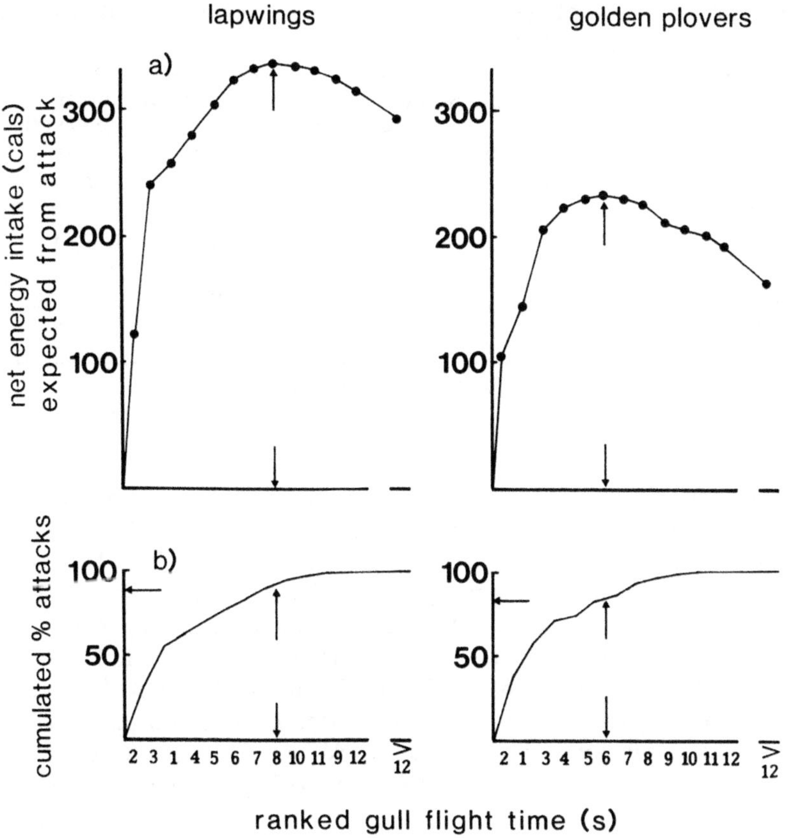

Figure 8.11: (a) The rate of energy intake expected from attacks over a range of distances from the most to the least profitable, see equation (8.2). (b) The cumulative percentage of attacks in relation to ranked distance. Arrows show the optimal range of distances over which to attack

more profitable worms and crouch duration correlates positively with worm size (see Figures 5.3d, 5.4d). This increased vulnerability is shown in Table 8.3, which compares the risk of attack for pecks preceded by and those not preceded by a crouch in lapwings and golden plovers. The apparent risk inherent in crouching is, however, explicable in at least two ways. Firstly, gulls may interpret crouching as indicating that a large worm is about to be procured and therefore direct their attention towards plovers which crouch. This would be helped by the positive correlation between crouch duration and worm size. Secondly, they may attack plovers which are handling large worms simply because their handling times are longer and more likely to be detected. Since crouching results in large worms, this would have the effect of making pecks preceded by a crouch appear more vulnerable. To see which is the case, we developed a simple model.

Table 8.3: The percentage of feeding postures assumed visible to gulls and the percentage attacked

	Lapwings		Golden plovers	
	Pecks only	Crouch then peck	Pecks only	Crouch then peck
% postures visible to gulls	61.9	38.1	11.3	88.7
% postures attacked by gulls	1.9	6.0**	0	2.1***
Total no. postures observed	645	396	260	2038

X^2 test comparing risk of attack for pecks with and without crouching.
** $p < .01$, *** $p < .001$.

The basic assumption of the model is that plovers are unable to detect approaching gulls while they are crouching or handling prey because visual attention is then directed mainly at the ground. This is likely to be an over-simplification since birds, even though they are not scanning, may still be able to detect gulls incidentally. Nevertheless, it is reasonable to suppose that gulls will be more successful if targets spend a long time crouching for and/or handling any worms they procure. We should thus be able to predict whether or not a gull will be successful over a given distance if we know the probability that the target will not scan during the attack flight. We can now formulate a model similar to that developed by Lendrem (1982) and Hart and Lendrem (1984) to cater for predators attacking prey on open ground. Following Hart and Lendrem (1984), the probability of a plover detecting an approaching gull with initial flight time t between gull and plover can be calculated as:

$$p_d = p(I_s < t) \tag{8.3}$$

where I_s is the interscan interval during which the target is non-vigilant. The probability that the gull will be successful is therefore simply:

$$p_s = 1 - p_d \tag{8.4}$$

An example of the relationship between t and p_d is shown in Figure 8.12b. The frequency distribution of handling times for lapwings in the presence of gulls is shown in Figure 8.12a. Obviously, as t increases, the probability that it will exceed I_s also increases and the gull is more likely to be detected. As expected from the response of plovers to the presence of gulls (Chapters 5 and 6), most interscan intervals are short and, since plovers tend to move away from gulls, are unlikely to exceed t.

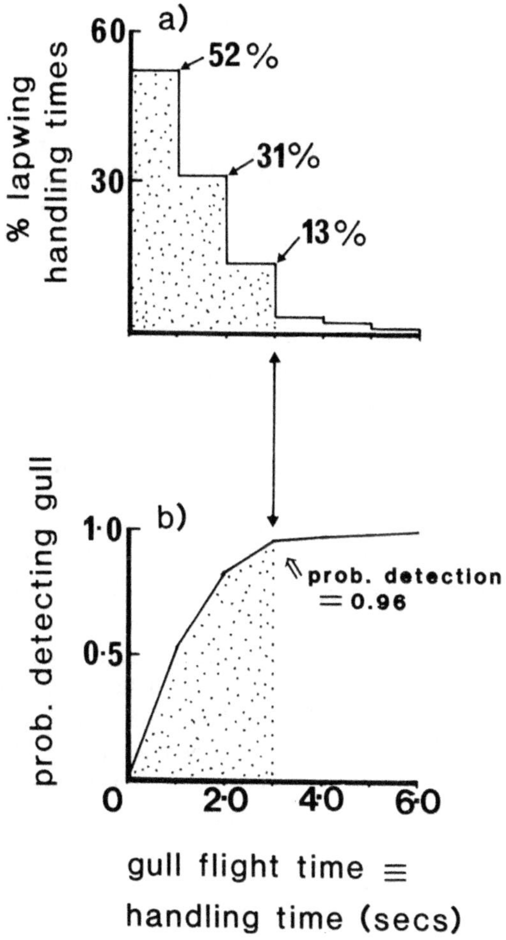

Figure 8.12: (a) The frequency distribution of handling time for lapwings in the presence of gulls; data for 168 observations. (b) The estimated probability of a gull avoiding detection as a function of attack distance, based on (a)

Source: Thompson and Lendrem (1985)

Given the relationship between t and p_d, when should a gull attack? If it attacks early, when the target begins to crouch, it risks attacking an unsuccessful bird. This will not only waste time and energy but may also drive away potentially fruitful targets. If it waits until a worm has been procured, however, it increases the likelihood that the target will scan before it completes its approach. To see how gulls resolve the problem, we compared the values of p_d measured in the field with those expected (a) if gulls wait until a target begins to handle prey and (b) if they begin their attack while targets are still crouching. Observed (pd_0) for any attack distance t was calculated as:

$$pd_o = ((nA_{d1} + nA_{d2})/NA)_t \tag{8.5}$$

where nA_{d1} and nA_{d2} are the number of Type 1 and Type 2 detected attacks respectively and NA is the total number of attacks. Figure 8.13 shows that pd_o increases in a sigmoid fashion with flight duration. If flights are short, gulls are unlikely to be detected, and the risk increases relatively slowly until flight times of about 3s against lapwings and 2s against golden plovers. The risk then rises steeply until, at a flight time of 12s, gulls are certain to be detected by both species. The increase, however, is much steeper against golden plovers so that, for any given flight time, gulls are more likely to be detected than if they attack lapwings (Kolmogorov-Smirnov two-sample test, D_{max} = .27, $p<.01$). This may be one of the reasons gulls prefer to attack lapwings. Also plotted in Figure 8.13 are the curves for expected p_d if gulls wait until targets are handling (dotted line) and if they attack at any time during crouching and handling (dashed line) (the *expected* curves were derived from the frequency distributions of handling/handling and crouching times recorded for plovers feeding at different distances from gulls). In both cases, the observed curve is closer to the curve expected if gulls attack while the target is crouching. Although the correspondence is closer for attacks against golden plovers, the difference between observed and expected curves is significant for both species (D_{max} against lapwings = .44, $p<.01$; against golden plovers = .21, $p<.05$). Both plover species therefore fail to detect gulls more often than expected. There are several possible explanations. It is likely, for instance, that plovers sometimes fail to detect approaching gulls even when they are scanning. If scanning is directional, birds may sometimes not scan in the direction of the gull before it reaches them. Another possibility is that *pd* depends on the size of worm procured by the target.

We showed earlier that gulls prefer to attack birds handling large worms (>32 mm). We have also shown that handling time in plovers increases with worm size (Figures 5.2c, 5.3c). Most of this increase, however, is due to increased extraction time rather than mandibulation time (e.g. extraction accounts for 54 % of handling time for class 1 and 2 worms, but 76 % for class 5 and 6 worms). Since extraction necessarily demands the focusing of attention on the ground, it is reasonable to suppose that plovers would be least able to spot an approaching gull at this time. By selectively attacking plovers which are procuring large worms, therefore, gulls may be able to reduce their risk of detection. We therefore recalculated expected *pd* for attacks against targets with worms over 32mm only (Figure 8.14a). Expected and observed curves are now much closer, and there is no significant difference between them for golden plovers. The fit is improved even more if we assume that gulls delay their attack for a short time to check whether a crouch is likely to result in prey. Many crouches are abortive and gulls might risk a fruitless flight if they attack as soon as a crouch begins. In Figure 8.14b we have assumed, arbitrarily, a delay of 1s between detecting a potential target and launching

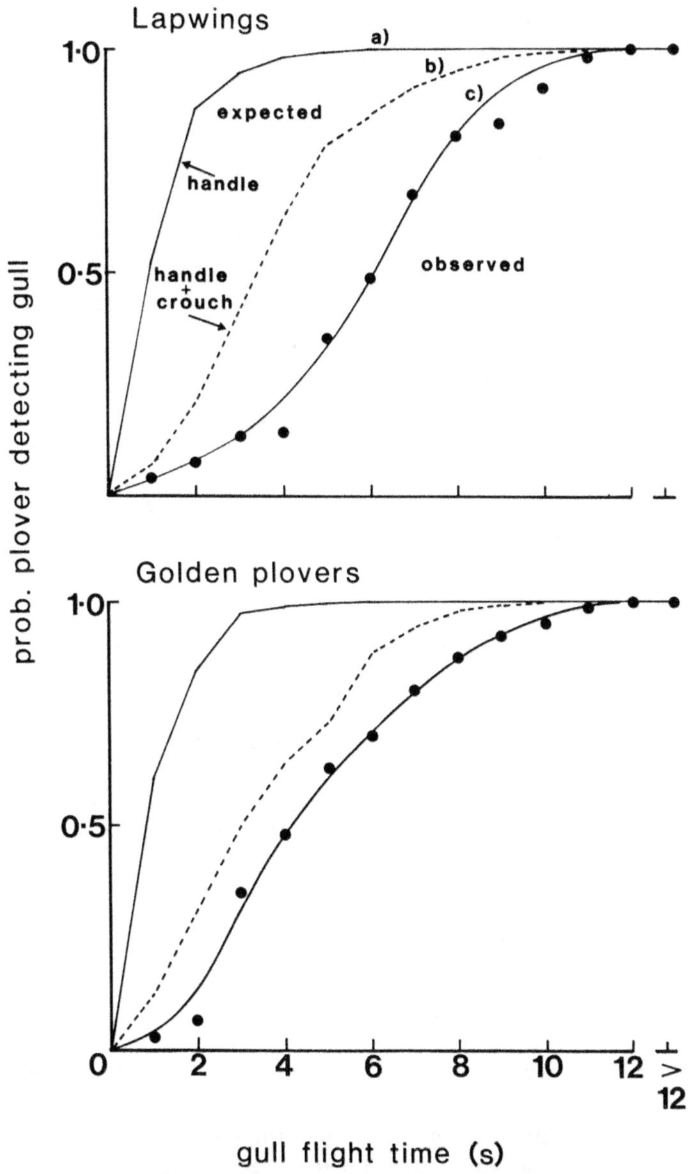

Figure 8.13: Observed and expected probability of a gull being detected over given attack distances. Data for lapwing and golden plover targets plotted separately. (a) The cumulative distribution of target handling time, (b) the cumulative distribution of target crouching + handling time, (c) the observed probability of detection. Number of observations in each case, for lapwings (a) 168, (b) 107, (c) 194; for golden plovers (a) 462, (b) 409, (c) 101

Source: Thompson and Lendrem (1985)

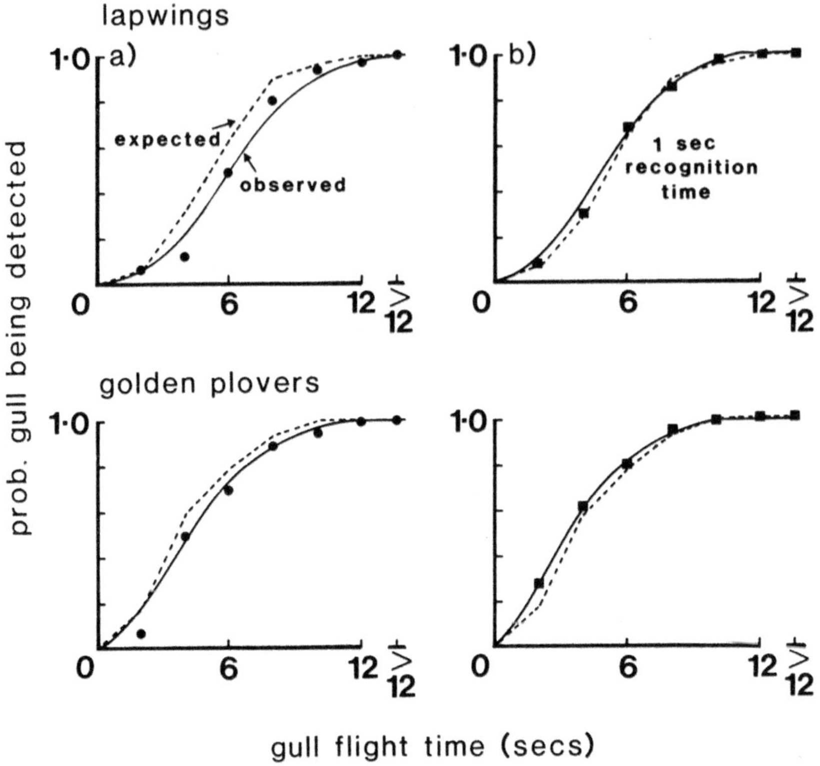

Figure 8.14: Same as Figure 8.13 but taking into account only targets handling large (> 32mm) worms. (a) Expected probability of detection (dotted line) based on crouching + handling times for large worms and observed probability (solid line); (b) as (a) but allowing a 1-second recognition time before attack. Kolmogorov-Smirnov one-sample test comparing observed and expected distributions for (a) lapwings, $D_{max} = .22$, $p < .01$, golden plovers, $D_{max} = .08$, n.s.; and for (b) lapwings, $D_{max} = .03$, n.s., golden plovers, $D_{max} = .09$, n.s.

Source: Thompson and Lendrem (1985)

an attack. The fit between expected and observed curves is now very close indeed.

It seems, therefore, that gulls pick their time to attack in a way which minimises their risk of detection and that they use crouching by a plover to indicate that the latter is likely to procure prey. The risk of detection is also reduced by selectively attacking plovers which find large worms. This, however, involves a further problem. It will pay gulls to specialise on large worms only if their availability is high. If the rate at which plovers procure large worms is low, it may pay gulls to attack for smaller worms (or move to another flock where the capture rate of large worms is high). If we assume that the cost of searching for a better flock is high (as it is likely to be, because worm supplies tend to vary similarly in pastures of similar type and most productive flocks have attendant gulls), we should be able to predict the mixture of worm sizes taken by gulls by means of an optimal-diet model similar to that in Chapter 5.

Host and Prey Selection by Gulls

To see whether gulls select worms as predicted by optimal foraging theory, we have first to identify the costs and benefits of worm-size selection.

Worm Size and Energy Content

As with plovers, we have assumed that gulls select worms on the basis of their energy content or some quality which correlates closely with it. The reasons for this are given in Chapter 5. Worms stolen by gulls were measured in relation to the bill length of the target lapwing or golden plover, or occasionally (if the worm was dumped on the ground as soon as it was extracted) in relation to the bill length of the gull. Worms were then allocated to the same six size classes as in Chapters 4 and 5. As before, the gross calorific value of each worm size was corrected to take account of the breakage of worms during extraction (see equation (5.1)). The relationship between corrected calorific value and the size of worm made available by lapwings and golden plovers is shown in Figure 8.15a. However, we also took into account the energetic cost of attack (see above) by subtracting it from the corrected calorific value for each worm size. Attack flight is so far the only foraging activity in gulls and plovers for which adequate estimates of energy expenditure are available. The corrected calorific value minus the cost of attack flight is shown for the average-sized worm in each class in Figure 8.15a. Two curves are shown for each target species. The upper line represents the gross value of worms, the lower line their value minus the cost of attack (net value).

Costs of Worm-size Selection

Again as with plovers, we measured the costs of worm-size selection in terms of time costs and the probability of obtaining prey. The profitability of different worm sizes was thus measured in terms of rate of energy intake.

Handling costs. Gulls take longer to mandibulate larger worms (Figure 8.15b), but their handling times for large worms are not disproportionately high as in plovers (cf. Figures 5.2c and 5.3c). They are also much shorter for any given worm size, because gulls do not have to extract worms from the ground and their larger bills allow more rapid mandibulation (see also Vernon 1972, Mudge and Ferns 1982, Curtis *et al.* 1985). There is no difference between the handling times for worms stolen from lapwings and those stolen from golden plovers.

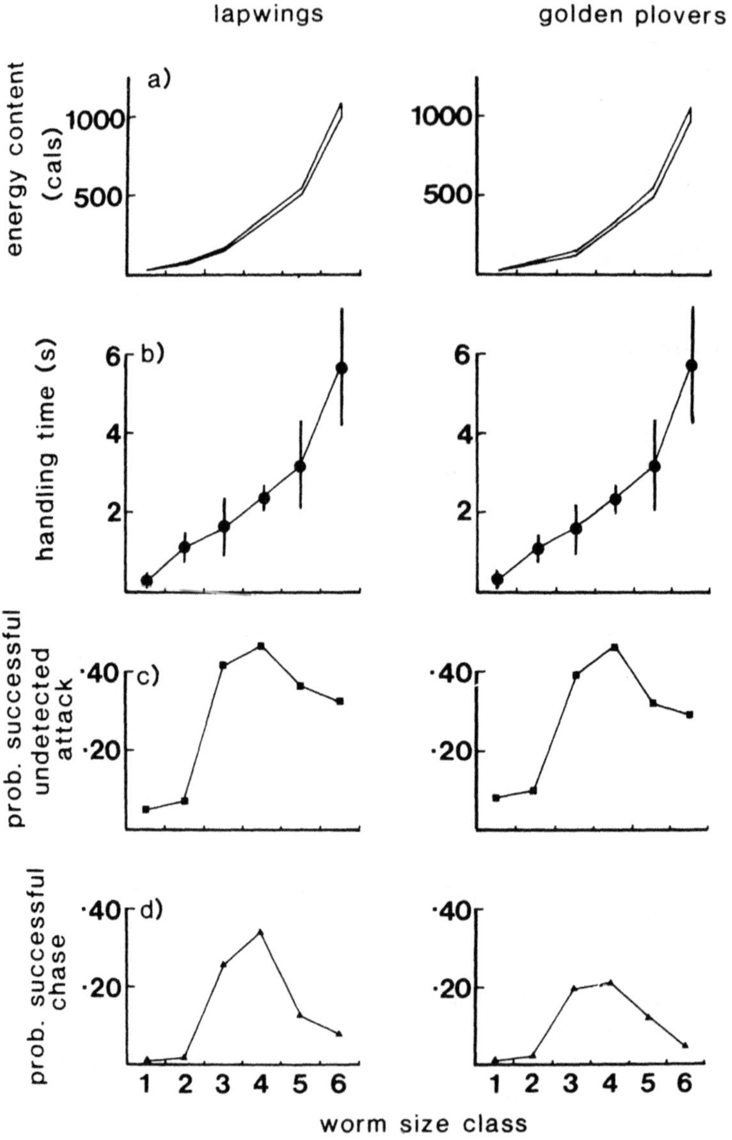

Figure 8.15: Costs and benefits of stealing different-sized worms from lapwings and golden plovers. (a) Energy content of the average-sized worm in each class (see text); upper line, gross energy content; lower line, gross energy content minus the estimated cost of attack. (b) Handling time for the average-sized worm in each class (data for 37 worms). (c) The probability of success in undetected attacks as a function of worm size (data for 68 attacks against lapwings and 39 against golden plovers). (d) The probability of success in detected attacks (data for 33 attacks against lapwings and 27 against golden plovers)

Costs of detection. From the discussion in the last section, we know that plovers spend longer crouching for and handling larger worms and that this reduces the likelihood of an approaching gull being detected. We should therefore expect the probability of successful attack to increase with worm size. Figure 8.15c,d shows that this is in fact the case for both detected and undetected attacks. The probability of success is less than 10 % for class 1 and 2 worms and rises steeply to nearly 50 % by class 4. As worm size increases further, however, success rate drops. In undetected attacks (Figure 8.15c), the most likely reason is that plovers tend to handle very large worms only when they are far away from gulls (Figure 8.9). Also, as we have seen, large worms, if they are extracted at all, are often discarded (Figures 5.2g, 5.3g) to reduce the risk of attack. Attacks against birds procuring very small class 1 and 2 worms are usually unsuccessful, because the handling times are so short that, except for targets close to a gull, they are likely to be shorter than the gull's attack-flight time. As might be expected, the success rate for any given-sized worm (at least above class 2) is lower against golden plovers than against lapwings. The shorter handling times of golden plovers (see Figures 5.2c, 5.3c) mean that they are more likely to detect approaching gulls. In detected attacks (Figure 8.15d), the probability of the gull's success combines risks of the target escaping with the worm and the risk of losing the worm to another gull. Now the decrease in success for large worms is due to the high risk of theft (8.3 % and 6.4 % of classes 5-6 worms stolen from lapwings and golden plovers respectively are lost to other gulls).

Worm Size and Profitability

Having quantified the apparent costs of attacking for different-sized worms, we can now make some predictions about worm-size selection by gulls. Because they are often available simultaneously to gulls, we have incorporated both lapwings and golden plovers in the calculations rather than treating them separately. Nevertheless, account is taken of the different costs and benefits accruing from attacking the two species.

Profitability in classical OFT models (*sensu* Krebs *et al.* 1983) has usually been calculated in terms of gross energy content and handling time. In our case, this can be calculated as:

$$E_{ri} = (E_{ai}/h_i)\, p_l + (E_{ai}/h_i)\, 1 - p_l \qquad (8.6)$$

where E_{ri} is the rate of energy intake from a worm of size i, E_{ai} is the corrected calorific value of worm size i minus attack cost, h_i the handling time for a worm of size i, and p_l the probability that the worm would be stolen from a lapwing ($1-p_l$ is therefore assumed to be the probability of theft from a golden plover). As before, however, we have also taken into account additional costs, in this case the size-dependent costs of detection and the energetic cost of attack. Profitability can now be calculated as:

$$E_{ri} = (((E_{ai}/h_i)\, ps_{ui}) + ((E_{ai}'/h_i)\, ps_{di})\, p_l) + (((E_{ai}/h_i)\, ps_{ui}) + ((E_{ai}'/h_i)\, ps_{di}(1-p_l))) \quad (8.7)$$

where E_n, h_i and p_l are as in equation (8.6), E_{ai} is the corrected calorific value minus attack-flight cost for worm size i in undetected attacks, E_{ai}' the corrected calorific value minus attack-flight cost for worm size i in detected attacks, ps_{ui} the probability of a gull making a successful undetected attack for worm size i, and ps_{di} the probability that it will make a successful detected attack for a worm of size i.

Figure 8.16 compares the observed distribution of worm sizes stolen from lapwings and golden plovers with those predicted from equations (8.6) and (8.7). If we take into account only handling-time costs (equation (8.6)), profitability increases with worm size and gulls should prefer size class 6 (Figure 8.16a). When the risk of detection is catered for (equation (8.7), Figure 8.16b), class 4 worms emerge as the most profitable with classes 1 and 2 predicted to yield almost nothing. The histogram in Figure 8.16c shows that the distribution of worm sizes taken by gulls most closely approximates to that predicted by equation (8.7). Gulls take mainly worms of size class 4 and virtually ignore classes 1 and 2.

Optimal-diet breadth and kleptoparasitism. So far we have looked at prey selection by gulls in terms of the relative profitability of individual size classes of worm. While the range of sizes taken reflects relative profitability, the comparison in Figure 8.16b,c indicates that gulls take more of the less profitable worm sizes than expected. Of course, Figure 8.16 assumes that gulls would do best by selecting for the single most profitable worm size. As we have seen in Chapter 5, this may not be the case. Birds may do better by taking a range of prey sizes. We therefore calculated expected rates of energy intake from different combinations of worm size.

Again, we calculated the rate of energy intake expected from taking only the most profitable worm size, then from taking the two most profitable, then the three most profitable and so on until all six size classes were included. We used a modified version of Charnov's (1976) equation to incorporate encounter rate with different-sized worms. Expected rate of energy intake was calculated as:

$$E_{r1\cdots n} = \frac{\sum\lambda_{1\ldots n} \cdot E_{a1\ldots n}}{1 + (\sum\lambda_{1\ldots n} + h_{1\ldots n})} \quad (8.8)$$

where $E_{r1\ldots n}$ is the rate of energy intake expected from taking the first to the

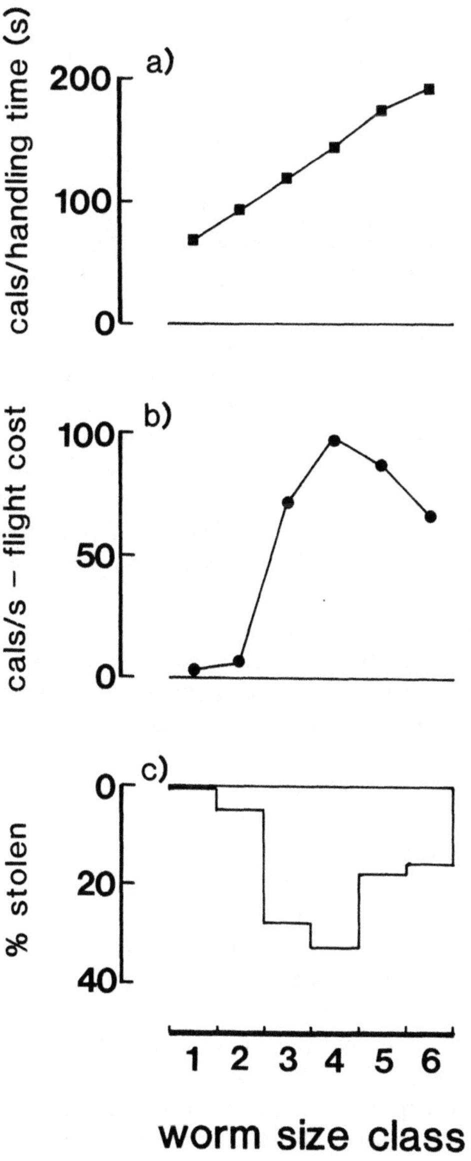

Figure 8.16: (a) The profitability (energy content/handling time) of the average-sized worm in each class, see equation (8.6). (b) The profitability of the average-sized worm in each class taking into account the risk of detection, see equation (8.7). (c) The contribution of different worm sizes to the diet. Data for 125 worms stolen from lapwings or golden plovers

nth most profitable worm size, $E_{a1\ldots n}$ is derived from E_{ai} or E_{ai}' in equation (8.7), $h_{1\ldots n}$ is the handling time for the average-sized worm of the first to nth classes, and $\lambda_{1\ldots n}$ is the gull's encounter rate with the first to nth size classes calculated as:

$$\lambda_{1\ldots n} = (Crl_{1\ldots n} \cdot \bar{n}_l) + (Crg_{1\ldots n} \cdot \bar{n}_g) \qquad (8.9)$$

where $Crl_{1\ldots n}$ is the capture rate of the first to nth classes by lapwings, $Crg_{1\ldots n}$ the capture rate by golden plovers, and $\bar{n}_l$ and $\bar{n}_g$ the average subflock size (number of birds) of lapwings and golden plovers respectively in flocks containing gulls.

The relationship between $E_{r1\ldots n}$ and diet breadth is shown in Figure 8.17a plotted against worm size class ranked in order of profitability (equation (8.7)). $E_{r1\ldots n}$ increases with the number of size classes included in the diet, reaching a peak when the fourth most profitable class is included. Beyond this, intake drops sharply. Gulls should therefore take worms of size classes 3-6 and ignore those of classes 1-2. When the size range of worms taken (Figure 8.16) is plotted cumulatively (Figure 8.17b), it turns out that 96 % of the gulls' intake is made up of classes 3-6 worms, closely approximating to the predictions of equation (8.7).

Gulls therefore follow the predictions of OFT at a qualitative level in that they prefer the most profitable range of prey. To test the predictions of OFT more rigorously, we examined prey selection in relation to encounter rate with profitable and unprofitable worm sizes. If gulls maximise their rate of energy intake, they should take profitable worms and ignore unprofitable worms when their encounter rate with the former is high.

We observed flocks containing gulls for periods of 15 to 20 minutes and recorded the kleptoparasitic activities of gulls and time budgets for a number of focal lapwings and golden plovers. Fifteen to 20 minutes was long enough to allow an adequate number of sample time budgets (mean duration = 88.9 ± 3.6 s) to be recorded from both species of plover, but short enough for their subflock sizes to remain constant or change only slightly. In this way the range of worm sizes taken by gulls and plovers was recorded over the same period, and selection by gulls could be examined in relation to changing apparent availability. Table 8.4 shows the relationships between the rate at which gulls capture profitable (classes 3-6) and unprofitable (classes 1-2) worms and the rate at which they are produced by plovers (encounter rate). The relationships are expressed as correlation coefficients. Capture rate of profitable worms increases significantly with their encounter rate, but there is no significant correlation between capture rate and availability for unprofitable worms. Capture rate of unprofitable worms is, however, negatively correlated with that of profitable worms; in other words, gulls are likely to take unprofitable worms only when they are taking few profitable ones. Nevertheless, there is no significant correlation between capture rate of

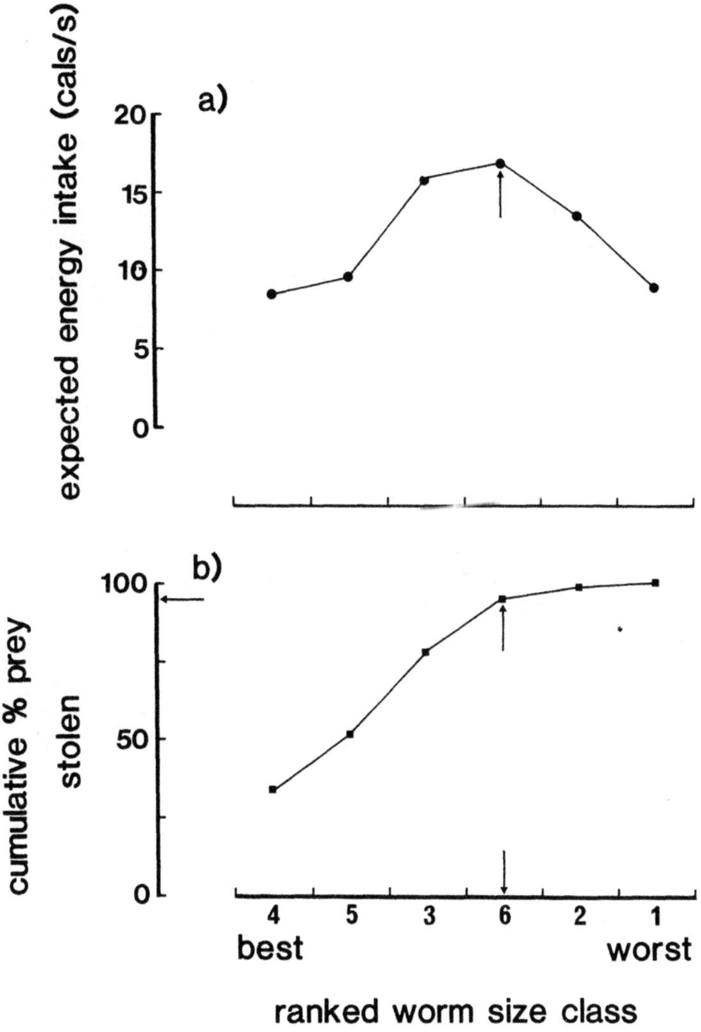

Figure 8.17: (a) Expected profitability of different ranges of worm size in the diet from taking just the most profitable size to taking all sizes, see equation (8.8). (b) The cumulative percentage of different-sized worms in the diet. Arrows indicate optimal-diet-breadth

unprofitable worms and *encounter rate* with profitable worms. In this respect, gulls follow the predictions of the optimal-diet model only weakly.

Choosing a Host

Throughout the preceding discussion, it is clear that lapwings and golden plovers represent very different kinds of potential target for gulls. Although they procure more or less the same type of prey, the greater speed and agility of golden plovers in the air increases the gull's energy expenditure and reduces its chances of obtaining food. In fact, the success rate of gulls against lapwings is almost exactly twice that against golden plovers (Table 8.5). Correspondingly, gulls show a two-fold preference for attacking lapwings (see Table 8.5). Similar figures have been reported on both counts for lapwings and golden plovers in Sweden (H. Källander pers. comm.). Does this mean that gulls do better the more lapwings there are in a flock? Figure 8.18a,b shows that capture rate by gulls increases with the number of lapwings in the flock, but not with the number of golden plovers. Interestingly, however, gull capture rate is significantly higher ($p<.001$) in flocks of lapwings only than in flocks containing both plover species (Figure 8.18c). This is despite the fact that there are on average more lapwings in mixed flocks and the rate at which they capture worms of size classes 3-6 is generally higher than in flocks of lapwings only.

Although there are more potential targets (of both plover species) to attack

Table 8.4: Correlation coefficients (r) for relationships between encounter rate with profitable (> 32mm) and unprofitable (< 32mm) worms and their capture rate by gulls

	Encounter rate with:		
Dependent variable	Profitable large worms	Unprofitable small worms	Rate of stealing profitable worms
Rate of stealing PROFITABLE (large) worms	+.541**	+.338	—
Rate of stealing UNPROFITABLE (small) worms	−.096	−.126	−.582**

** $p < .01$, — not included in analysis.
Data for 32 flocks.

Table 8.5: Frequency (mean ± se) and success rate (%) of gulls against lapwings and golden plovers

	Targets		
	Lapwings	Golden plovers	
No. attacks/plover/ gull/minute	0.012 ± .0023	0.006 ± .0021	$t = 3.08$, $p < .05$
% attacks successful	74	36	$X^2 = 5.89$, $p < .05$
n (attacks)	87	46	

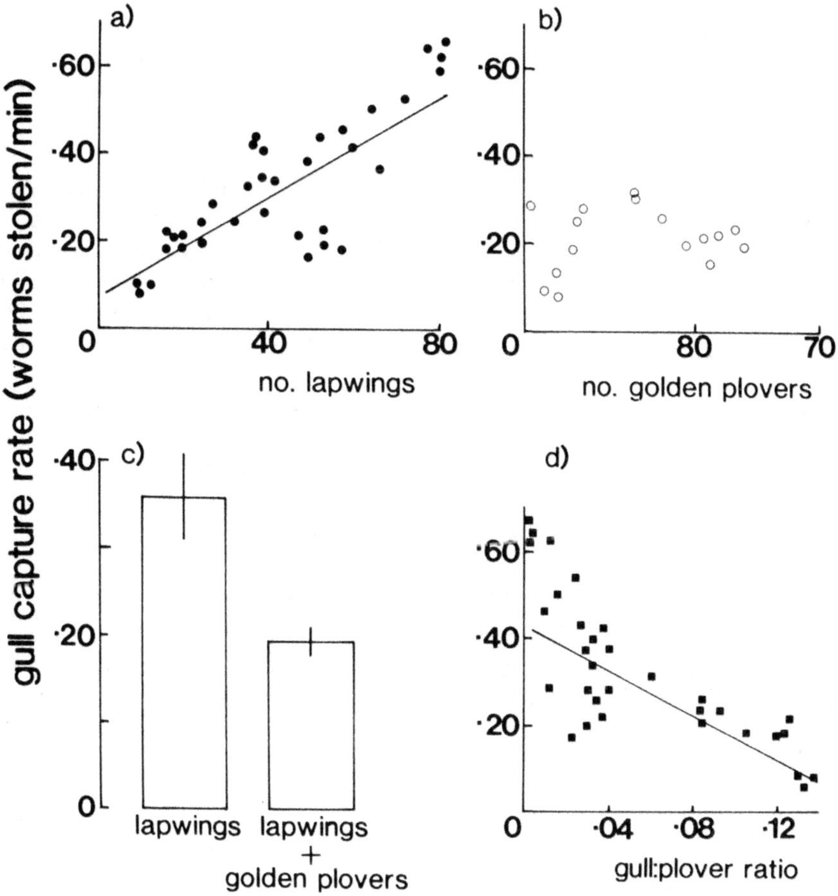

Figure 8.18: Factors affecting capture rate by gulls. (a) The relationship with lapwing subflock size, r = .79, p < .001, d.f. = 31; (b) relationship with golden plover subflock size, r = .25, n.s., d.f. = 14; (c) comparison between capture rates in flocks of lapwings only and mixed flocks; (d) the relationship with the gull:plover ratio, r = −.60, p < .001, d.f. 30. All lines fitted by regression

in mixed flocks, it may be that gulls are less able to capitalise on large worms procured by golden plovers. An analysis of the relationship between the rate at which gulls capture size class 3-6 worms and the rate at which lapwings and golden plovers procure them supports this ($r = .64$, $p < .005$, d.f. = 18 for flocks of lapwings; and $r = .51$, $p < .05$, d.f. = 14 for mixed plover flocks). As expected from the preceding discussion, therefore, feeding efficiency in gulls is apparently reduced by stealing from golden plovers. We have already mentioned, however, that gulls show a two-fold bias in their attacks towards lapwings. Could they expect to do better by preferring lapwings than by attacking lapwings and golden plovers as encountered? Table 8.6 suggests that they could. Gulls do better by attacking lapwings in single-species flocks than by attacking plovers at random. They do not, however, do better with a two-fold bias in mixed flocks than when attacking lapwings in single-species flocks.

Although the greater difficulty of stealing from golden plovers may be one reason for the reduced feeding efficiency of gulls in mixed flocks, increased competition from other gulls is another. There tend to be more gulls in mixed flocks, so that competition between gulls is likely to be more severe (see Table 8.7). The gull:plover ratio is significantly greater in mixed flocks ($t = 2.73$, $p < .05$) and there is an inverse relationship between gull:plover (both species) ratio and capture rate by gulls (Figure 8.18d). Furthermore, the proportion of variance in gull capture rate explained by the relationship between *rates* at which class 3-6 worms are procured by plovers and stolen by gulls is lower in mixed flocks ($r^2 = .41$, $p < .005$, d.f. = 17 in flocks of lapwings; $r^2 = .26$, $p < .05$, d.f. = 13 in mixed flocks). The efficiency with which large worms are attacked in mixed flocks is therefore reducd ($F = 3.62$, $p < .01$). Presumably, this partly explains why gull success rate against lapwings tends to be lower in mixed flocks (% attacks successful = .568 ± .078 in mixed flocks, and .748 ± .048 in lapwing/gull flocks; $t = 1.96$, n.s., $n = 14$ flocks). Reasons for the latter two differences probably include the dietary shifts by plovers when gulls are present and the increased intraspecific competition discussed above and in Chapter 6 (see also Thompson 1984).

Kleptoparasitism and Daily Energy Demand

Although at first sight, kleptoparasitism appears to be a convenient time- and labour-saving means of obtaining food, there are clearly substantial costs involved. Given the various factors which limit its profitability, can gulls live by kleptoparasitism alone? Källander (1977) estimated that black-headed

Table 8.6: Estimates of the rate of energy intake by gulls using different attack strategies

Attack strategy	Calories/second
(a) Random attacks at L and GP	14.09 ± 0.44
(b) Attacks at L and GP in observed ratio of 2:1	15.76 ± 0.67*
(c) Attacks at L only in mixed flocks	15.62 ± 1.13
(d) Attacks at GP only	13.26 ± 0.97
(e) Attacks at L in single-species flocks	16.47 ± 0.53**

* $p < .05$, ** $p < .01$; t-test comparing (a) against (b)-(e), L = lapwings, GP = golden plovers.

Data for 24 mixed plover flocks and 18 flocks of lapwings only.

Values were obtained using equation (8.7). See text for details.

Table 8.7: Intraspecific aggression between gulls in relation to the number of gulls in the flock

	Attacks/gull/min	
Number of gulls in flock	**At immigrants**	**At neighbours**
1 gull	0.28 ± .06	—
2-3 gulls	0.46 ± .03*	0.70 ± .24
> 3 gulls	0.51 ± .09*	1.09 ± .27

* $p < .05$; t-test comparing rates of aggression in flocks with 1 gull and flocks with > 1 gull.
Data for a total of 36 flocks (n = 16, 12, 8).

gulls could meet their daily energy requirement solely by stealing earthworms from plovers. Källander based his estimate on basal metabolic requirement (BMR). We also estimated existence energy requirement (EER) (see Chapter 6 for details of calculation), because this provides a better measure of the effects of environmental conditions on energy needs (Kendeigh *et al.* 1977, Walsberg 1983).

We calculated the costs of different maintenance and other activities performed by gulls during the day as multiples of BMR (e.g. King and Farner 1974, Maxson and Oring 1980, Puttick 1980). BMR was calculated from the following equation, modified after Aschoff and Pohl (1970):

$$\mathrm{BMR} = 0.0808 \cdot W^{0.734} \tag{8.10}$$

where W is the average weight (285 g) of black-headed gulls (from Cramp and Simmons 1983). Following Maxson and Oring (1980) and Puttick (1980), we costed activities as follows: roosting = 1.4 × BMR; flight from roost to feeding sites, between feeding sites and relocation flights within flocks = 12.0 × BMR; scanning plovers in feeding flocks = 4.0 × BMR; kleptoparasitic attacks = 15.0 × BMR. Daily costs for each activity were estimated by multiplying the number and fraction of hours spent performing it by the appropriate multiple of BMR.

We observed gulls stealing from plovers over complete daylight periods, and recorded all activities they performed from arriving at a feeding site first thing to returning to the roost at the end of the day. Since gulls do not forage on pastureland at night, daytime feeding represents their intake for the 24-hour period. Assuming an assimilation efficiency of 80 % (see Källander 1977; and for other bird species King and Farner 1974), we calculated daily energy intake as:

$$E_d = (M \cdot N)\left(\sum (E_{1\ldots 6} \cdot p_{1\ldots 6})\right) \tag{8.11}$$

where E_d is the total energy intake over the day, M the average time spent in flocks of plovers, N the number of worms captured per unit time in flocks of plovers, $E_{1...6}$ the calorific value of worms of each size class (1-6, see Chapter 5), and $p_{1...6}$ the probability that worms in the gulls' diet would be from each of size classes 1-6. Estimates of energy spent and gained during a day are given in Table 8.8. Kleptoparasitism was estimated to account for about 32 % of total daily energy expenditure; 22 % of this (and 6 % of daily expenditure) was attributable to attack flights. The table also shows that gulls marginally exceed their daily energy requirement by stealing from plovers, though only by the equivalent of two to five worms.

Kleptoparasitism and Feeding Efficiency: Résumé

Although kleptoparasitism differs in many important ways from the other foraging strategies considered in this book, kleptoparasites nevertheless face the same kinds of problem as other predators in maximising their feeding efficiency. The availability of different types of prey is determined by what host individuals turn up. This, of course, reflects their own needs rather than those of the kleptoparasite, so usually only a proportion of procured items is suitable for stealing. In the case of the gulls, it is the larger of the range of

Table 8.8: Estimates of daily energy expenditure and intake (kJ/24 hours) by gulls stealing earthworms from plovers

	Average EXPENDITURE (KJ/day)	Average INTAKE (KJ/day)
24-HOUR ACTIVITY BUDGET		
Roosting	129.8 ± 4.86	—
Flights between roost and feeding site	33.5 ± 1.34	—
Kleptoparasitism		
watching plovers	84.5 ± 5.07	—
attacking plovers	18.8 ± 4.21	330.3 (92 ± 7)*
Flights within and between fields	57.8 ± 5.19	—
TOTAL	324.4	—
BALANCE	—	+ 5.9 KJ/day

* Bracketed value is the mean ± se number of worms stolen per day, obtained from observations of gulls throughout the day. Average energy INTAKE was estimated using equation (8.11).

BMR/day (basic metabolic requirement, 132.7 KJ = 31.7 kcals).

Temperature range = 4.8°C-8.6°C (mean daylight temperature).

Assumed assimilation efficiency of 80%.

worm sizes taken by plovers that are worth stealing. Even the relative profitability of these, however, is determined by a complex range of factors involving the gulls' detection capabilities, the time and energy costs of attack, the chances of success in attacks and various anti-theft responses employed by plovers.

There are striking differences in the success rate of gulls against lapwings and golden plovers. Attacks against lapwings are more successful and result in greater feeding efficiency over a greater range of attack distances. There are several reasons for this. Probably the most important is the greater flight speed and agility of golden plovers. Lapwings tend to spend longer crouching and handling worms and are more likely to miss approaching gulls. Not surprisingly, therefore, gulls prefer to attack lapwings.

This difference between species is also reflected in the performance of gulls in different types of flock. Gulls do better in flocks of lapwings than in mixed flocks. Surprisingly, however, they tend to be most numerous in mixed flocks. There are at least three possible explanations. Firstly, as we have seen in Chapter 7, gulls often arrive in flocks with golden plovers because these may lead them to profitable feeding areas; secondly, defence by gulls which are already present may make landing in lapwing flocks more difficult; and, thirdly, gulls may gain some anti-predator benefit from being in large, mixed flocks. We explore the last of these explanations in the following Chapter.

A major problem facing a gull, whichever species it attacks, is to get to a successful bird before being spotted. Gulls thus do better against birds which are close by and if they launch their attack early in the target's period of prey procurement. The fact that plovers have to spend time locating worms which are concealed underground provides gulls with a useful cue to prey procurement. Larger worms are deeper down and take longer to target. Long targeting postures (crouches), during which plovers are unlikely to be vigilant, allow gulls additional time to make an undetected approach. For this reason, attacks are often launched before a worm has actually been caught. The distribution of attack timings, however, suggests that gulls wait a short time before attacking to make sure a crouch is not abortive.

Not surprisingly, plovers are under pressure to reduce the impact of gulls on their feeding efficiency. They appear to do this in a number of ways. The most conspicuous is the tendency to move away from the gull's vantage point so that flock density in the vicinity of gulls rapidly decreases. After a while, gulls make relocation flights within the flock to find higher-density areas. The tendency to make relocation flights, however, depends on how well gulls have been doing. The better they have done, the more likely they are to move. We can thus talk in terms of giving-up times for plover patches which are analogous to those of non-kleptoparasitic predators in other patchy environments. Plovers also increase the distance over which gulls have to attack by turning their backs on gulls which are nearby and by handling large, vulnerable worms only when gulls are far away. Although gulls certainly defend areas of plover flocks against incoming and resident gulls, caution is needed in interpreting the restriction of attack ranges at high gull densities as being territorial. A simpler explanation is that, in productive flocks, gulls can be supported by fewer plovers. Thus, more gulls operate over smaller distances. When the energetic cost of attack and the time and probability costs incurred by plover anti-theft responses are taken into account, gulls appear to select worms as predicted by a simple optimal-diet model similar to that applied to plovers. Selective stealing means that gulls do better than they would by being unselective and slightly exceed their estimated daily energy requirement.

Summary

1. An important factor influencing the cost and success of kleptoparasitic attacks by gulls is the risk of detection by target plovers.

2. The risk of detection increases with distance, and several responses of plovers to the presence of gulls tend to increase the distance over which attacks have to be made.

3. Gulls usually attack over those distances which are predicted to be the most profitable.

4. A simple vigilance model successfully predicts the probability of gulls making undetected attacks over given distances. To increase their chances of reaching targets undetected, gulls attack when the targets are crouching.

5. The range of worm sizes stolen by gulls conforms to the predictions of an optimal-diet model and gulls do better by foraging selectively than by stealing worms at random.

6. Gulls prefer to attack lapwings rather than golden plovers because the cost of attacking the latter is greater.

7. Gulls appear to obtain sufficient worms by kleptoparasitism to meet their energy requirement.

Chapter 9

Vigilance, Alarm Responses and an Early Warning System

Most animals must not only successfully exploit limited resources, they must also avoid becoming the exploited limited resource of predators and competitors. Both finding limited resources and avoiding predation and exploitation, however, may require considerable investment of time and energy in scanning the environment, because resource availability and encounters with predators and competitors tend to be unpredictable in time and space. The need to scan can seriously constrain an animal's ability to carry out important maintenance activities such as feeding and sleeping (e.g. Powell 1974, Kacelnik *et al.* 1981, Barnard and Brown 1981, Lendrem 1982, 1984). Factors which enable animals to modify their investment in scanning are therefore likely to influence their behavioural efficiency and thus reproductive potential. One which we have already seen has an important effect on time and energy budgeting, especially with regard to vigilance, is grouping behaviour (see Chapters 1 and 6, also Barnard 1985). The relationships between grouping behaviour, scanning and responsiveness to alarm stimuli, however, are not necessarily straightforward.

Social Behaviour and Vigilance for Predators

Group Size and the Allocation of Time to Scanning

The potential relationship between grouping behaviour and individual investment in scanning for predators was first formalised in a simple mathematical model by Pulliam (1973). Pulliam envisaged feeding flocks of birds in which individuals reduced their allocation of time to scanning as a simple function of increasing flock size (the number of birds in the flock). His model assumed that (a) scans ('head-cocks' in his terminology) were more or less instantaneous actions, (b) they occurred randomly and (c) individual birds scanned independently. The model also assumed that, if one bird in the flock detected an approaching predator, it alerted all the others. If o is the time a predator takes to make its final uncovered dash to attack the flock and λ_n is the mean rate of scanning per bird in a flock of n birds, then the probability P_n that any bird in the flock scans in a time less than or equal to o (and therefore detects the approaching predator) is:

$$P_n = 1 - e^{-n \cdot \lambda_n \cdot \theta} \tag{9.1}$$

If λ_n and θ remain constant, the probability that a predator will be detected increases with flock size. As a corollary, a bird joining a large flock could reduce its own scanning rate without reducing P_n and therefore its own risk of predation.

Although there are difficulties with the assumptions of Pulliam's simple model (see e.g. Elgar and Catterall 1981), several field and laboratory studies have borne out the broad prediction of decreasing individual investment in scanning with increasing flock size (e.g. Lazarus 1972, 1978, Powell 1974, Barnard 1980a, Sullivan 1984, Lendrem 1984, and see Chapter 1).

Evolutionarily stable scanning. Exactly how this negative relationship arises is interesting, because, at first sight, it seems that birds scanning in large flocks risk being exploited by cheats who opt out and spend all their time feeding. One way to avoid this would be to monitor the scanning behaviour of neighbours and adjust investment accordingly. Pulliam *et al.* (1982) have examined the dependence of individual investment in scanning and the scanning behaviour of companions using a games theory approach. The degree of investment in scanning can be viewed as an optimisation problem, since time and energy spent scanning cannot be spent in other important activities. The problem, however, is that several optimal scanning strategies are possible. This is because the optimal scanning rate for any one bird depends on how frequently other birds are scanning. The optimal strategy might, for instance, be for all birds in the flock to co-operate and scan at the rate that maximises survival for all co-operators. This, however, is open to cheats who scan at less than the co-operative rate. An alternative optimum might therefore be the rate which, if adopted by all flock members, has the property that any individual deviating from it has a lower probability of survival (i.e. an evolutionarily stable strategy (ESS) *sensu* Maynard Smith 1974). Pulliam *et al.* (1982) calculated the expected pay-off (probability of surviving the day) accruing to 'co-operative' and 'selfish' scanners and compared the different optima with actual scanning rates in flocks of yellow-eyed juncos. To their surprise, the observed relationship between scanning rate and flock size was close to the co-operative optimum.

A possible explanation is that Pulliam *et al.* left out of their calculations the fact that individuals within junco flocks are likely to meet more than once, because these birds live in small, closed populations (e.g. Caraco 1979b). Axelrod and Hamilton (1981) have shown that, under certain conditions where alternative strategies can meet again and again, the most stable strategy seems to be to do whatever the other individual has just done ('tit-for-tat'). In the case of scanning, this could be equivalent to the rule 'co-operate as long as everyone else does, but defect if they don't'. The optimal solution may therefore be to play a conditional strategy: 'scan if, and only if, other members of the flock are scanning'. Pulliam *et al.* (1982) referred to this as the 'judge' strategy. A 'judge' thus scans at the co-operative rate if it is associating with co-operators or other judges, but at the selfish rate if it detects a cheat. In simulations, judge does as well as co-operator in flocks of co-operators and/or judges, and better than co-operator in flocks containing selfish strategists. Judge also does well as a selfish strategist under the latter conditions, and better when flocking with another judge than either of two selfish strategists associating together. Of the three strategies considered, therefore, only judge emerges as an ESS.

At first sight, it might seem that judge could be stable only in small flocks, since a bird is unlikely to be able to monitor all companions in a large flock. A judge should stop scanning at the co-operative rate if it notices just one bird in the flock cheating. Very quickly, therefore, all birds in a large flock will stop scanning if a cheat is discovered. Consequently, a selfish strategist will have a lower chance of survival than a judge and judge emerges as an ESS, which is independent of flock size. It may be, therefore, that Pulliam *et al.*'s juncos scanned at the co-operative rate because judge has gone to fixation in the population. There is, however, an alternative explanation: there may be an advantage in being the first bird to spot an approaching predator (see Charnov and Krebs 1975). If so, then birds would benefit by scanning at a higher rate than the selfish equilibrium (Pulliam *et al.* 1982).

Group Size and Corporate Vigilance

The modulation of individual scanning rate as a function of flock size is clearly likely to affect the chances of an approaching predator being spotted. An important consideration, therefore, is how flock-size-related changes in individual scanning behaviour affect the probability that at least one bird in the flock will spot the predator. Depending on the relationship between group size and the probability of group detection by a predator (and hence individual risk) (Vine 1971, Treisman 1975), we might expect the probability to remain constant, so that individual risk does not increase with reduced scanning rates, or to increase where larger groups are more vulnerable (because they are more easily detected). Several studies have attempted to test this by using various indices of group or *corporate* vigilance.

The simplest index of corporate vigilance is that used by Barnard (1980a), which calculates the total number of scans per unit time (V_c) in a given flock size as:

$$V_c = S \cdot F \tag{9.2}$$

where S is the scanning rate of a focal bird and F is the number of birds in the flock. This has proved quite successful at predicting responsiveness to approaching apparent predators and will be used again later. Barnard (1980a) recorded the responsiveness of different-sized sparrow flocks to an approaching human observer. Responsiveness was measured as the distance from the observer at which flocks took off (the *flight distance*: Altmann 1958, Owens 1977, Byrkjedal and Kålås 1983, Thompson and Thompson 1985). From the relationship between corporate vigilance and flock size, Barnard predicted that the latter would not influence flight distance. Multivariate analysis controlling for the correlation between flock size and food density showed that, indeed, flock size did not itself influence flight distance. Instead, flight distance decreased when birds were feeding on higher food densities and spending less time scanning.

The index of corporate vigilance in equation (9.2) assumes that the probability of detecting a predator is a simple function of the rate at which flock members are scanning. Other indices have considered more specific relationships between the distribution of inter-scan intervals (when birds are presumed to be unable to detect an approaching predator) and the attack strategy of predators needing to make a final, uncovered dash to secure prey (Lendrem 1982, Hart and Lendrem 1984). A model along the lines developed by Lendrem is applied to the time-budgeting and attack strategies of plovers and gulls in Chapter 8. The most important point arising from these more sophisticated models is that the benefit of being in a flock varies with the attack strategy of the predator (see Hart and Lendrem 1984).

Flocking Behaviour and Modulating Responsiveness to Predators

So far, we have assumed that birds will respond in some way to having detected a predator. Indeed, empirically, some kind of response by birds is usually the only indication that detection has occurred. Despite the fact that field studies have often equated alarm responses (e.g. calling, taking-off) with predator detection (e.g. Siegfried and Underhill 1975, Barnard 1980a), there is now good evidence that the two may be temporally separate events and that an identifiable alarm response may occur some time after detection of the predator. The best evidence comes from Lazarus's (1979a) study of alarm responses in weaver-birds.

Lazarus (1979a) found that the number of birds responding to an alarm stimulus (a small light appearing in the cage) in flocks of different size generally did not differ from that expected if birds were responding to detecting the light rather than to the spread of alarm through the group (see Treherne and Foster 1981, Byrkjedal and Kålås 1983). Analysis of the *type* of response they showed, however, revealed a pronounced effect of flock size. Lazarus distinguished three categories of response: *orientation* (head movements orientating the birds' visual field towards the stimulus), *flight intention plus flight*, and *orientation plus flight intention/flight (mixed response)* (hopping, running and flying were all regarded as 'flight responses'). As flock size increased, birds were more likely to orientate and less likely to show flight intention/flight responses. The reduced tendency to flee in large flocks can be understood in terms of reduced individual risk (Hamilton 1971, Foster and Treherne 1981). Because they are safer, birds can afford to check out detected stimuli (by orientating) to see whether they really constitute danger. In this way, the risks of wasteful false alarms can be reduced. Greig-Smith (1981) found a similar separation of detection and response in feeding barred ground doves. Here, however, birds tended to take flight sooner in large flocks. Greig-Smith attributed this to the greater likelihood of a large flock containing a particularly nervous individual.

Early warning. Lazarus and Greig-Smith were concerned with responsiveness in relation to flock size in single-species aggregations. Greig-Smith's suggestion that birds may be responding to a particular kind of individual, however, raises an interesting possibility. If certain individuals are, for some reason, more likely to detect and/or respond to approaching danger, they can be exploited by other flock members. If their enhanced responsiveness does not result in unnecessarily frequent interruptions to other important activities, companions may be able to reduce their responsiveness to false alarms and perhaps even their commitment to scanning. In certain cases, one or a few individuals may effectively act as 'sentinels' or early warners, spending a high proportion of their time scanning, often from a prominent vantage point (e.g. Murton and Isaacson 1962, Byrkjedal and Kålås 1983). Other individuals in the group can then feed, rest and so on until the sentinel provides a warning. Sentinel behaviour is, however, an extreme. In mixed associations, at least, individuals are likely to differ in size, morphology, time budgeting and other characteristics which could influence their likelihood of detecting danger. In these cases, species-typical behaviour may allow individuals of one species to capitalise on the vigilance properties of another. As we mentioned in Chapter 1, this may be one reason why certain species form mixed aggregations. Examples are known among waders. Dunlin, for instance, associate with golden plovers on their breeding grounds (Oakes 1948, Ratcliffe 1976). Golden plovers are taller than dunlin and the latter appear to benefit from their greater vigilance (Byrkjedal and Kålås 1983, Thompson and Thompson 1985). In some places, the association between the two species is so consistent that the dunlin has been referred to as the 'plover's page' (in England: Oakes 1948) and the 'loupraell' (= plover's parasite, in Iceland: Byrkjedal and Kålås 1983).

Interestingly, Byrkjedal and Kålås showed that, in single-species flocks, dunlin scan more often but have shorter flight distances than when associating with golden plovers. In a recent study, Thompson and Thompson (1985) found that both species had longer flight distances in mixed associations. Furthermore, in every case observed, dunlins took off after golden plovers and *only* when golden plovers took off (Figure 9.1, see later).

So far, we have seen how flock size might relate to individual scanning behaviour and how a relationship between flock size and individual commitment to scanning might evolve. We have also looked at the potential effects of flock-size-related scanning on the probability of at least one individual spotting approaching danger, and how being in a flock might allow individuals to vary their response to alarm stimuli. Our investigation of plovers and gulls in the preceding chapters has focused on the consequences of single- and mixed-species association for individual feeding efficiency. Since flock size and composition affects individual time budgeting, however, it is possible that birds

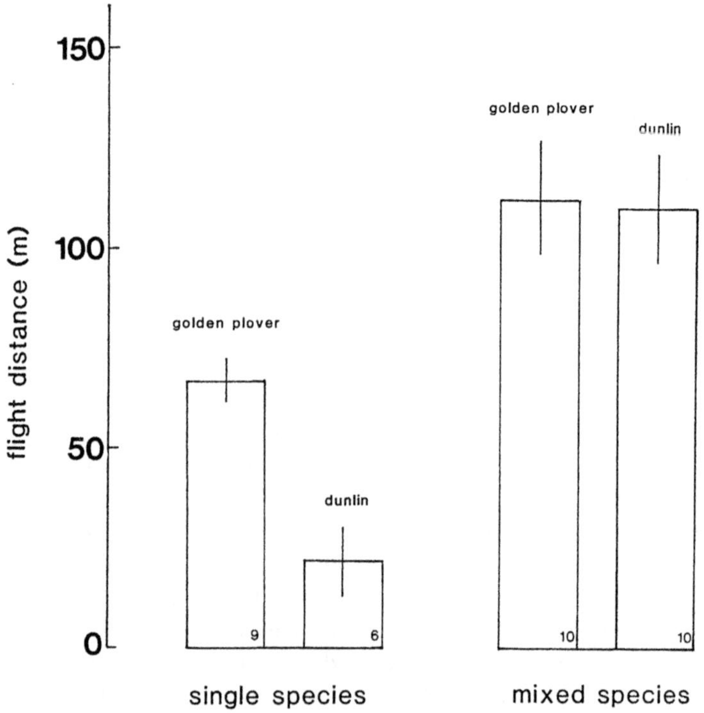

Figure 9.1: Flight distances (m) for golden plovers and dunlin in single- and mixed-species associations. One-way analysis of variance shows a significant effect of flock composition on flight distance ($F = 6.37$, $p < .01$, d.f. 3,31). Bars represent standard errors

Source: from data in Thompson and Thompson (1985)

also gain anti-predator benefits from the association. In this chapter, we examine the interrelationships between flock composition, environmental factors, time budgeting and alarm responses. In particular, we look at the effects of mixed association on responsiveness in plovers.

Vigilance and Responsiveness in Gulls and Plovers

As in several other flocking species, the amount of time plovers spend scanning changes with flock size and composition (Chapter 6). From casual observations, it also appears that the ease with which flocks can be alarmed by humans and potential predators and the number of birds which respond by taking off vary considerably from flock to flock. In our case, natural predators comprise kestrels (*Falco tinnunculus*), sparrowhawks (*Accipiter nisus*), rooks and carrion crows, all of which we have seen skimming and alarming flocks and, in the case of carrion crows, killing lapwings (K.R. Futter pers. obs.). For further evidence of predation by raptors and corvids, see e.g. Tinbergen (1953b), Franck (1955), Yalden (1980). From what we know already, both flock composition and physical environmental factors are likely to have important effects on responsiveness. In the first part of this chapter, we test the hypothesis that the responsiveness of birds to a standard alarm stimulus is determined by those aspects of flock composition which affect individual scanning behaviour and thus, presumably, vigilance.

Observations were made in the Hill Farm area during the winters of 1980/1981 and 1981/1982. As in several other field studies where vulnerability to avian predators has been considered (e.g. Davis 1973, Page and Whitacre 1975, Barnard 1980a, Metcalfe 1984a, b), the number of attacks actually observed was very small. In 950 hours of observation, we saw only five unsuccessful attacks (two by kestrels, three by rooks). All were against lapwings in mixed flocks. Both attacks by kestrels were directed at peripheral birds, but the rook attacks were directed at targets in different parts of the flock. To standardise the alarm stimulus (Markgren 1960), one of us alarmed birds by making himself conspicuous from the perimeter of the field for a period of 3s. The intensity of alarm was standardised by considering only those flocks for which the distance from the observer to the nearest peripheral bird was approximately 40m. We considered three categories of response: (i) no birds took flight; (ii) some or all birds took flight but resettled in the same field within 30s; and (iii) some or all birds took flight and departed (occasionally a few birds returned to the same field after flying away for some distance).

The total number of birds of each species resettling or departing expressed as a proportion of their species subflock size provided an estimate of the probability that an individual of each species would respond to alarm (= individual responsiveness). For each presentation, we also recorded the size and composition of the flock, pasture area and the usual climatic and temporal variables (see earlier chapters).

Flock Composition, Corporate Vigilance and Individual Responsiveness

To see whether individual responsiveness within a given flock is predictable from the effects of flock size and composition on scanning behaviour, we first used a modified form of equation (9.2) to calculate corporate vigilance for species subflocks (*subflock vigilance*). Subflock vigilance (V_{sci}) was calculated as:

$$V_{sci} = S \cdot F_{si} \tag{9.3}$$

where F_{si} is the number of birds in the subflock of species *i* and S is as in equation (9.2). This provides an estimate of the probability that one or more birds of species *i* will be scanning at any given moment. The assumption that scanning (as defined in Chapter 5) reflects vigilance for danger was supported by an analysis of what focal birds were doing when they took off in response to an approaching predator during time-budget recordings. On each of 11 occasions when a rook or carrion crow flew low over a flock, causing birds to take off, focal birds (three lapwings and eight golden plovers) were scanning immediately prior to departure.

How does subflock vigilance vary with flock composition? In particular, how does it vary with those aspects influencing individual scanning behaviour? Table 9.1 shows the results of stepwise partial regression analysis relating subflock vigilance to flock composition: subflock vigilance in lapwings increases with lapwing subflock size, golden plover subflock density and the number of gulls. Earlier analyses (Barnard and Stephens 1981) suggested that individual scanning rate decreased with lapwing subflock size and increased with the number of gulls (but see Table 6.3a, b, which takes a wider range of flock composition and environmental variables into account). Table 6.3a also shows that the proportion of time spent scanning by lapwings decreases with the total density of plovers in the flock. Subflock vigilance therefore tends to increase with subflock size, despite a loose tendency for individual scanning rate to decrease (see also Bertram 1980). In golden plovers, subflock vigilance increases with conspecific subflock size, the ratio of golden plovers to gulls and the number of gulls. Table 6.4a, b shows that the number of gulls correlates with a reduction in individual scanning rate in golden plovers, while conspecific subflock density correlates with an increase in scanning rate, at least when gulls are present.

From the relationships between subflock vigilance and flock composition in lapwings and golden plovers, and on the assumption that birds would respond as soon as an alarm stimulus was detected, we made some simple predictions about responsiveness of birds of each species in mixed flocks:

Table 9.1: Beta values from stepwise partial regression analysis for the relationship between plover subflock vigilance and flock composition

	Independent variables				
Subflock vigilance	**GP**	**GP/ha**	**L**	**B-HG**	**GP/B-HG**
Lapwings	ns	.324*	.383**	.271*	ns
Golden plovers	.326***	ns	ns	.214*	.265**

* $p < .05$, ** $p < .01$, *** $p < .001$; see Table 9.2 for details.
Data for 54 observations of lapwings and 263 of golden plovers.

(i) Lapwings should be more responsive (more likely to take off) when there are more lapwings and gulls in the flock and more golden plovers per hectare.
(ii) Golden plovers should be more responsive when there are more golden plovers per gull and more gulls.
(iii) On the assumption that gulls are sensitive to the responsiveness of other flock members, gulls should be more responsive when more lapwings, golden plovers or gulls are present.

Table 9.2 shows the effects of flock composition on individual responsiveness in each species. Predictions (i) and (ii) are largely borne out in that responsiveness in lapwings increases significantly with the density of golden plovers and also increases (though not significantly, $r = .224$, $p > .05 < .08$) with lapwing subflock size. Responsiveness in golden plovers increases with the number of golden plovers per gull and golden plover density (though not golden plover subflock size as predicted). This last discrepancy is perhaps not surprising, since subflock size and density are positively correlated over the range considered in the analysis of time budgets (see Chapter 6). As might be expected from the discussion of flock dynamics in Chapter 7, golden plovers tend to respond more cohesively than lapwings (a mean of $54 \pm 3.0\%$ responded per alarm, compared with $18 \pm 3.0\%$ in lapwings). They also leave the site more often after being alarmed. Nevertheless, despite these differences, mean responsiveness in the two plover species is very similar within flock types (Table 9.3). In general, therefore, those aspects of flock composition which influence subflock vigilance also influence individual responsiveness to a standard alarm. Two predicted relationships which are not borne out, however, are those between the responsiveness of lapwings and golden plovers and the number of gulls. Although subflock vigilance increases with the number of gulls in both species, responsiveness *decreases*. How might this apparent paradox arise?

Casual observation of take-offs after disturbance is enough to suggest that gulls respond to alarm more quickly than the two plover species, and that responses by gulls are usually quickly followed by lapwing and then golden

Table 9.2: Beta values for the relationship between flock composition, environmental variables and responsiveness in lapwings, golden plovers and gulls (see text)

Probability of response	Independent variables GP/ha	GP/L	B-HG	GP/B-HG	L/B-HG	Max °C	Daylength	Field size
Lapwing (n = 232)	.17*	ns	−.22*	ns	ns	.18*	.28**	−.36**
GP (n = 98)	.18*	ns	−.19*	.18*	ns	.24*	ns	—
Gull (n = 33)	ns	.19*	ns	ns	.23*	ns	ns	−.17*

GP = no. of golden plovers, L = no. of lapwings, BHG = no. of black-headed gulls, ha = hectare.

* $p < .05$, ** $p < .01$; significance levels for F-ratios associated with beta values; ns = not significant relationships; — variable not included in analysis.

Table 9.3: Mean (± se) probability of a bird of each species responding to a standard alarm in flocks of different composition

	Probability of response L only	L + GP	L + B-HG	L + GP + B-HG
Lapwing	.47 ± .03	.39 ± .06	.27 ± .14	.27 ± .09
Golden plover	—	.36 ± .08	—	.26 ± .12
Gull	—	—	.67 ± .32	.25 ± .13

Initials as in Table 9.2.
Data for 232 flocks. — not included.

Source: Thompson and Barnard (1983)

plover responses. It may be that the almost continuous scanning by gulls, and their conspicuousness during take-off and flight, provide an early-warning system for the other species. Birds might use the responsiveness of those individuals most likely to detect danger first (in this case, gulls) as a guide to when it is worth taking off. If gulls remain, then danger is unlikely to be close and birds can continue feeding; if they take off, danger has probably been spotted. This is a further response modulation to those considered by Lazarus (1979a) and Greig-Smith (1981) (see above). To test this, we carried out a further series of experiments.

Gulls as Early Warners

To see whether gulls really are the first birds in mixed flocks to respond to alarm, and whether lapwings and golden plovers are able to capitalise on their departure, we recorded the time elapsing between making ourselves conspicuous and the first bird of each species taking off (response time). Table 9.4 shows mean response times for all three species in flocks containing gulls and for lapwings and golden plovers in flocks without gulls. Gulls respond almost immediately (with mean response times of less than a second). The response times of lapwings and golden plovers, however, depend on whether or not gulls are present. When they are, lapwings tend to respond after just over a second and golden plovers after about 2.5 s. When they are not, the response times of both species are significantly longer. Gulls therefore respond first — and very quickly — and the other species follow suit, but sooner after the alarm if gulls are present. This is strong suggestive evidence that gulls have an early-warning effect.

Since the presence of gulls does appear to reduce the response time of plovers and hence, potentially, their vulnerability, we might expect plovers to be less likely to respond to a standard alarm when gulls are present. Table 9.3 suggests that they are. When all four types of flock are compared, lapwings and golden plovers turn out to be least responsive in flocks with gulls and, at least in lapwings, most responsive in single-species flocks. Interestingly, gulls are more likely to respond in flocks of lapwings and gulls than in flocks containing all three species (no flocks of gulls and golden plovers only were observed). Although there is no suggestion in Table 9.2 that responsiveness in gulls decreases with their number in the flock, three-species flocks tend to contain more than one gull (mean number per flock = 1.8 ± .66, range 1-9, compared with a mean of 1.0 ± .02, range 1-3 in lapwing/gull flocks). It may be, therefore, that gulls also modify their responsiveness when other gulls are present. It is important to point out that reduced responsiveness in the presence of gulls is not due simply to gulls joining only large flocks, in which the subflock denominator in the calculation of species responsiveness is large. In our sample flocks there was, on average, the same number of lapwings in lapwing/golden plover flocks (mean number = 34.5 ± 5.2) as in three-species

Table 9.4: The mean (± se) time elapsing between presentation of a standard alarm and the first bird of each species responding. Data for flocks with and without gulls

	Response times (secs)		
	Gulls	Lapwings	Golden plovers
With gulls	0.87 ± .44	1.36 ± .19	2.55 ± .33
Without gulls	—	3.64 ± .70	6.82 ± .68
t-test		2.86 $p < .02$	3.94 $p < .001$

Data from a total of 26 flocks. — not included.

Source: Thompson and Barnard (1983)

flocks (mean number = 35.4 ± 4.4) (see Table 9.3 for mean responsiveness in different types of flock).

If lapwings and golden plovers use gulls as early warners, we might expect their responsiveness to be greater if gulls respond than if they do not. This seems to be the case (Table 9.5). Birds of both species are significantly more likely to respond if a gull takes off and the probability of response is, in both cases, greater than in flocks without gulls ($t = 2.26$, $p < .05$ for lapwings; $t = 2.42$, $p < .02$ for golden plovers). Conversely, the probability of response when gulls do not take off is significantly *lower* than in flocks without gulls ($t = 8.06$, $p < .001$ for lapwings; $t = 4.27$, $p < .001$ for golden plovers). The fact that responsiveness in plovers is lower when gulls do not respond than when they are not present at all is strong evidence that plovers are using gulls as a simple means of deciding when to take off.

Early warning and flight distance. So far, we have considered responsiveness solely in relation to a standard alarm stimulus given at a fixed distance. If gulls provide early warning of danger, however, we might also expect their presence to increase the *distance* from danger at which birds respond (flight distance, see above). Because they scan all the time, gulls are likely to detect an approaching object sooner than plovers and, if they respond, we should also expect lapwings and golden plovers to respond (see Table 9.5). To test this, one of us again acted as an alarm stimulus, but this time walked at a constant pace (one pace/sec) towards flocks. Only flocks giving unobscured approaches of >200 m from the periphery were used.

As expected, Figure 9.2 shows that both lapwings and golden plovers have significantly greater flight distances when gulls are present. Lapwings also have greater flight distances than golden plovers. Mean distances for each condition are shown in the legend to Figure 9.2. Flight distances for each species are also plotted against lapwing and golden plover subflock size to control for any confounding effect with the number of conspecifics. While there are some suggestive trends, only for golden plovers in the presence of gulls is there a significant effect of conspecific subflock size (Figure 9.2). Again,

Table 9.5: Mean (± se) probability of a plover responding to a standard alarm when gulls do or do not respond

	Probability of response in	
	Lapwings	Golden plovers
Gulls responding	0.70 ± .11	0.82 ± .17
Gulls not responding	0.13 ± .03	0.01 ± .01
t-test	4.39 $p < .001$	4.67 $p < .001$

Data for 33 flocks.

Source: Thompson and Barnard (1983)

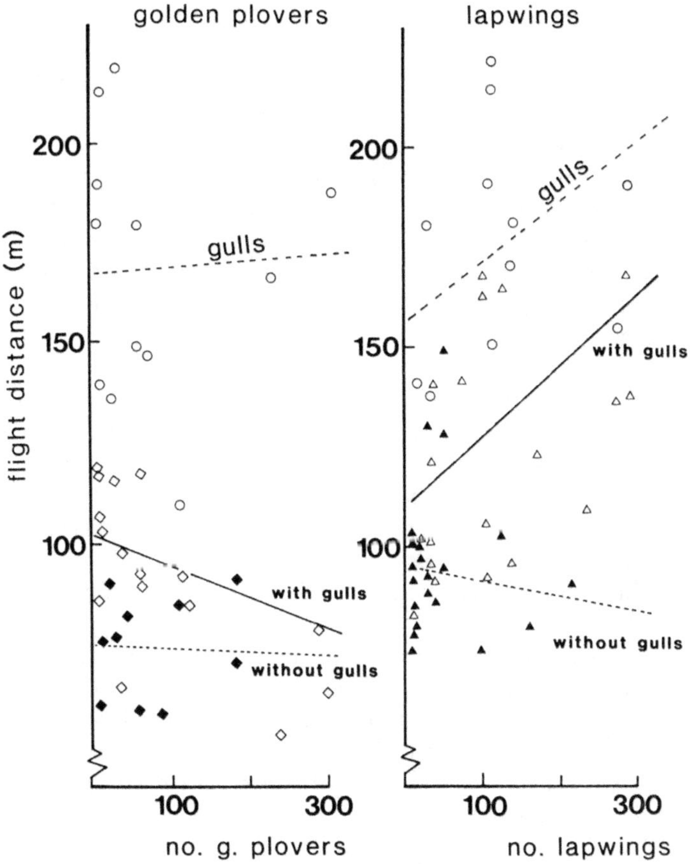

Figure 9.2: Flight distances for lapwings, golden plovers and gulls in relation to plover flock/subflock size. Distances with and without gulls plotted separately for lapwings and golden plovers. Golden plovers: open diamonds, with gulls; closed diamonds, without gulls. Lapwings: open triangles, with gulls; closed triangles, without gulls. The relationship between flight distance and flock/subflock size is significant only in golden plovers with gulls ($r = -.58$, $p < .02$). All lines fitted by regression. Mean (± se) flight distances for each condition are given below (data for 44 flocks):

		Flight distance (m)	
	Gulls	*Lapwings*	*Golden plovers*
With gulls	*169.0 ± 8.62*	*121.2 ± 5.82*	*93.7 ± 4.44*
Without gulls	—	*96.4 ± 4.61*	*75.7 ± 4.45*
t-test		*3.38* *p < .002*	*2.42* *p < .05*

Source: in part from Thompson and Barnard (1983)

therefore, the presence of gulls emerges as the best predictor of alarm responses.

As we saw in Chapters 2 and 8, black-headed and other gulls associate with a wide range of species, either on their feeding grounds or in mixed breeding colonies. If enhanced incidental vigilance is a widespread consequence of the opportunistic feeding habits of gulls, it would be surprising if other species had not capitalised on it. Burger (1984) has discovered that Rolland's grebes (*Rollandia rolland*) and silver grebes (*Podiceps occipitalis*) obtain early warning of predation by nesting in colonies with brown-hooded gulls (*Larus maculipennis*). Although the context is rather different from our own, it is worth detailing.

Grebes and gulls. Burger (1984) studied colonies of the two grebe species in Argentinian tule marshes and compared the behaviour and reproductive success of birds nesting in association with the gulls with those nesting just with grebes (about 50 % of grebes nested with gulls). She found that birds nesting with gulls lost fewer eggs to predators (mainly caracaras, *Milvago chimango* and *Polyborus plancus*, and weasels, *Grison vittatus*), and had a greater hatching success. She attributed this to a form of early warning provided by the gulls.

In most seabird colonies, at least some birds are usually flying above the colony at any given time. These individuals can spot approaching predators and, in many species, emit characteristic warning cries. These may be acted upon not only by conspecifics, but also by birds of other species. Neuchterlein (1981), for instance, showed that western grebes (*Aechmophorus occidentalis*) respond to the warning cries of Forster's terns (*Sterna forsteri*) as well as those of Franklin's gulls (*Larus pipixcan*) (see Burger 1984). Similarly, in Burger's study, Rolland's and silver grebes responded to the gulls' warning cries by covering their eggs when they left the nest. This was not done in the grebe colonies without gulls. Caracaras ignored covered nests and ate eggs only from nests which were not covered. Advance warning was also important in reducing adult mortality among incubating grebes. Over 90 % of adult predation occurred in grebe colonies without gulls.

The grebe example is one of many where birds appear to capitalise on so-called protector species within mixed associations (see e.g. Fuchs 1977 for black-headed gulls, Gotmark and Andersson 1980, Drycz *et al.* 1981). Burger's study is particularly interesting because it is the first to examine the effects of 'protector species' on reproductive success. The black-headed gulls in our study may provide an example of a protector species in a different context. As in the case of mixed breeding colonies, their 'protective' function is an incidental spin-off of their opportunistic (and, in this case, exploitative) feeding habits. Although lapwings and golden plovers may not be able to avoid gulls joining flocks, the interruptions and losses sustained through their attacks may be partly offset by their early-warning effect.

Vulnerability, Feeding Priority and Responsiveness to Alarm

Vulnerability and Responsiveness

An important factor which we might expect to influence an animal's responsiveness to alarm is its assessment of its vulnerability to attack. If vulnerability is assessed as being high, then we should expect responsiveness to any given alarm to be high. For animals feeding in exposed areas, vulnerability to attack may be influenced by proximity to cover. In birds, at least, the effect of distance to cover on time budgeting and responsiveness differs depending on how the cover is used. Some species, especially small passerines, use hedges bordering fields as a refuge between feeding bouts. Barnard (1980a) found that scanning rate in house sparrows feeding in newly-sown barley fields increased with distance from perimeter hawthorn hedges. Furthermore, birds fed for shorter periods and in larger flocks the further they were from cover. The higher rate of scanning in the open fields resulted in a doubling of the flight distance from an approaching human observer compared with distances in covered cattlesheds. As well as providing refuge from predators (e.g. Barnard 1979, Cramp and Simmons 1980), however, cover may provide predators with a concealed approach (e.g. Brown 1976) and thus constitute a danger. Perhaps for this reason, some species tend to feed away from cover. This seems to be the case in fieldfares. Barnard and Stephens (1983) found that, when feeding with redwings, fieldfares tended to scan more often and for longer periods the *nearer* they were to hedges. They fed close to cover only when few redwings (which they appeared to use to find food, see Chapter 1) were present. As in the house sparrows, there was a positive correlation between redwing subflock size and distance of feeding away from cover. Pulliam and Mills (1977) found that vesper sparrows differed in their flight responses to alarm according to the distance at which they fed from cover. If cover was close by, they fed in flocks and returned to cover when alarmed. When they fed away from cover, they tended to feed solitarily and went to ground if alarmed. This made sense in that vulnerability to resident aerial predators was likely to be increased by a long, exposed, flight.

From what we have seen of field choice by plovers so far, it seems that both species, but especially golden plovers, prefer to feed in larger fields (see Chapter 3). Although this is at least partly due to several of the old, worm-rich pastures being large, it may be that birds avoid feeding near perimeter hedges to reduce the risk of surprise attack. If so, we might expect birds to be more responsive when they are feeding in small fields and are thus constrained to be nearer cover.

Table 9.2 includes analyses of the effects of field size on responsiveness. Both lapwings and gulls are less likely to respond when they are feeding in large fields, with a high surface area:perimeter ratio. Although a significant negative relationship with field size did not emerge for golden plovers, Figure 9.3 shows that there is a tendency for golden plovers to be less responsive on large fields. Another measure of response strength might be the tendency for birds to leave a site altogether after being alarmed. Again, birds might be less willing to return to a small field after a disturbance because it is likely to be more difficult to tell whether the source of the disturbance has gone. Figure 9.4 bears this out. Both lapwings and golden plovers when feeding on small fields are more likely to leave the field, rather than resettle, if they are alarmed. Of course, an alternative explanation for the effect of field size on departure might be that, on small fields, birds are unable to settle outside their (perhaps extended) post-alarm flight distance (e.g. Pulliam and Mills 1977, Greig-Smith 1981), but this seems unlikely because field size does not directly influence the probability of resettling (Figure 9.4). In plovers, therefore, field size affects both quantitative and qualitative aspects of individual alarm response. There is no direct effect of field size on the tendency for gulls to depart. They depart whatever the conditions at the time of alarm (100 % of observed gull responses resulted in departure, compared with 39 % in lapwings and 74 % in golden plovers).

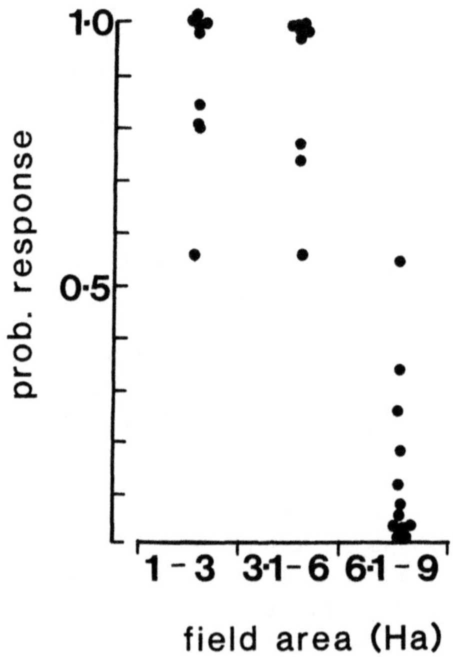

Figure 9.3: The relationship between field area and responsiveness in golden plovers. Data for 30 flocks

Source: Thompson and Barnard (1983)

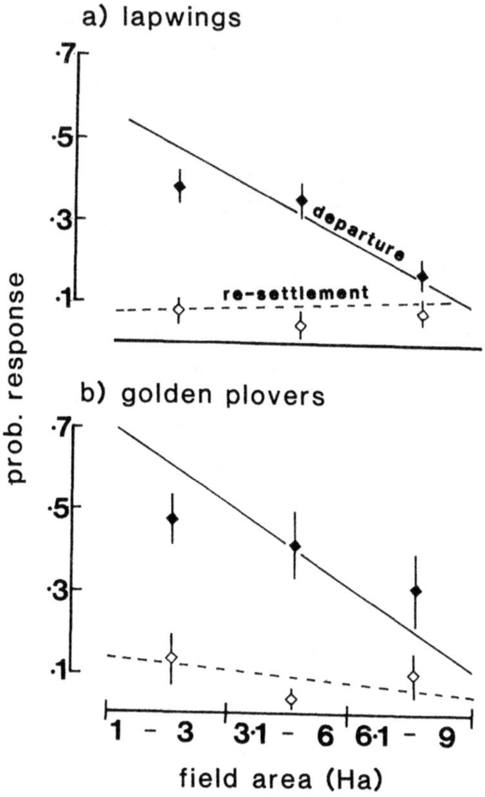

Figure 9.4: The relationship between field area and (a) the probability of lapwings departing ($r = -.35$, $p < .001$) or resettling ($r = .03$, n.s.), data for 186 flocks, and (b) the probability of golden plovers departing ($r = -.29$, $p < .01$) or resettling ($r = .04$, n.s.); data for 54 flocks. Means plotted with standard errors (bars). All lines fitted by regression

Source: Thompson and Barnard (1983)

If plovers prefer feeding in large fields because they can keep away from perimeter hedges, we might expect flock density to be high when they feed in small fields as birds aggregate in the middle of the field. We could test this only in lapwings, however, because golden plovers tended to avoid small fields altogether. Figure 9.5 shows firstly that the number of lapwings feeding in a field increases with field size (Figure 9.5a, see also Crooks and Moxey 1966) and, secondly, that flock density is indeed higher on small fields. Taken together, the two trends in Figure 9.5 imply that fewer birds are more densely packed and thus concentrated in a disproportionately small area of the field. Why golden plovers show a more pronounced aversion to small fields than lapwings is not clear, but Thompson (1983b) suggests that it may be because many of the smaller fields in the study area happen to be close to farm buildings. Golden plovers are the warier of the two species and may be more reluctant to feed near habitation.

Feeding Priority and Responsiveness

In Chapter 6, we saw that plovers foraging during the short, cold days of midwinter achieve relatively low rates of energy intake and undershoot their requirement. We might expect conditions that affect the urgency of food intake (feeding priority) also to affect responsiveness to alarm, because behaviours associated with feeding (e.g. crouching, pecking and stepping, see Chapters 5 and 6) reduce the time available for scanning (Tables 6.2, 6.4) and the likelihood that an alarm stimulus will be detected. Temperature is clearly one factor which is likely to affect feeding priority. In several species, like the lapwings and golden plovers above, reduced temperature increases feeding

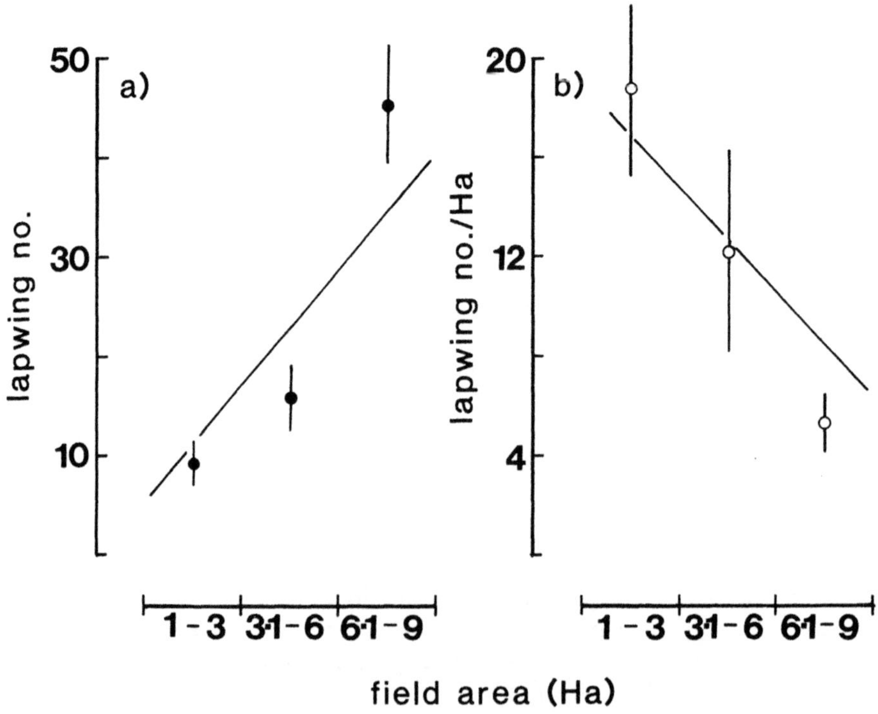

Figure 9.5: The effect of field area on flock/subflock size and density in lapwings. (a) $r = .56$, $p < .001$, d.f. 175, (b) $r = -.30$, $p < .005$, d.f. 175. Means plotted with standard errors (bars). Lines fitted by regression. Relationships best described by linear rather than higher-order regression

priority (e.g. Caraco 1979a, Barnard 1980a, Pienkowski 1982, Chapter 6). Similarly, feeding priority tends to decrease with increasing daylength (Barnard 1980a, Chapter 6). We might thus predict greater responsiveness on warm and/or long days, when feeding priority is apparently low and more time can be devoted to scanning. Generally, this seems to be the case (Table 9.2). Both lapwings and golden plovers are more responsive on days for which the maximum recorded daylight temperature is high, and responsiveness in lapwings increases with daylength. Responsiveness in gulls does not seem to be influenced significantly by climatic factors. This is not, however, surprising, since gulls tend to be present only in flocks which are productive in terms of turning up large worms (Chapters 6 and 8) and may thus be buffered to a large extent from the effects of variation in temperature and daylength (adverse conditions simply reduce the number of productive flocks which are available). Furthermore, the gulls' feeding strategy demands almost continuous vigilance whatever the climatic conditions, so responsiveness is unlikely to be affected in the same way as in plovers.

Anti-predator Qualities of Mixed-species Association: Résumé

Throughout the book, we have underlined the negative effect of kleptoparasitism by gulls on plover feeding efficiency. Gulls have been seen as exploiting the foraging effort of lapwings and golden plovers, particularly the former. That there is a cost to the plovers from attacks by gulls is clear from the investigation of feeding efficiency and the range of defensive strategies adopted when gulls are present (see Chapters 5, 6 and 8). In this chapter, however, we suggest at least one reason why plovers might actually *benefit* from the presence of gulls.

While plovers generally respond to alarm stimuli in ways which are predictable from the effects of flock size and composition on individual scanning behaviour and corporate vigilance, responsiveness does not increase as expected in relation to the number of gulls in the flock. Instead, responsiveness decreases in both species. A major reason for the decrease appears to be the apparent early warning of alarm provided by gulls. Gulls are usually the first birds to respond to alarm in mixed flocks and, if lapwings or golden plovers respond, they respond sooner to a standardised alarm when gulls are present. Overall, however, both species of plover are *less* likely to respond in flocks with gulls than in flocks without. Moreover, their tendency to respond to alarm when gulls are present depends on whether the gulls themselves respond. If they do, responsiveness in plovers is high; if they do not, it is lower than in flocks without gulls, suggesting that plovers use the responses of gulls as a simple guide as to when it is worth responding themselves. The early-warning effect of gulls is also shown in their effect on flight distances. Gulls depart sooner than plovers when an observer walks towards the flock. As a result, both species of plover have significantly longer flight distances in flocks with gulls.

It has been suggested (D.W. Lendrem pers. comm.) that, rather than gulls providing early warning, plovers might be responding to an increased risk of predation when gulls take off. Gulls taking flight are highly conspicuous and thus might attract the attention of a predator who would otherwise have missed the flock. This, however, seems highly unlikely, because gulls are conspicuous even when stationary and they also frequently take flight to steal prey. Taking flight in response to alarm is therefore unlikely to render gulls significantly more conspicuous to predators.

The early-warning benefit may be one explanation for the otherwise curious fact that lapwings sometimes join one or more gulls which are not feeding with plovers (authors' pers. obs.). This is difficult to account for if gulls have only a negative effect on plover fitness, although it is possible that lapwings are simply making a mistake. The conspicuous gulls may provide an easy means of spotting profitable feeding sites at a distance. As Källander (1977) points out, a human observer can identify fields containing flocks of lapwings from several kilometres simply by the characteristic distribution of gulls. It may be difficult to detect plovers until the field is almost reached. The early-warning effect may also explain why plovers tend to face towards gulls when they are at a relatively safe distance (see Figure 8.8).

Summary

1. Being in a flock may allow birds to reduce their commitment to scanning for predators and modify their responses to alarm. The way a relationship between scanning behaviour and flock size might evolve can be predicted from simple mathematical models.
2. Responsiveness to alarm in lapwings and golden plovers correlates with those social and environmental factors which influence individual scanning behaviour and corporate vigilance.
3. Responsiveness also varies in a predictable way with apparent vulnerability and feeding priority.
4. Black-headed gulls appear to provide plovers with early warning of alarm. Responsiveness in plovers depends on whether or not gulls respond, and plovers have longer flight distances in flocks containing gulls. Gulls may thus act as one form of 'protector species' as hypothesised in many mixed breeding colonies.

Chapter 10

Gulls and Plovers: an Overview

In the preceding chapters we have looked in detail at a number of different aspects of the association between black-headed gulls and two species of plover, from the function of pre-foraging flocks and the aggregation of birds on different types of pasture to the effects of flock size and composition on time budgeting and feeding efficiency and the kleptoparasitic activities of gulls. While the important points raised in each chapter are emphasised in a résumé and summary, we feel that the results and interrelationships between chapters are sufficiently complex to merit a brief synthesis. The aim of this short, final chapter is to provide an overview of the study as a whole and to highlight some of its main conclusions.

The Build-up of Foraging Flocks: Choosing Where to Feed

Like many other predators, overwintering gulls and plovers are choosy about where they feed. The criteria they use, however, vary between species. Lapwings almost always arrive to feed first and, having a range of fields to choose from, opt for older pasture. This turns out to be where earthworms, the chief prey over winter, are most abundant.

The location of rich feeding areas within fields is also more predictable in old pasture, and lapwings selectively land in locally-rich areas. The cues used to do this are not clear, but may include some measure of vegetation density and quality.

Golden plovers seldom feed in single-species flocks, but instead use the presence of lapwings in deciding where to land. By choosing fields in which lapwings are already feeding, golden plovers end up at good feeding sites. They fine-tune their decision by going for the densest lapwing flocks, because, on average, these are feeding on the areas of highest worm density. Gulls steal worms from plovers, so, not surprisingly, tend to land where the density of plovers is high. Their choice is largely unaffected by climatic or habitat variables. Gulls not only appear to reduce their foraging costs by stealing worms, they also use golden plovers as a quick means of locating profitable flocks in the first place by selectively associating with them as they leave pre-foraging flocks. Pre-foraging flocks, however, appear to act as assembly points rather than information centres as the latter term was originally conceived.

Of course, not all birds can feed at the most preferred sites. Numbers in old pasture eventually reach a limit and birds spill over onto lower-quality fields. As the number of birds trying to feed in the area increases, flock size in the best fields stays more or less constant, but a greater number of lower-quality fields are occupied. Numbers on less-preferred fields also increase with temperature as worms become more available and the difference in quality between fields is reduced.

Earthworms in pasture are a renewing source of food, and several aspects of foraging behaviour in plovers suggest that their pattern of movement within fields is geared to that of worm replenishment. Birds tend to make circuits of fields so that they do not end up back at a previously-depleted area until worm density has recovered to at least its pre-predation level. Local worm density also influences where birds land after disturbance or other movements within fields. Birds are less easy to disturb and more likely to return to the same place when local worm density is high.

Time Budgeting, Feeding Efficiency and Flock Dynamics

As in many other socially-feeding species, the way individual lapwings and golden plovers allocate time to different activities varies with flock size and affects feeding efficiency. In this case, however, the effects of flocking are complicated by variation in species composition.

Broadly speaking, lapwings tend to do better when conspecific density is high, while feeding efficiency in golden plovers correlates more closely with the combined density of both plover species. The way time budgeting affects individual feeding efficiency also differs between species. In golden plovers, there appears to be a direct trade-off between time spent in prey assessment (as suggested by crouching behaviour) prior to pecking and presumed vigilance for predators or kleptoparasites (scanning). In situations where scanning commitment is low (when gulls are absent and plover density high) and/or feeding priority low, golden plovers spend more time crouching. In lapwings there is a much less obvious relationship between vigilance and crouching. Crouching tends to increase with worm density when travel time between potential prey is low and birds can afford to be fussy.

Crouching appears to be a means of targeting deeper worms rather than assessing size *per se*. The most profitable, intermediate-sized worms tend to be deeper down (within the 3 cm soil horizon available to plovers) and harder to locate. Plovers appear to maximise their feeding efficiency by using the simple rule of thumb 'go for the deepest', even though this means more time spent targeting and a reduced success rate. When crouching is inhibited (in the presence of gulls), plovers are less able to locate deep worms and end up taking less profitable items from the soil surface.

Selective foraging increases feeding efficiency in terms of energy intake per unit time. Although large worms contain more energy than small worms, several factors reduce their profitability to lapwings and golden plovers. These include increased time costs of handling and crouching, increased risk of breakage and decreased peck success rate. Large worms are more likely to be stolen by gulls because they are more visible and take longer to extract and mandibulate. They are also more likely to be discarded after extraction, presumably because of the high risk of attack. The presence of gulls alters the relative profitability of different worm sizes, and plovers shift their preference accordingly. Plovers do best, however, by taking a range of worm sizes rather than simply the most profitable size. When this is taken into account, both plover species follow closely the predictions of a simple optimal-diet model.

The impact of gulls is much greater on lapwings than on golden plovers, apparently because lapwings respond to attack less quickly and are less agile in the air. Gulls are twice as successful in attacks against lapwings compared with golden plovers and are twice as likely to attack them. The presence of gulls is an important factor reducing daytime feeding efficiency in lapwings. During mid-winter, both plover species fail to meet their estimated daily energy requirement by daytime feeding alone, and feed into the night as daytime feeding efficiency decreases.

Although the kleptoparasitic activities of gulls have a negative effect on plover feeding efficiency, there appear to be positive consequences of having gulls in the flock. Gulls provide a form of early-warning system. Because they spend most of their time in the flock scanning for potential targets, gulls are usually the first to respond to approaching danger. Intriguingly, responsiveness to alarm in lapwings and golden plovers decreases as the number of gulls in the flock increases. If they do respond, however, they respond sooner when gulls are present than when they are absent. Furthermore, whether or not they respond depends on the response of gulls. If gulls respond (by taking off), responsiveness in plovers is high; if they do not, it is lower than in flocks without gulls. In other words, plovers appear to use the behaviour of gulls as a simple guide to when it is worth responding themselves. This is also manifested in flight distances. Gulls depart sooner than plovers as a potential alarm stimulus approaches, and both plover species have longer flight distances when gulls are present. Plovers may thus capitalise on the greater vigilance of gulls by reducing their responsiveness to innocuous 'alarm' stimuli and minimising the amount of time and energy wasted in unnecessary flights.

Plovers are often aggressive towards one another, more commonly within than between species. Aggression takes the form of both chasing and kleptoparasitism and is most intense when feeding priority is likely to be high and the size and composition of the flock unfavourable to the attacker. Aggression appears to be related to increased prey depletion when the density of birds around the attacker is high. Successful foragers tend to attack copiers, especially when turning up the most profitable, intermediate-sized worms.

Changes in individual time budgeting and feeding efficiency with flock composition are reflected in the dynamics of foraging flocks. Arrival and departure rates at flocks are not random, but vary with the costs and benefits expected to accrue to birds of each species by joining or remaining in a flock. Arrival and departure of golden plovers and gulls correlate with flock composition and physical environmental factors which affect time budgeting and feeding efficiency. Although gulls use golden plovers as a means of locating good foraging flocks, they are more likely to depart as the number of golden plovers, the less preferred target, feeding in the flock increases. The pattern of arrival in lapwings, however, does not seem to correlate with flock composition. This is in keeping with flock composition, except for the presence or absence of gulls, having little effect on lapwing time budgeting and feeding efficiency. Nevertheless, departure rates in lapwings are influenced by flock composition and birds are especially likely to leave when pressure from kleptoparasitic gulls is high. As might be expected, the presence of gulls has much less effect on departure by golden plovers.

There is a tendency for arrivals and departures of plovers to be concentrated near the edge of the flock, apparently because worm density and patchiness increase towards the flock periphery. Coupled with this, lapwings tend to occur at the edge and golden plovers in the centre of established foraging flocks, mainly because lapwings displace golden plovers in disputes over locally-rich worm supplies.

Changes in individual time budgeting and feeding efficiency as a consequence of changing flock composition in turn feed back to influence flock composition. The effect of newly-arriving gulls and plovers on changes in flock size and composition is predictable from the effects of their arrival on birds already present. Thus, lapwing subflock size decreases when gulls arrive because of the negative effect of kleptoparasitism on lapwing feeding efficiency. On the other hand, it decreases when golden plovers arrive only if gulls are present. When lapwings are already under pressure from gulls, additional competition from golden plovers becomes important. In the absence of gulls, golden plovers have no significant effect on lapwing time budgeting and feeding efficiency. In golden plovers, by contrast, the arrival of gulls has no consistent effect on subflock size, while additional lapwings cause a reduction in subflock size only when there are no gulls. Gulls have little direct effect on golden plover feeding efficiency, but their preference for lapwings means that the latter interfere with golden plovers less when under pressure from kleptoparasitism. The different patterns of arrival and departure shown by lapwings and golden plovers, especially the tendency for golden plovers to move between flocks (of all types) in a wide range of group sizes, mean that the two species show different subflock dynamics. In particular, golden plover subflocks are less likely than those of lapwings to reach an equilibrium size.

Kleptoparasitism

Black-headed gulls join flocks of plovers to steal food. At first sight, kleptoparasitism appears to be an obvious way of reducing searching and handling-time costs during foraging. Like any other foraging strategy, however, kleptoparasitism has its disadvantages as well as its advantages.

One of the most serious problems facing gulls is to reach a target bird without being spotted, at least until it is too late to affect the probability of stealing the target's worm without a protracted chase. Gulls are therefore more successful against birds which are nearby and when they launch an attack early in the target's period of prey procurement. The crouching posture adopted by plovers prior to pecking provides gulls with a useful cue in timing attacks. Since larger worms are deeper down and take longer to locate, crouch duration tends to correlate with the size of worm procured. Plovers may also be less likely to spot an approaching gull while their attention is focused on the ground. It may thus pay gulls to launch attacks during the crouching phase, even though this carries the risk of the target not capturing a worm.

Kleptoparasitism is costly to plovers. Not surprisingly, therefore, they take a number of steps to reduce loss to gulls. The most obvious are (a) the tendency to move away from gulls when they arrive in the flock and (b) the tendency to face away from gulls when they are nearby, thus obscuring prey-handling activities. Both these responses increase the distance over which gulls have to attack and, therefore, the time and energy costs of attack and the risk of detection. To counter this, gulls periodically move to new vantage points within the flock at a rate which is at least partly dependent on their current success rate. When the time, energy and risk costs of attacks over different distances are taken into account, gulls appear to launch attacks and select target worms as predicted by a simple optimal-diet model. By attacking selectively, they do better than they would on the basis of a random-attack strategy and marginally exceed their estimated daily energy requirement through kleptoparasitism alone.

References

Abramson, M. 1979. Vigilance as a factor influencing flock formation among curlews (*Numenius arquata*). *Ibis 121*: 213-16.

Altmann, M. 1958. The flight distance of free ranging big mammals. *J. Wild. Mgmt 22*: 207-9.

Altmann, S.A. and Altman, J. 1970. *Baboon Ecology: African field research.* Chicago: Chicago University Press.

Andersson, M., Gotmark, F. and Wiklund, C. 1981. Food information in the black-headed gull, *Larus ridibundus. Behav. Ecol. Sociobiol. 9*: 195-200.

_____ and Wiklund, C.C. 1978. Clumping versus spacing out: experiments on nest predation in fieldfares (*Turdus pilaris*). *Anim. Behav. 26*: 1207-12.

Aschoff, J. and Pohl, H. 1970. Der Ruheumsatz von als funktion der Togeszeit und der Korpergrosse. *J. Orn. 111*: 38-47.

Ashmole, P.N. 1971. Seabird ecology and the marine environment. In D.S. Farner and J.R. King (eds.), *Avian Ecology*, Vol. I: 223-86. New York: Academic Press.

Avery, M.I. and Krebs, J.R. 1984. Temperature and foraging success of great tits (*Parus major*) hunting for spiders. *Ibis 126*: 33-8.

Axelrod, R. and Hamilton, W.D. 1981. The evolution of co-operation. *Science 211*: 1390-6.

Baker, J.M. 1981. Winter feeding rates of redshank (*Tringa totanus*) and turnstone (*Arenaria interpres*) on a rocky shore. *Ibis 123*: 85-7.

Baker, M.C. 1973. Stochastic properties of the foraging behaviour of six species of migratory shorebirds. *Behaviour 45*: 242-70.

_____ and Baker, A.E.M. 1973. Niche relationships among 6 species of shorebirds on their wintering and breeding ranges. *Ecol. Monogr. 43*: 193-212.

Balph, D.F. and Balph, M.H. 1979. Behavioural flexibility in pine siskins in mixed species foraging groups. *Condor 81*: 211-12.

Barnard, C.J. 1978. Aspects of winter flocking and food-fighting in the house sparrow (*Passer domesticus* L.). D.Phil. thesis, University of Oxford.

_____ 1979. Interactions between house sparrows and sparrow hawks. *Brit. Birds 72*: 569-73.

_____ 1980a. Flock feeding and time budgets in the house sparrow (*Passer domesticus* L.). *Anim. Behav. 28*: 295-309.

_____ 1980b. Factors affecting flock size mean and variance in a winter population of house sparrows (*Passer domesticus* L.). *Behaviour 74*: 114-27.

_____ 1980c. Equilibrium flock size and factors affecting arrival and departure in feeding sparrows. *Anim. Behav. 28*: 503-11.

_____ 1982. Social mimicry and interspecific exploitation. *Am. Nat. 120*: 411-13.

_____ 1984a. The evolution of food-scrounging strategies within and between species. In C.J. Barnard (ed.), *Producers and Scroungers: strategies of exploitation and parasitism*: 95-126. London: Croom Helm.

_____ 1984b. Snap decisions for survival. *New. Sci. 102* (1411): 24-7.

_____ and Brown, C.A.J. 1981. Prey size selection and competition in the common shrew (*Sorex araneus* L.). *Behav. Ecol. Sociobiol. 8*: 239-43.

_____ and Sibly, R.M. 1981. Producers and scroungers: a general model and its application to feeding flocks of house sparrows. *Anim. Behav. 29*: 543-50.

_____ and Stephens, H. 1981. Prey size selection by lapwings in lapwing/gull

association. *Behaviour 77*: 1-22.
_____ and _____ 1983. Costs and benefits of single and mixed species flocking in fieldfares (*Turdus pilaris*) and redwings (*T. iliacus*). *Behaviour 84*: 91-123.
_____ Thompson, D.B.A. and Stephens, H. 1982. Time budgets, feeding efficiency and flock dynamics in mixed species flocks of lapwings, golden plovers and gulls. *Behaviour 80*: 44-69.
Barnes, B.T. and Ellis, F.B. 1982. The effects of different methods of cultivation and direct drilling and contrasting methods of straw disposal on populations of earthworms. *Soil Science* (in press).
Bateson, P.P.G. (ed.) 1983. *Mate Choice.* Cambridge: Cambridge University Press.
_____ Lotwick, W. and Scott, D.R. 1980. Similarities between the faces of parents and offspring in Bewick's swan and the differences between mates. *J. Zool. Lond. 191*: 61-74.
Bayer, R.D. 1982. How important are bird colonies as information centres? *Auk 99*: 31-40.
Beer, J.R. 1961. Winter feeding patterns in the house sparrow. *Auk 78*: 63-71.
Begon, M., Harper, J.L. and Townshend, C.R. 1985. *Ecology: organisms, populations and communities.* Oxford: Blackwell.
Belovsky, G.E. 1978. Diet optimization in a generalist herbivore: the moose. *Theor. popul. Biol. 14*: 105-34.
Bengtson, S-A., Nilsson, A. and Nordstrom, S. 1978. Selective predation on lumbricids by golden plovers (*Pluvialis apricaria*). *Oikos 31*: 164-8.
_____, _____, _____ and Rundgren, S. 1976. Effects of bird predation on lumbricid populations. *Oikos 27*: 9-12.
Bertram, B.C.R. 1978. Living in groups: predators and prey. In J.R. Krebs and N.B. Davies (eds.), *Behavioural Ecology: An Evolutionary Approach*: 64-96. Oxford: Blackwell Scientific Publications.
_____ 1980. Vigilance and group size in ostriches. *Anim. Behav. 28*: 278-86.
Beuchat, C.A., Chaplin, S.B. and Morton, M.L. 1979. Ambient temperature and the daily energetics of two species of hummingbirds, (*Calypte anna* and *Selasphorus rufus*). *Physiol. Zool. 52*: 280-95.
Bibby, C.J. and Green, R.E. 1980. Foraging behaviour of migrant pied flycatchers (*Ficedula hypoleuca*) on temporary territories. *J. Anim. Ecol. 49*: 507-21.
Birkhead, T.R. 1974. Utilization of guillemot (*Uria aalge*) colonies by jackdaws (*Corvus monedula*). *Ornis Scand. 5*: 71-81.
Boswall, J. 1970. The association of the northern carmine bee-eater (*Merops n. nubicus*) with mammals, birds and motor vehicles in Ethiopia. *Bull. Brit. Orn. Club. 90*: 92-6.
Bradbury, J. 1977. Social organisation and communication. In W. Wimstatt (ed.), *Biology of Bats.* New York: Academic Press.
Brockmann, H.J. and Barnard, C.J. 1979. Kleptoparasitism in Birds. *Anim. Behav. 27*: 497-514.
Brown, C.A.J. 1983. Prey abundance of the European badger (*Meles meles* L.) in North-East Scotland. *Mammalia 47*: 81-6.
Brown, L. 1976. *Birds of Prey: their biology and ecology.* London: Hamlyn.
Bryant, D.M. 1979. Effects of prey density and site character on estuary usage by overwintering waders (Charadrii). *Est. Coast. Mar. Sci. 9*: 369-84.
_____ and Leng, J.M. 1975. Feeding distribution and behaviour of shelduck in relation to food supply. *Wildfowl 26*: 20-30.
_____ and Turner, A.K. 1982. Central place foraging by swallows (Hirundidae): the question of load size. *Anim. Behav. 30*: 845-57.
Buckley, F.G. and Buckley, P.A. 1974. Comparative feeding ecology of adult and wintering juvenile ringed terns (Aves: Laridae). *Ecology 55* 1053-63.

Burger, J. 1984. Grebes nesting in gull colonies: protective associations and early warning. *Am. Nat. 123*: 327-37.

_____ and Gochfeld, M. 1979. Age differences in ring-billed gull kleptoparasitism on starlings. *Auk 96*: 806-8.

_____ and _____ 1981. Age related differences in piracy behaviour of four species of gulls (*Larus* spp.). *Behaviour 77*: 242-67.

_____ Hahn, C. and Chase, J. 1979. Aggressive interactions in mixed species flocks of migrating shorebirds. *Anim. Behav. 27*: 459-69.

Burgess, J. 1974. Kleptoparasitism in black-headed gulls. *Bird Study 24*: 279.

Burton, P.J.K. 1974. *Feeding and feeding apparatus in waders.* British Museum (Natural History), London.

Buskirk, W.H. 1974. Social system in tranquil forest avifauna. *Am. Nat. 110*: 293-310.

Butler, P.J. 1979. The use of radio telemetry in the studies of diving and flying in birds. In C.J. Almaner and D.W. Macdonald (eds.), *A Handbook on Biotelemetry and Radio Tracking*: 569-77. Oxford: Pergamon Press.

_____ West, N.H. and Jones, D.R. 1977. Respiratory and cardiovascular responses of the pigeon to sustained level flight in a wind-tunnel. *J. Exp. Biol. 71*: 1-26.

_____ and Woakes, A.J. 1980. Heart rate, respiratory frequency and wingbeat frequency of free-flying barnacle geese (*Branta leucopsis*). *J. Exp. Biol. 85*: 213-26.

Buxton, N.E. 1981. The importance of food in the determination of the winter flock sites of the shelduck. *Wildfowl 32*: 79-87.

Byrkjedal, I. and Kålås, J.A. 1983. Plover's page turns into plover's parasite: a look at the dunlin/golden plover association. *Ornis Fenn. 60*: 10-15.

Campbell, J.W. 1935. Notes on the food of some British birds. *Brit. Birds. 29*: 183.

_____ 1946. Notes on the food of some British birds. *Brit. Birds. 39*: 371-3.

Caraco, T. 1979a. Time budgeting and group size: a test of theory. *Ecology 60*: 618-27.

_____ 1979b. Time budgeting and group size: a theory. *Ecology 60*: 611-17.

_____ 1980. Stochastic dynamics of avian foraging flocks. *Am. Nat. 115*: 262-76.

_____ 1981. Risk-sensitivity and foraging groups. *Ecology 62*: 527-31.

_____ Martindale, S. and Pulliam, H.R. 1980a. Avian flocking in the presence of a predator. *Nature 285*: 400-1.

_____, _____ and _____ 1980b. Avian time budgets and distance to cover. *Auk 97*: 872-5.

Charnov, E.L. 1976. Optimal foraging attack strategy of a mantid. *Am. Nat. 110*: 141-51.

_____ and Krebs, J.R. 1975. The evolution of alarm calls: altruism or manipulation. *Am. Nat. 109*: 107-12.

_____ Orians, G.H. and Hyatt, K. 1976. The ecological implications of resource depression. *Am. Nat. 110*: 247-59.

Clegg, W.E. 1944. Predatory habits of black-headed gulls. *Brit. Birds. 38*: 57.

Cody, M.L. 1971. Finch flocks in the Mohave desert. *Theor. Popul. Biol. 2*: 142-58.

Collinge, W.E. 1924-27. *The Food of some British Birds.* York: Focal Press.

Cottam, C., Williams, C.S. and Sooter, C.A. 1942. Co-operative feeding in white pelicans. *Auk 59*: 444-5.

Coulson, J.C. 1959. Observations on the Tipulidae (Diptera) of the Moor House nature reserve, Westmorland. *Trans. R. Ent. Soc. Lond. 111*: 157-74.

_____ 1962. The biology of *Tipula subnodicornis* Zeiterstedt, with comparative observations on *Tipula paludosa* Meigen. *J. Anim. Ecol. 31*: 1-21.

_____ and Whitacker, J.B. 1977. The ecology of moorland animals. In O.W. Heal and D. Perkins (eds.), *The Ecology of some British Moors and Mountain Grasslands.* Springer-Verlag: Berlin.

Cox, F.E.G. 1968. Parasites of British earthworms. *J. Biol. Educ. 2*: 151-64.
Cramp, S., Bourne, W.R.P. and Saunders, D. 1974. *The Seabirds of Britain and Ireland.* London: Collins.
_____ and Simmons, K.E.L. (eds.) 1983. *The Birds of the Western Palearctic.* Vol III. Oxford: Oxford University Press.
Crook, J.H. 1964. The evolution of social organisation and vocal communication in weaver birds (Ploceinae). *Behav. Monogr.* No. 10.
Crooks, S.E. and Moxey, P.A. 1966. Study of wintering lapwing population in N.W. Middlesex. *London Bird Report 30*: 60-79.
Curtis, D.J., Galbraith, C.G., Smyth, J.C. and Thompson, D.B.A. 1985. Prey selection and feeding efficiency in estuarine black-headed gulls (*Larus ridibundus*). *Est. Coast. Shelf. Sci.* (in press)
_____ and Thompson, D.B.A. 1985. Spacing and foraging in black-headed gulls (*Larus ridibundus*) in an estuary. *Ornis Scand.* (in press)
Cvitonic, A. and Novak, P. 1966. The diet of golden plovers (*Pluvialis apricaria*) in Yugoslavia. *Larus 20*: 80-100.
Davidson, N.C. 1981. Survival of shorebirds (Charadrii) during severe weather: the role of nutritional reserves. In N.V. Jones and N.J. Wolff (eds.), *Feeding and Survival Strategies of Estuarine Organisms*: 231-50. London: Plenum Press.
Davies, N.B. and Green, R.E. 1976. The development and ecological significance of feeding techniques in the reed warbler (*Acrocephalus scirpaceus*). *Anim. Behav. 24*: 213-29.
_____ and Houston, A.I. 1981. Owners and satellites: the economics of territory defence in the pied wagtail, (*Motacilla alba*). *J. Anim. Ecol. 50*: 157-80.
_____ and _____ 1984. Territory economics. In J.R. Krebs and N.B. Davies (eds.), *Behavioural Ecology: an evolutionary approach* (2nd edn): 148-69. Oxford: Blackwell.
_____ and Krebs, J.R. 1978. Introduction: ecology, natural selection and social behaviour. In J.R. Krebs and N.B. Davies (eds.), *Behavioural Ecology: an evolutionary approach*: 1-18. Oxford: Blackwell.
Davis, J.M. 1973. Habitat preferences and competition of wintering juncos and golden-crowned sparrows. *Ecology 54*: 174-80.
De Groot, P. 1980. A study of the acquisition of information concerning resources by individuals in small groups of red-billed weaver birds (*Quelea quelea*). Unpublished Ph.D. thesis, University of Bristol.
Dittberner, H. and Dittberner, W. 1981. Aspects of behaviour in lapwings. *Orn. Mitt. 33*: 231-3.
Dobbs, A. 1979. *The Birds of Nottinghamshire: past and present.* Newton Abbot: David and Charles.
Draulans, D. 1981. Foraging and size selection of mussels by the tufted duck (*Aythya fuligula*). *J. Anim. Ecol. 51*: 943-56.
_____ and Vessem, J.V. 1982. Flock size and feeding behaviour of migratory whinchats (*Saxicola rubetra*). *Ibis 124*: 347-57.
Drent, R. and Daan, S. 1980. The prudent parent: energetic adjustments in avian breeding. *Ardea 68*: 225-52.
_____ and van Eerden, M. 1980. Goose flocks and food exploitation: how to have your cake and eat it. *Proc. XVII Int. Ornith. Congr. Berl*: 800-6.
_____ and Swierstra, R. 1977. Goose flocks and food finding: field experiments with barnacle geese in winter. *Wildfowl 28*: 15-20.
Drycz, A., Witkowski, J. and Okulewicz, J. 1981. Nesting of 'timid' waders in the vicinity of 'bold' ones as an anti-predator adaptation. *Ibis 123*: 542-5.
Dugan, P.J. 1982. Seasonal changes in patch use by a territorial grey plover: weather-dependent adjustments in foraging behaviour. *J. Anim. Ecol. 51*: 849-57.

_____ Evans, P.R., Goodyer, L.R. and Davidson, N.C. 1981. Winter fat reserves in shorebirds: disturbance of regulated levels by severe weather conditions. *Ibis 123*: 359-63.

Dummigan, K.A. 1977. Food piracy by Iceland gull on oystercatchers. *Brit. Birds. 70*: 392.

Edwards, C.A. and Lofty, J.R. 1972. *Biology of Earthworms.* London: Chapman and Hall.

_____ and _____ 1982. The effect of direct drilling and minimal cultivation on earthworm population. *J. Appl. Ecol., 19*: 723-34.

Edwards, K.C. 1966. *Nottingham and its region.* Nottingham: British Association for the Advancement of Science.

Eiserer, L.A. 1980. Effects of grass length and mowing on foraging behaviour of the American robin (*Turdus migratorius*). *Auk 97*: 576-80.

Ekman, J.B. and Askenmo, C.E.H. 1984. Social rank and habitat use in willow tit groups. *Anim. Behav. 32*: 508-14.

_____ Cederholm, G. and Askenmo, C.E.H. 1981. Spacing and survival in winter groups of willow tit (*Parus montanus*) and crested tit (*P. cristatus*) — a removal study. *J. Anim. Ecol. 50*: 1-9.

Elder, W.H. and Elder, W.L. 1970. Social groupings and primate associations of the brushbuck (*Tragelaphus scriptus*). *Mammalia 34*: 356-62.

Elgar, M.A. and Catterall, C.P. 1981. Flocking and predator surveillance in house sparrows: test of an hypothesis. *Anim. Behav. 29*: 868-72.

Emlen, J.T. and Ambrose, H.W. 1970. Feeding interactions in snowy egrets and red breasted mergansers. *Auk 87*: 164-5.

Emlen, S.T. and Demong, N.J. 1975. Adaptive significance of synchronized breeding in a colonial bird: a new hypothesis. *Science 188*: 1029-31.

Ens, B.J. and Goss-Custard, J.D. 1984. Interference among oystercatchers (*Haematopus ostralegus*) feeding on mussels (*Mytilus edulis*) on the Exe Estuary. *J. Anim. Ecol. 53*: 217-31.

Erichsen, J.T., Krebs, J.R. and Houston, A.I. 1980. Optimal foraging and cryptic prey. *J. Anim. Ecol. 49*: 271-6.

Evans, A.C. and Guild, W.J. MCL. 1948. Studies on the relationships between earthworms and soil fertility. V. Field populations. *Ann. Appl. Biol. 35*: 485-93.

Evans, P.R. 1976. Foraging behaviour and optimal foraging strategies in shorebirds: some implications for their distributions and movements in the non-breeding season. *Ardea 64*: 117-39.

_____ 1979. Adaptations shown by foraging shorebirds to cyclical variations in the activity and availability of their invertebrate prey. In E. Naylor and R.G. Hartnoll (eds.), *Cyclical Phenomena in Marine Plants and Animals*: 357-66. Oxford: Pergamon Press.

_____ Herdson, D.M., Knights, P.J. and Pienkowski, M.W. 1979. Short-term effects of reclamation of part of Seal Sands, Teesmouth, on wintering waders and shelduck. 1. Shorebird diets, invertebrate distribution and the impact of predation on invertebrates. *Oecologia 41*: 183-206.

_____ and Pienkowski, M.W. 1982. Behaviour of shelducks (*Tadorna tadorna*) in a winter flock: does regulation occur? *J. Anim. Ecol. 51*: 241-62.

_____ and Smith, P.C. 1975. Studies of shorebirds at Lindisfarne, Northumberland. 2. Fat and pectoral muscle as indicators of body condition in the bar-tailed godwit. *Wildfowl 25*: 64-76.

Evans, R.M. 1982a. Foraging flock recruitment at a black-billed gull (*Larus bulleri*) colony: implications for the information center hypothesis. *Auk 99*: 24-30.

_____ 1982b. Flock size and formation in black-billed gulls (*Larus bulleri*). *Can. J. Zool. 60*: 1806-11.

Evans, W. 1908. The black-headed gull as a persecutor of the lapwing. *Ann. Scot. Nat. Hist.* (1908): 255.
Fallet, M. 1962. Über Bodenvögel und ihre terricolen Beutetiere. Technik der Nahrungssuche-Population Dynamik. *Zool. Anz. 168*: 187-212.
Feare, C.J. 1984. *The Starling.* Oxford: Oxford University Press.
_____ and Inglis, I.R. 1979. The effect of reduction in feeding space on the behaviour of captive starlings (*Sturnus vulgaris*). *Ornis Scand. 10*: 42-7.
Ficken, M.S. 1981. Food finding in black-capped chickadee flocks: altruistic communication. *Wilson Bull. 93*: 393-4.
Fiscus, C.H. and Niggol, K. 1965. Observations of cetaceans off California, Oregon and Washington. *Fish. Wildl. Serv. 498*: 1-27.
Fitzherbert, A. 1534. *Book of Husbandry* (cited in R.K. Murton 1971, *Man and Birds.* Collins: London).
Foster, W.A. and Treherne, J.E. 1981. Evidence for the dilution effect in the selfish herd from fish predation on a marine insect. *Nature 293*: 466-7.
Franck, D. 1955. Hooded crows attacking black-headed gulls. *Brit. Birds. 68*: 332-3.
Fuchs, E. 1977. Predation and anti-predator behaviour in a mixed colony of terns (*Sterna* sp.) and black-headed gulls (*Larus ridibundus*) with special reference to the sandwich tern (*Sterna sandvicensis*). *Ornis Scand. 8*: 17-32.
Fuller, R.J. 1982. *Bird Habitats.* Calton: Poyser.
_____ and Lloyd, D. 1981. The distribution and habitats of wintering golden plovers in Britain, 1977-78. *Bird Study 28*: 169-85.
_____ and Youngman, R.E. 1979. The utilization of farmland by golden plovers wintering in Southern England. *Bird Study 26*: 37-46.
Gerard, B.M. 1964. Lumbricidae (Annelida). Synopses of the British Fauna. No. 6 (Linnaean Society of London).
_____ and Hay, R.K.M. 1979. The effect on earthworms of ploughing, timed cultivation, direct drilling and nitrogen in a barley monoculture system. *J. Agric. Sci. Camb. 93*: 147-55.
Gibb, J.A. 1960. Populations of tits and goldcrests and their food supply in pine plantations. *Ibis 102*: 163-208.
Gilbert, H.A. and Brook, A. 1924. *Secrets of Bird Life.* London.
Gilbert, L.E. 1975. Ecological consequences of a co-evolved mutualism between butterflies and plants. In L.E. Gilbert and P.R. Raven (eds.), *Coevolution of Animals and Plants*: 210-240. Austin, Texas: Texas University Press.
Glutz von Blotzheim, U.N., Baer, K.M. and Bezzel, E. 1975. *Handbuch der Vögel Mitteleuropas*, VI. Akademische Verlagsgesellschaft.
Gochfeld, M. and Burger, J. 1984. Age differences in foraging behaviour of the American robin (*Turdus migratorius*). *Behaviour 88*: 227-39.
Goodyer, L.R. 1976. Lapwing weights and moult. *Wader Study Group Bull.*, *18*: 9-12.
Gosling, L.M. and Petrie, M. 1981. Economics of social organisation. In C.R. Townsend and P. Calow (eds.), *Physiological Ecology: an evolutionary approach to resource use*: 315-45. Oxford: Blackwell.
Goss-Custard, J.D. 1969. Winter feeding ecology of redshank (*Tringa totanus* L.). *Ibis 111*: 338-56.
_____ 1970a. Feeding dispersion in some overwintering wading birds. In J.H. Crook (ed.), *Social Behaviour in Birds and Mammals*: 3-35. London: Collins.
_____ 1970b. Factors affecting the diet and feeding rate of the redshank (*Tringa totanus* L.). In A. Watson (ed.), *Animal Populations in Relation to their Resources*: 101-10. Oxford: Blackwell.
_____ 1976. Variations in the dispersion of Redshank (*Tringa totanus* L.) on their winter feeding grounds. *Ibis 118*: 257-63.

_____ 1977a. Optimal foraging and the size selection of worms by redshank (*Tringa totanus* L.) in the field. *Anim. Behav. 25*: 10-29.
_____ 1977b. Predator responses and prey mortality in redshank (*Tringa totanus* L.) and preferred prey (*Corophium volutator*). *J. Anim. Ecol. 46*: 21-35.
_____ 1977c. The ecology of the Wash. III. Density related behaviour and the possible effects of a loss of feeding grounds on wading birds (Charadrii). *J. Appl. Ecol. 14*: 721-39.
_____ 1980. Competition for food and interference among waders. *Ardea 68*: 31-52.
_____ 1983. Spatial and seasonal variations in the food supply of waders (Charadrii) wintering in the British Isles. *Dansk Orn. Foren. Tidsskr. 3*: 85-96.
_____ Clarke, R.T. and Durell, S.E.A. le V dit. 1984. Rates of food intake and aggression of oystercatchers (*Haematopus ostralegus*) on the most and least preferred mussel (*Mytilus edulis*) beds. *J. Anim. Ecol. 53*: 233-45.
_____ and Durell, S.E.A. le V dit. 1983. Individual and age differences in the feeding ecology of oystercatchers (*Haematopus ostralegus L.*). *Anim. Behav. 30*: 917-28
_____, _____ and Ens, B.J. 1982a. Individual differences in aggressiveness and food stealing among wintering oystercatchers (*Haematopus ostralegus* L.). *Anim. Behav. 30*: 917-28.
_____, _____, McGrorty, S. and Reading, C.J. 1982b. Use of mussel (*Mytilus edulis* L.) beds by oystercatchers (*Haematopus ostralegus* L.) according to age and population size. *J. Anim. Ecol. 51*: 543-54.
_____ and Durell, S.E.A. le V dit. 1983. Individual and age differences in the feeding ecology of oystercatchers (*Haematopus ostralegus*) wintering on the Exe estuary, Devon. *Ibis 125*: 155-71.
Gotmark, F. and Andersson, M. 1980. Breeding associations between common gull (*Larus canus*) and arctic skua (*Stercorarius parasiticus*). *Ornis Scand. 11*: 121-4.
Greig, S.A. 1984. The feeding behaviour of *Larus argentatus* and other *Larus* gulls at refuse tips. Unpubl. Ph.D. thesis, University of Durham.
Greig-Smith, P.W. 1981. Responses to disturbance in relation to flock size in foraging groups of barred ground doves (*Geopelia striata*). *Ibis 123*: 103-6.
Guild, W.J. McL. 1951. The distribution and population density of earthworms (Lumbricidae) in Scottish pasture fields. *J. Anim. Ecol. 20*: 88-97.
Hale, W.G. 1980. *Waders*. London: Collins.
Halliday, J.B., Curtis, D.J., Thompson, D.B.A., Smyth, J.C. and Bignal, E.M. 1982. The status and distribution of shorebirds on the Clyde Estuary. *Scot. Birds. 12*: 65-72.
Hamilton, W.D. 1971. Geometry for the selfish herd. *J. theor. Biol. 31*: 295-311.
Hart, A. and Lendrem, D.N. 1984. Vigilance and scanning patterns in birds. *Anim. Behav. 32*: 1216-24.
Hart, J.S. and Berger, M. 1972. Energetics, water economy and temperature regulation during flight. *Proc. 15th Int. Ornith. Congr.* 189-99.
Harvey, P.H. and Greenwood, P.J. 1978. Anti-predator defence strategies: some evolutionary problems. In J.R. Krebs and N.B. Davies (eds.), *Behavioural Ecology: an evolutionary approach*: 129-58. Oxford: Blackwell.
Heatwole, H. 1965. Some aspects of the association of cattle egrets with cattle. *Anim. Behav. 13*: 79-83.
Heppleston, P.B. 1971a. The feeding ecology of the oystercatchers (*Haematopus ostralegus*) in winter in N.E. Scotland. *J. Anim. Ecol. 40*: 651-72.
_____ 1971b. Feeding techniques of the oystercatcher. *Bird Study 18*: 15-20.
Heppner, F. 1975. Sensory mechanisms and environmental cues used by the American robin in locating earthworms. *Condor 67*: 247-56.
Herrera, C.M. 1979. Ecological aspects of heterospecific flock formation in a Mediterranean passerine bird community. *Oikos 33*: 85-96.

Hilgarth, N. and Kear, J. 1981. Diseases of perching ducks in captivity. *Wildfowl 32*: 156-63.
HMSO. 1980. *Agricultural Statistics for the United Kingdom.* London: MAFF/ HMSO.
Hoffman, W., Heineman, D. and Wiens, J.A. 1981. The ecology of seabird feeding flocks in Alaska, USA. *Auk 98*: 437-56.
Hogstad, O. 1976. Sexual dimorphism and divergence in winter foraging behaviour of the three-toed woodpecker (*Picoides tridactylus*). *Ibis 118*: 41-50.
Holling, C.S. 1959. Some characteristics of simple types of predation. *Can. Entomol. 91*: 385-98.
Hoogland, J.L. and Sherman, P.W. 1976. Advantages and disadvantages of bank swallow (*Riparia riparia*) coloniality. *Ecol. Monogr. 46*: 33-58.
Horn, H.S. and Rubenstein, D.I. 1984. Behavioural adaptation and life history. In J.R. Krebs and N.B. Davies (eds.), *Behavioural Ecology: an evolutionary approach* (2nd edn.): 279-98. Oxford: Blackwell.
Houston, A.I., Krebs, J.R. and Erichsen, J.I. 1980. Optimal prey choice and discrimination time in the great tit (*Parus major* L.). *Behav. Ecol. Sociobiol. 6*: 169-75.
Hughes, R.N. 1979. Optimal diets and the energy maximization premise: the effects of recognition time and learning. *Am. Nat., 113*: 209-11.
Hulscher, J.B. 1974. An experimental study of the food intake of the oystercatcher (*Haematopus ostralegus*) in captivity during the Summer. *Ardea 62*: 155-71.
_____ 1982. The oystercatcher (*Haematopus ostralegus*) as a predator of the bivalve (*Macoma balthica*) in the Dutch Wadden Sea. *Ardea 70*: 89-152.
Hurlbert, S.H. 1978. The measurement of niche overlap and some relatives. *Ecology 59*: 67-77.
Inglis, I.R. and Lazarus, J. 1981. Vigilance and flock size in brent geese: the edge effect. *Z. Tierpsychol. 57*: 193-200.
Ireland, M. 1977. Heavy worms. *New Sci. 76*: 486-7.
Jennings, T. and Evans, S.M. 1980. Influence of position in the flock and flock size on vigilance in the starling (*Sturnus vulgaris*). *Anim. Behav. 28*: 634-5.
Johnstone, F.J. 1945. Herring-gulls robbing lapwings. *Brit. Birds 38*: 278.
Kacelnik, A., Houston, A.I. and Krebs, J.R. 1981. Optimal foraging and territorial defence in the great tit (*Parus major*). *Behav. Ecol. Sociobiol. 8*: 35-40.
Källander, H. 1977. Piracy by black-headed gulls on lapwings. *Bird Study 16*: 45-52.
_____ 1979. Skrattmåsen (*Larus ridibundus*) som kleptoparasit på tofsvipa (*Vanellus vanellus*). *Fauna och Flora 75*: 200-7.
Kamil, A.C. 1978. Systematic foraging for nectar by the amakihi (*Loxops virens*). *J. Comp. Psychol. 92*: 388-96.
Kendeigh, S.C. 1949. Effect of temperature and season on energy resources of the English sparrow. *Auk 66*: 111-27.
_____ 1970. Energy requirements for existence in relation to size of bird. *Condor 72*: 60-5.
_____ and Blem, C.R. 1974. Metabolic adaptation to local climate. *Comp. Biochem. Physiol. 48A*: 175-87.
_____ Dol'nik, V.R. and Gavrilov, V.M. 1977. Avian energetics. In J. Pinowski and S.C. Kendeigh (eds.), *Granivorous Birds in Ecosystems*: 127-204. Cambridge: Cambridge University Press.
Kenward, R.E. 1978. Hawks and doves: factors affecting success and selection in goshawk attacks on woodpigeons. *J. Anim. Ecol. 47*: 449-60.
Keymer, I.F. 1972. Diseases of birds of prey. *Vet. Rec.* 1972: 579-94.
_____ Rose, J.H., Beesley, W.N. and Davies, S.E.M. 1962. A survey and review of parasitic diseases of wild and game birds in Great Britain. *Vet. Rec. 74*: 887-94.

King, J.R. and Farner, D.S. 1974. Seasonal allocation of time and energy resources in birds. In R.A. Paynter (ed.), *Avian Energetics*: 4-85. Cambridge, MA: Nuttal Ornithological Club. No. 15.

Kirkwood, R.S.M. 1986. Models and experiments in foraging group dynamics. Ph.D. thesis, University of Nottingham.

Kistyakiviski, O.B. 1957. The diet of lapwings (*Vanellus vanellus*) in the USSR. *Fauna Ukraini* Vol. *4* (Kiev).

Klomp, H. 1954. De terreinkeus van de Kievit. *Ardea 42*: 1-139.

Krebs, J.R. 1973. Social learning and the significance of mixed-species flocks of chickadees (*Parus* spp). *Can. J. Zool. 51*: 1275-88.

——— 1974. Colonial nesting and social feeding as strategies for exploiting food resources in the great blue heron (*Ardea herodias*). *Behaviour 51*: 99-134.

——— 1978. Optimal foraging: decision rules for predators. In J.R. Krebs and N.B. Davies (eds.), *Behavioural Ecology: an evolutionary approach*: 23-63. Oxford: Blackwell.

——— 1979. Foraging strategies and their social significance. In P. Marler and J.G. Vandenbergh (eds.), *Handbook of Behavioural Neurobiology. Volume 3. Social Behaviour and Communication*: 225-70.

——— and Davies, N.B. 1981. *An Introduction to Behavioural Ecology*. Oxford: Blackwell.

——— Erichsen, J.T., Webber, M.I. and Charnov, E.L. 1977. Optimal prey selection in the great tit (*Parus major*). *Anim. Behav. 25*: 30-8.

——— MacRoberts, M.H. and Cullen, J.M. 1972. Flocking and feeding in the great tit (*Parus major*): an experimental study. *Ibis 114*: 507-30.

——— and McCleery, R.H. 1984. Optimization in behavioural ecology. In J.R. Krebs and N.B. Davies (eds.), *Behavioural Ecology: an evolutionary approach* (2nd edn): 91-121. Oxford: Blackwell.

——— Ryan, J.C. and Charnov, E.L. 1974. Hunting by expectation or optimal foraging? A study of patch use by chickadees. *Anim. Behav. 22*: 953-64.

——— Stephens, D.W. and Sutherland, W.J. 1983. Perspectives in optimal foraging. In G.A. Clark and A.H. Bruch (eds.), *Perspectives In Ornithology*: 165-221. Cambridge: Cambridge University Press.

Kroll, J.C. and Fleet, R.R. 1979. Impact of woodpecker predation on over-wintering within-tree populations of the southern pine beetle (*Dendroctomus frontalis*). In J.G. Dickson, R.N. Connor, R.R. Fleet, J.A. Jackson and J.C. Kroll (eds.), *The Role of Insectivorous Birds in Forest Ecosystems*. London: Academic Press.

Kushlan, J.A. 1978. Non rigorous foraging by robbing egrets. *Ecology 59*: 649-53.

Lack, D. 1954. *The Natural Regulation of Animal Numbers*. Oxford: Clarendon.

Laidlaw, T.G. 1908. Food of the black-headed gull. *Ann. Scot. Nat. Hist.* 1908: 139-41.

Lange, G. 1968. Über Nahrung, Nahrungsaufnahme und Verdauungstrakt mitteleuropäischer Limikolen. *Beitr. Vogelkunde 13*: 225-334.

Lasiewski, R.C. and Dawson, W.R. 1967. A re-examination of the relation between standard metabolic rate and body weight in birds. *Condor 69*: 12-23.

Lazarus, J. 1972. Natural selection and the function of flocking in birds: a reply to Murton. *Ibis 114*: 556-8.

——— 1978. Vigilance, flock size and domain of danger in the white-fronted goose. *Wildfowl 29*: 135-45.

——— 1979a. The early warning function of flocking in birds: an experimental study with captive *Quelea*. *Anim. Behav. 27*: 855-65.

——— 1979b. Flock size and behaviour in captive red-billed weaverbirds (*Quelea quelea*): implications for social facilitation and the functions of flocking. *Behaviour 71*: 127-45.

_____ and Inglis, I.R. 1978. The breeding behaviour of the pink-footed goose. Parental care and vigilant behaviour during the fledging period. *Behaviour 65*: 62-88.
Lee, D.L. 1958. A note on a species of *Syngamus* found in the intestine of (*Turdus pilaris*). *Parasitology 48*: 121-3.
Lendrem, D.W. 1982. Vigilance in birds. D.Phil. thesis, University of Oxford.
_____ 1984. Flocking, feeding and predation risk: absolute and instantaneous feeding rates. *Anim. Behav. 32*: 298-9.
Lind, A.H. 1957. Territorial opforsel hos vibe (*Vanellus vanellus* L.) om efteretaret. *Dansk Orn. Foren. Tidsskr. 51*: 22-9.
Lockie, J.D. 1956. The food and feeding behaviour of the jackdaw, rook and carrion crow. *J. Anim. Ecol. 25*: 421-8.
Macdonald, D.W. and Henderson, D.G. 1977. Aspects of the behaviour and ecology of mixed-species bird flocks in Kashmir. *Ibis 119*: 481-91.
MacFadyen, A. 1962. Soil Arthropod Sampling. In J.B. Cragg (ed.), *Advances in Ecological Research*. No. 1. London.
Madon, P. 1935. Contribution à l'étude du régime des oiseaux aquatiques. *Alauda 7*: 178-97, 382-401.
Markgren, M. 1960. Fugitive reactions in avian behaviour. *Acta Vertebratica 2*: 1-161.
Maxson, S.J. and Oring, L.W. 1980. Breeding season time and energy budgets of the polyandrous spotted sandpiper. *Behaviour 74*: 200-63.
Maynard Smith, H. 1974. The theory of games and the evolution of animal conflicts. *J. theor. Biol. 47*: 209-21.
_____ 1982. *Evolution and the Theory of Games*. Cambridge: Cambridge University Press.
McLellan, C.R. 1958. Role of woodpeckers in control of the Codling Moth in Nova Scotia. *Can. Ent. 90*: 18-22.
McLennan, J.A. 1979. The formation and function of mixed species wader flocks in fields. Ph.D. thesis, University of Aberdeen.
McIlhenny, E.A. 1939. Feeding habits of black vulture. *Auk 56*: 472-4.
McVean, D.N. and Lockie, J.D. 1969. *Ecology and Land Use in Upland Scotland*. Edinburgh: Oliver and Boyd.
Metcalfe, N.B. 1984a. The effects of mixed species flocking on the vigilance of shorebirds: who do they trust? *Anim. Behav. 32*: 930-7.
_____ 1984b. Prey detection by intertidally feeding lapwing. *Z. Tierpsychol. 67*: 45-57.
_____ 1984c. The effects of habitat on the vigilance of shorebirds: is visibility important? *Anim. Behav. 32*: 925-9.
Milinski, M. 1977. Experiments on the selection by predators against spatial oddity of their prey. *Z. Tierpsychol. 43*: 311-25.
_____ and Heller, R. 1978. Influence of a predator on optimal foraging behaviour of sticklebacks (*Gasterosteus aculeatus*). *Nature 275*: 642-4.
Miller, R. 1922. The significance of the gregarious habit. *Ecology 3*: 122-6.
Milsom, T.D. 1984. Diurnal behaviour of lapwings in relation to moonphase during winter. *Bird Study 31*: 117-20.
Morse, D.H. 1967. Foraging relationships of brown-headed nuthatches and pine warblers. *Ecology 48*: 94-103.
_____ 1968. A quantitative study of foraging of male and female spruce-woods warblers. *Ecology 49*: 779-84.
_____ 1970. Ecological aspects of some mixed species foraging flocks of birds. *Ecol. Monogr. 40*: 119-68.
_____ 1978. Structure and foraging patterns of flocks of tits and associated species

in an English woodland during the Winter. *Ibis 120*: 295-312.
Moynihan, M. 1955. Some aspects of reproductive behaviour in the black-headed gull (*Larus ridibundus* L.) and related species. *Behaviour 5*: 58-80.
_____ 1968. Social mimicry: character convergence versus character displacement. *Evolution 22*: 315-31.
Mudge, G.P. and Ferns, P.N. 1982. The feeding ecology of five species of gulls (Aves: Larinii) in the inner Bristol Channel. *J. Zool. Lond. 197*: 497-510.
Murdie, G. and Hassell, M.P. 1973. Food distribution, searching success and predator-prey models. In M.S. Bartlett and R.W. Hirons (eds.), *The Mathematical Theory of the Dynamics of Biological Populations.* London: Academic Press.
Murphy, R.C. 1936. *Oceanic Birds of South America.* New York: Amer. Mus. Nat. Hist.
Murton, R.K. 1971a. Why do some birds feed in flocks? *Ibis 113*: 534-6.
_____ 1971b. The significance of a specific search image in the feeding behaviour of the woodpigeon. *Behaviour 39*: 10-42.
_____ 1971c. *Man and Birds.* London: Collins.
_____ and Isaacson, A.J. 1962. The functional basis of some behaviour in the wood pigeon (*Columba palumbus*). *Ibis 104*: 503-21.
Neill, S.R. and Cullen, J.M. 1974. Experiments on whether schooling by their prey affects the hunting behaviour of cephalopod and fish predators. *J. Zool. Lond. 172*: 549-69.
Neub, M. 1970. *Orn. Mitt. 22*: 31-5 (cited in Cramp and Simmons 1983).
Neuchterlein, G.L. 1981. Information parasitism in mixed colonies of western grebes and Forster's terns. *Anim. Behav. 29*: 985-9.
Newton, I. 1967. The adaptive radiation and feeding ecology of some British finches. *Ibis 109*: 33-99.
_____ 1972. *Finches.* London: Collins.
_____ 1979. *Population Ecology of Raptors.* Berkhamsted: Poyser.
Nie, N.H., Hull, C.H., Jenkins, J.G., Steinbrenner, K. and Bent, D.H. 1975. *Statistical Packages for the Social Sciences.* 2nd Edition. New York: McGraw Hill.
Nilsson, S.G. 1979. Seed density, cover, predation and the distribution of birds in a beech wood in southern Sweden. *Ibis 121*: 177-85.
Nordstrom, S. and Rundgren, S. 1974. Association of Lumbricids in Southern Sweden. *Pedobiologia 13*: 301-26.
Norton-Griffiths, M. 1967. Some ecological aspects of the feeding behaviour of the oystercatcher (*Haematopus ostralegus*) on the edible mussel (*Mytilus edulis*). *Ibis 109*: 412-24.
Oakes, C. 1948. 'Plover's page' behaviour of dunlin. *Brit. Birds 41*: 226-8.
Ogilvie, M.A. 1975. *Ducks of Britain and Europe.* Berkhamsted: Poyser.
Orians, G.H. and Pearson, N.E. 1979. On the theory of central place foraging. In D.F. Horn (ed.), *Analysis of Ecological Systems*: 157-77. Columbus: Ohio State University.
Owen-Smith, N. and Novellie, E.P. 1982. What should a clever ungulate eat? *Amer. Nat. 119*: 151-78.
Owens, N.W. 1977. Responses of overwintering brent geese to human disturbance. *Wildfowl 28*: 5-14.
Page, G. and Whitacre, D.F. 1975. Raptor predation on wintering shorebirds. *Condor 77*: 73-83.
Parker, G.A. 1982. Phenotype-limited evolutionarily stable strategies. In King's College Sociobiology Group (eds.), *Current Problems in Sociobiology*: 173-201. Cambridge: Cambridge University Press.
Parks, J.M. and Bressler, S.L. 1963. Observation of joint feeding activities of certain

fish-eating birds. *Auk 80*: 198-9.
Parr, R. 1980. Population study of golden plovers *(Pluvialis apricaria)* using marked birds. *Ornis Scand. 11*: 179-89.
Parslow, J. 1974. *Breeding Birds of Britain and Ireland.* Berkhamsted: Poyser.
Patterson, I.J. 1975. Aggressive interactions in flocks of rooks (*Corvus frugilegus* L.). In G. Baerends, C. Beer and A. Manning (eds.), *Function and Evolution in Behaviour*: 169-83. Oxford: Clarendon Press.
_____ 1977. Aggression and dominance in winter flocks of shelduck (*Tadorna tadorna* L.). *Anim. Behav. 25*: 447-59.
_____ 1982. *The Shelduck: a study in behavioural ecology.* Cambridge: Cambridge University Press.
_____ Dunnet, G.M., and Fordham, R.A. 1971. Ecological studies of the rook (*Corvus frugilegus*) in NE Scotland. Dispersion. *J. Appl. Ecol. 8*: 815-33.
Payne, R.B. and Howe, H.F. 1976. Cleptoparasitism by gulls of migrating shorebirds. *Wilson Bull. 88*: 349-50.
Perrins, C.M. 1979. *British Tits.* London: Collins.
Perry, R. 1945. *A Naturalist in Lindisfarne.* London: Country Life.
Piearce, T.G. 1972. The calcium relations of selected Lumbricidae. *J. Anim. Ecol. 41*: 167-88.
Pienkowski, M.W. 1981. How foraging plovers cope with environmental effects on invertebrate behaviour and availability. In N.V. Jones and W.J. Wolff (eds.), *Feeding and Survival Strategies of Estuarine Organisms*: 179-92. London: Plenum Press.
_____ 1982. Diet and energy intake of grey and ringed plovers (*Pluvialis squatarola*) and (*Charadrius hiaticula*) in the non-breeding season. *J. Zool. Lond. 197*: 511-50.
_____ 1983a. Changes in the foraging pattern of plovers in relation to environmental factors. *Anim. Behav. 31*: 244-64.
_____ 1983b. Surface activity of some intertidal invertebrates in relation to temperature and the foraging behaviour of their shorebird predators. *Mar. Ecol. Prog. Ser. 11*: 141-50.
_____ Lloyd, C.S. and Minton, C.D.T. 1979. Seasonal and migrational weight changes in dunlins. *Bird Study 20*: 134-8.
Powell, G.V.N. 1974. Experimental analysis of the social value of flocking by starlings (*Sturnus vulgaris*) in relation to predation and foraging. *Anim. Behav. 22*: 501-5.
_____ 1980. Mixed species flocking as a strategy for neotropical birds. *Actis. XVII. Congr. Int. Ornith.*: 813-19. Berlin: Deutsche Ornithologen-Gesellschaft.
Prater, A.J. 1982. *Estuary Birds.* Berkhamsted: Poyser.
Prins, H.H.Th, Ydenberg, R.C. and Drent, R.H. 1980. The interaction of Brent geese (*Branta bernicla*) and sea plantain (*Plantago maritima*) during spring: field observations and experiments. *Acta Bot. Neerl. 29*: 585-96.
Pulliam, H.R. 1973. On the advantages of flocking. *J. theor. Biol. 38*: 419-22.
_____ and Caraco, T. 1984. Living in groups: is there an optimal group size? In J.R. Krebs and N.B. Davies (eds.), *Behavioural Ecology: an evolutionary approach* (2nd edn): 122-47. Oxford: Blackwell.
_____ and Millikan G.C. 1982. Social organisation in the non reproductive season. In D.S. Farner, J.R. King and K.C. Parkes (eds.), *Avian Biology.* Vol. VI: 169-97. London: Academic Press.
_____ and Mills, G.S. 1977. The use of space by wintering sparrows. *Ecology 58*: 1393-9.
_____ Pyke, G.H. and Caraco, T. 1982. The scanning behavior of juncos: a game theoretical approach. *J. theor. Biol. 95*: 89-104.

Puttick, G.M. 1980. Energy budgets of curlew sandpipers at Longebaan Lagoon, S. Africa. *Est. Coast. Mar. Sci. 11*: 207-15.

_____ 1981. Sex related differences in foraging behaviour of curlew sandpipers. *Ornis Scand. 12*: 13-17.

Ralph, C.L. 1957. Persistent rhythms of activity and energy consumption in the earthworm. *Physiol. Zool. 30*: 41-55.

Rand, A.L. 1954. Social feeding behaviour of birds. *Fieldiana Zool. 36*: 5-71.

Rands, M.R.W. and Barkham, J.P. 1981. Factors controlling within flock feeding densities in three species of wading birds. *Ornis Scand. 12*: 28-36.

Rasa, O.E.A. 1981. Raptor recognition: an interspecific tradition? *Naturwissenschaften 68*: 151.

_____ 1983. Dwarf mongoose and hornbill mutualism in the Taru Desert, Kenya. *Behav. Ecol. Sociobiol. 12*: 181-90.

Ratcliffe, D.A. 1976. Observations on the breeding of the golden plover in Great Britain. *Bird Study 23*: 63-116.

_____ (ed.) 1977. *A Nature Conservation Review.* Cambridge: Cambridge University Press.

_____ 1980. *The Peregrine Falcon.* Calton: Poyser.

Raw, F. 1959. Estimating earthworm populations by using formalin. *Nature 184*: 1661-2.

Reading, C.J. and McGrorty, S. 1978. Seasonal variations in the burying depth of *Macoma balthica* (L.) and its accessibility to wading birds. *Est. Coast. Mar. Sci. 6*: 135-44.

Recher, H.F. and Recher, J.A. 1969. Some aspects of the ecology of migrational shorebirds. II. Aggression. *Wilson Bull. 84*: 140-54.

Rechten, C., Avery, M. and Stevens, A. 1983. Optimal prey selection: why do great tits show partial preferences? *Anim. Behav. 31*: 576-84.

Richford, A.S. 1978. The effect of jackdaws on the breeding of auks on Skomer Island. *Nature in Wales 16*: 135-44.

Rohwer, S. and Ewald, P.W. 1981. The cost of dominance and advantage of subordination in a badge signalling system. *Evolution 35*: 441-54.

Rubenstein, D.I., Barnett, R.J., Ridgely, R.S. and Klopfer, P.H. 1977. Adaptive advantages of mixed-species feeding flocks among seed-eating finches in Costa Rica. *Ibis 119*: 10-29.

Ryabov, V.F. and Mosalova, N.I. 1966. The food of lapwings in the USSR. *Zool. Zh. 45*: 910-18.

Sage, B.L. 1963. Gulls parasitizing ducks and other birds. *Ann. Rep. Wildfowl Trust. 14*: 173-4.

Satchell J.E. (ed.). 1983. *Earthworm Ecology: from Darwin to vermiculture.* London: Chapman and Hall.

Schaller, G.B. 1972. *The Serengeti Lion: a study of predator-prey relations.* Chicago: University of Chicago Press.

Schluter, D. 1982. Seed and patch selection by Galapagos ground finches: relation to foraging efficiency and food supply. *Ecology 63*: 1106-20.

Selous, E. 1927. *Realities of Bird Life.* London: Constable.

Sharrock, J.T.R. (comp.) 1979. *The Atlas of Breeding Birds in Britain and Ireland.* BTO, IWC; T. and A.D. Poyser, Berkhamsted.

Sherry, D. 1984. Food storage by black-capped chickadees: memory for the location and contents of caches. *Anim. Behav. 32*: 451-64.

Sibly, R.M. 1983. Optimal group size is unstable. *Anim. Behav. 31*: 947-8.

Siegfried, W.R. and Underhill, L.G. 1975. Flocking as an anti-predator strategy in doves. *Anim. Behav. 23*: 504-8.

Sigg, H. and Stolba, A. 1981. Home range and daily march in a Hamadryas baboon troop. *Folia Primatol.* *36*: 40-75.

Simmons, K.E.L. 1961. Further observations on foot-movements in plovers and other birds. *Brit. Birds.* *54*: 418-22.

Simms, E. 1974. *British Thrushes.* London: Collins.

Smith, J.N.M. 1974a. The food searching behaviour of two European thrushes. I. Description and analysis of search paths. *Behaviour* *48*: 276-302.

_____ 1974b. The food searching behaviour of two European thrushes. II. The adaptiveness of search patterns. *Behaviour* *49*: 1-61.

_____ 1977. Feeding rates, search paths and surveillance for predators in great tailed grackle flocks. *Can. J. Zool.* *55*(6): 891-8.

_____ and Sweatman, H.P.A. 1974. Food searching behaviour of titmice in patchy environments. *Ecology* *55*: 1216-32.

Smith, P.C. 1975. A study of the winter feeding ecology and behaviour of the bar-tailed godwit (*Limosa lapponica*). Ph.D. thesis, University of Durham.

_____ and Evans, P.R. 1973. Studies of shorebirds at Lindisfarne, Northumberland. I. Feeding ecology and behaviour of the bar-tailed godwit. *Wildfowl* *24*: 135-9.

Smith, S.M. 1967. An ecological study of winter flocks of black-capped and chestnut-backed chickadees. *Wilson Bull.* *79*: 200-7.

Solomon, M.E. 1949. The natural control of animal populations. *J. Anim. Ecol.* *18*: 1-35.

Southwood, T.R.E. 1978. *Ecological Methods.* London: Chapman and Hall.

Spencer, K.G. 1953. *The Lapwing In Britain.* London: Collins.

Stinson, C.H. 1980. Flocking and predator avoidance: models of flocking and observations on the spatial dispersion of foraging shorebirds. *Oikos* *34*: 35-43.

Sullivan, K.A. 1984. The advantages of social foraging in downy woodpeckers. *Anim. Behav.* *32*: 16-32.

Sutherland, W.J. 1982. Do oystercatchers select the most profitable cockles? *Anim. Behav.* *30*: 857-61.

_____ and Koene, P. 1982. Field experiments of the strength of interference between oystercatchers (*Haematopus ostralegus*). *Oecologia* *55*: 108-9.

Swennen, C., Heessen, H.J.L. and Hocker, A.W.M. 1979. Occurrence and biology of trematodes (*Cotylurus erraticus*, *C. variegatus* and *C. platycephalus* (Digenea: Strigeida)) in the Netherlands. *Neth. J. Sea Res.* *13*: 161-91.

Taklor, L.R. 1961. Aggregation, variance and the mean. *Nature* *189*: 732-5.

Thompson, D.B.A. 1981. Feeding behaviour of wintering shelduck (*Tadorna tadorna* L.) on the Clyde Estuary. *Wildfowl* *32*: 88-98.

_____ 1982. The abundance and distribution of intertidal invertebrates, and an estimation of their selection by shelduck. *Wildfowl* *33*: 151-8.

_____ 1983a. Winter gold. *Birds* *9*(8): 32-3.

_____ 1983b. Prey assessment by plovers (Charadriidae): net rate of energy intake and vulnerability to kleptoparasites. *Anim. Behav.* *31*: 1226-36.

_____ 1984. Foraging economics in flocks of lapwings (*Vanellus vanellus*), golden plovers (*Pluvialis apricaria*) and black-headed gulls (*Larus ridibundus*). Ph.D. thesis, University of Nottingham.

_____ and Thompson, M.L.P. 1985. Early warning in mixed species associations: the 'plover's-page' revisited. *Fbis* (in press).

_____ and Barnard, C.J. 1983. Anti-predator responses in mixed species associations of lapwings, golden plovers and gulls. *Anim. Behav.* *31*: 585-93.

_____ and _____ 1984. Prey selection by plovers: optimal foraging in mixed species groups. *Anim. Behav.* *32*: 554-63.

_____ and Lendrem, D.W. 1985. Gulls and plovers: host vigilance, kleptoparasite success and a model of kleptoparasite detection. *Anim. Behav.* *33* (in press).

Thompson, W.A., Vertinsky, I. and Krebs, J.R. 1974. The survival value of flocking

in birds: a simulation model. *J. Anim. Ecol. 43*: 785-820.
Thorpe, W.H. 1963. *Learning and Instinct in Animals.* London: Methuen.
Tinbergen, L. 1960. The natural control of insects in pinewoods. I. Factors influencing the intensity of predation by songbirds. *Archs. Neerl. Zool. 13*: 265-336.
Tinbergen, N. 1951. *The Study of Instinct.* Oxford: Oxford University Press.
_____ 1953a. Carrion crow striking lapwing in the air. *Brit. Birds. 46*: 377.
_____ 1953b. *The Herring Gull's World.* London: Collins.
_____ and Norton-Griffiths, M. 1964. Oystercatchers and mussels. *Brit. Birds. 57*: 64-70.
Torre-Bueno, J.R. and La Rochelle, J. 1978. The metabolic cost of flight in unrestricted birds. *J. Exp. Biol. 75*: 223-9.
Townshend, D.J. 1981. The importance of field feeding to the survival of wintering male and female curlews (*Numenius arquata*) on the Tees Estuary. In N.W. Jones and W.J. Wolff (eds.), *Feeding and Survival Strategies of Estuarine Organisms*: 261-74. London: Plenum Press.
_____ Dugan, P.J. and Pienkowski, M.W. 1984. The unsociable plover: use of intertidal areas by grey plovers. In P.R. Evans, J.D. Goss-Custard and W.G. Hale (eds.), *Coastal Waders and Wildfowl in Winter.* Cambridge: Cambridge University Press.
Treherne, J.E. and Foster, W.A. 1981. Group transmission of predator avoidance in a marine insect: the Trafalgar effect. *Anim. Behav. 29*: 911-17.
Treisman, M. 1975. Predation and the evolution of gregariousness. I. Models for concealment and evasion. *Anim. Behav. 23*: 779-800.
Trivers, R.L. 1971. The evolution of reciprocal altruism. *Q. Rev. Biol. 46*: 35-7.
Tucker, V.A. 1968. Respiratory exchange and evaporative water loss in the flying budgerigar. *J. Exp. Biol. 48*: 67-87.
_____ 1969. The energetics of bird flight. *Sci. Am. 220* (May): 70-8.
_____ 1972. Metabolism during flight in the laughing gull (*Larus atricilla*). *Am. J. Physiol. 222*: 237-45.
Tukey, J.W. 1977. *Exploratory Data Analysis.* Reading: Addison-Wesley.
Turner, A.K. 1982. Optimal foraging by the swallow (*Hirundo rustica*): prey selection. *Anim. Behav. 30*: 862-72.
van Eerden, M. and Keij, P. 1979. Counting golden plovers (*Pluvialis apricaria*) on passage: some results of two country-wide surveys in the Netherlands. *Wader. Study. Group Bull. 27*: 25-7.
Vaughan, R. 1980. *Plovers.* Lavenham, Surrey: Terence Dalton.
Vepsäläinen, K. 1968. The effect of the cold spring of 1966 upon the lapwing (*Vanellus vanellus*) in Finland. *Ornis Scand. 45*: 33-47.
Vernon, J.D.R. 1970. Feeding habits and food of the black-headed and common gulls. Part I. Feeding habits. *Bird Study 17*: 287-96.
_____ 1972. Feeding habits and food of the black-headed and common gulls. Part II. Food. *Bird Study 19*: 173-86.
Vickery, W.L. 1984. Optimal diet models and rodent food consumption. *Anim. Behav. 32*: 340-8.
Vine, I. 1971. Risk of visual detection and pursuit by a predator and the selective advantage of flocking behaviour. *J. theor. Biol. 30*: 405-22.
_____ 1973. Detection of prey flocks by predators. *J. theor. Biol. 40*: 207-10.
Vines, G. 1980. Spatial consequences of aggressive behaviour in flocks of oystercatchers (*Haematopus ostralegus* L.). *Anim. Behav. 28*: 1175-83.
Vollrath, F. 1984. Kleptobiotic interactions in invertebrates. In C.J. Barnard (ed.), *Producers and Scroungers: strategies of exploitation and parasitism*: 61-94. London: Croom Helm.

Waite, R.K. 1981. Local enhancement for food finding by rooks (*Corvus frugilegus*) foraging on grassland. *Z. Tierpsychol. 57*: 14-36.

_____ 1983. Some issues in corvid behaviour and ecology. Unpublished Ph.D. thesis, University of Keele.

_____ 1984a. Sympatric corvids: effects of social behaviour, aggression and avoidance on feeding. *Behav. Ecol. Sociobiol. 15*: 55-9.

_____ 1984b. Winter habitat selection and foraging behaviour in sympatric corvids. *Ornis Scand. 15*: 55-62.

Wallace, D.I.M. 1983. Lapwings robbing golden plovers. *Brit. Birds. 76*: 452-3.

Walsberg, G.E. 1983. Avian ecological energetics. In D.S. Farner, J.R. King and K.C. Parkes (eds.), *Avian Biology*. Vol. VII: 161-220. London: Academic Press.

Ward, P. and Zahavi, A. 1973. The importance of certain assemblages of birds as 'information centres' for food finding. *Ibis 115*: 517-34.

Washburn, S.L. and DeVore, I. 1961. The social life of baboons. *Sci. Am. 204*: 62-71.

Williamson, P. and Gray, L. 1975. Foraging behaviour of the starling (*Sturnus vulgaris*) in Maryland. *Condor 77*: 84-9.

Willis, E.O. 1967. The behaviour of bicolored antbirds. *Univ. Calif. Publ. Zool. 79*: 1-127.

Witherby, H.F., Jourdain, F.C.R., Ticehurst, N.F. and Tucker, B.W. 1940. *The Handbook of British Birds*. Vol 4. London: Witherby.

Wittenberger, J.F. 1981. *Animal Social Behavior*. Boston: Duxbury Press.

Wynne Owen, R. and Pemberton, L. 1962. Helminth infection of the starling (*Sturnus vulgaris*) in northern England. *Proc. Zool. Soc. Lond. 139*: 557-87.

Yalden, D.W. 1980. Notes on the diet of urban kestrels. *Bird Study 27*: 235-8.

Ydenberg, R.C. and Prins, H.H. Th. 1981. Spring grazing and the manipulation of food quality by barnacle geese. *J. Appl. Ecol. 18*: 443-53.

_____ and _____ 1984. Why do birds roost communally in winter. In P.R. Evans, J.D. Goss-Custard and W.G. Hale (eds.), *Coastal Waders and Wildfowl in Winter*: 121-39. Cambridge: Cambridge University Press.

_____, _____ and van Dijk, J. 1983. Post roost gatherings and information centres. *Ardea 71*: 125-32.

Zach, R. and Falls, J.B. 1977. Influence of capturing prey on subsequent search in the ovenbird (Aves: Parulidae). *Can. J. Zool. 55*: 1958-69.

_____ and _____ 1979. Foraging and territoriality of male ovenbirds (Aves: Parulidae) in a heterogeneous habitat. *J. Anim. Ecol. 48*: 33-52.

Zwarts, L. 1976. Density related processes in feeding dispersion and feeding activity of teal (*Anas crecca*). *Ardea 64*: 192-209.

_____ 1978. Intra- and inter-specific competition for space in estuarine bird species in a one prey situation. *Proc. XVII. Int. Orn. Congr. Berlin*: 1045-50.

_____ and Drent, R.H. 1981. Prey depletion and regulation of predator density: oystercatchers (*Haematopus ostralegus*) feeding on mussels (*Mytilus edulis*). In N.V. Jones and W.J. Wolff (eds.), *Feeding and Survival Strategies of Estuarine Organisms*. London: Plenum Press.

Index

aerial chase 220-2, 224, 228; *see also* attack; kleptoparasitism
age 47; and competitive ability 191; effects on foraging behaviour 185-6; and foraging skill 186; and vulnerability 185
aggregative response 16, 29, 37, 143, 167
aggression 18, 22, 30, 31-3, 34, 37, 42, 44, 58, 165-74, 198, 212, 278; back-up display 186; effects of environmental conditions 165-74; intrasexual 187; parallel walk 187-8; redirected pecking 187
alarm call 27; warning cry 268
alarm response: to approaching predator 258-9; duration 265-8; and field size 269-72; flight intention 259; in gulls and plovers 255-74; mixed 259; orientation 259; and vulnerability 269-73
alarm stimulus 260-1, 268, 274, 278; responsiveness to 255, 258, 274, 277
Allolobophora: caliginosa 73, 79, 125; *chlorotica* 73, 79, 125; *longa* 79; *rosea* 73, 79
alternative strategy 257
antcatchers (*Gymnopithys* and *Alethe* spp.) 25
anti-predator strategies: avoiding detection 2; confusion effect 3; deterrence 4-7; early detection 3, 4-7; reduced individual risk 4-7
area-restricted searching 172
Arenicola 144
arms race 20
assimilation 117, 177-9, 251-2
attack: choosing distance 232-4; detection of 219-24, 228, 234, 236, 238, 243-4; effective distance of 231; efficiency in single versus mixed species flocks 250; energetic cost of 218-19, 224, 241, 279; flight 218-54; optimal breadth of attack time 232-4; optimal distance of 232; range 254; success 224, 232, 243, 248, 250, 253, 277, 279; undetected 219-21, 224, 243; *see also* kleptoparasitism
auditory detection (of prey) 157
auk (Alcidae) 13, 25; guillemot (*Uria aalge*) 25

baboon 27; hamadryas (*Papio hamadryas*) 206
badger (*Meles meles*) 84
basal metabolic rate 175-9, 251-2
bat 27
beating effect 22, 25, 26
bee-eater: carmine (*Merops nubicus*) 25
behavioural ecology 1, 117
bird feeding hours: and pasture age 103-6
bivalve (*Macoma balthica*) 134
blackbird (*Turdus merula*) 26, 134
breeding: lapwing 47, 57-8; nesting 57; in plovers 53-7; plumage 184
bushbuck (*Tragelaphus scriptus*) 27
bustard (*Ardeotis* spp.) 25

calcium 125
caracara: *Milvago chimango* 268; *Polyborus plancus* 268
carrying capacity 194
cephalopod 3, 14
cetacean 27
cheat 256-7
cheetah (*Acinonyx jubatus*) 27
chickadee: black-capped (*Parus atricapillus*) 11, 12, 24-5; chestnut-backed (*P. rufescens*) 11, 24
cloud cover: and prey activity 144
cockle (*Cerastoderma edule*) 120; handling time 120
competition 2, 34, 39, 44, 91, 107, 118, 194, 199, 278; between gulls 250-1; and intrasexual aggression 187; intraspecific 166; limited responses and 255
competitive ability 22, 185, 187; dominance 18, 22, 30, 32, 34, 198
competitive exclusion 24
conditional strategy 257
contact calls 2

co-operative foraging 22
co-operative scanning 256-7; versus selfish 256-7
coot (*Fulica atra*) 6
'copier' strategy 18
copying behaviour: area-copying 17, 18, 20, 133, 205, 278, *see also* flocking; and aggression 172-4; intraspecific 26; perch copying 17; and prey profitability 172, 175, 190
corporate vigilance; flock composition and 262-3; group size and 257-8, 262-8, 273
courtship 65
creepers (*Certhia* spp.) 11
crow: carrion (*Corvus corone*) 13, 25, 84, 112, 261-2
curlew (*Numenius arquata*) 6, 10, 58, 88

daily energy balance: and feeding efficiency 174-84
Daphnia 3, 14
daylength: and choice of feeding site 67-92; effect on flight over flock 201-2; and energy demand 94; and flock size variance 196-8; and gull's food requirement 207; effect on foraging behaviour 143; effect on prey size taken 146; effect on time budgeting 145-65
dead organic matter 73
decision rules 116
diet: black-headed gull 52; golden plover 49, 52, 57; lapwing 47-8; selection 94, 116-42; *see also* optimal diet selection
dietary opportunism 26, 62, 204
dietary shift 231, 250
digger wasp, *Sphex ichneumoneus* 26
display flight: lapwing 47
diver (loon) 25
domain of danger 14
double benefit of grouping 15
dove: barbary (*Streptopelia risoria*) 5, 16; ground (*Geopelia striata*) 6, 9, 259
ducks: Cairinini 9
dung 100, 115; indicating local worm availability 100, 115
dunlin (*Calidris alpina*) 12, 56, 259-60

early warning system 255-74, 277; and flight distance 266-8; gulls as 264-74
edge effect 15, 202-3, 278
egg predation: by black-headed gull 62; in golden plover 57
egrets: cattle (*Bubulcus ibis*) 22, 26; snowy (*Egretta thula*) 25
enclosure experiment 81-4, 139
encounter rate 120, 135-42, 246-7
energy conservation 176-7
energy deficit 179-84, 190; and time spent feeding 179-82
energy intake 117-42, 217-54, 272, 277; daily 251-2
energy requirement: daily 58, 85, 86, 174-9, 198, 272, 277, 279; and environmental conditions 252; existence 174-9, 217, 250-4; in kleptoparasites 250-2, 254; *see also* temperature
energy resource 183-4
equilibrium flock size *see* flock dynamics
evolutionarily stable scanning strategy 256-7
exploitation 30, 204; of flock members 259-61
exposure 105

Falconiformes 10
fat reserve 183
feeding efficiency 2, 17, 18, 22, 29, 30, 78, 93-4, 143-91, 205, 217, 224, 227 275-9; and flock dynamics 210-16; and food availability 161; crouching and 159-61; effects of environmental conditions 145-65; effects of flock composition 145-65; effects of golden plovers on lapwings 146-65; effects of gulls on plovers 147-65, 273-4; of golden plovers 150-91; of gulls 217-54; of lapwings 146-91
feeding priority 28, 32, 149, 165, 198, 277, 278; and responsiveness to alarm 270-2
ferret (*Mustela furo*) 16
fieldfare (Turdus pilaris) 3, 5, 12, 28-30, 84, 134, 269
fitness 33
flight: energetic cost 218-19; wingbeat frequency 218
flight distance 258, 260, 274, 278; post-alarm 270

flock (group) dynamics 30-45, 96, 192-216, 217, 263, 276-9; arrival and departure 34-44, 192-216, 278, 279; assessment of flock size 199; constant arrival rate 34-44; continuously growing flocks 37-44; equilibrium flock size 34-44, 81, 193, 195-212, 215-16, 278-9; and feeding efficiency 192-9, 199-201; and field size 195-6, 200, 202; and effect of golden plovers and lapwings 210-16; and effect of gulls on plovers 210-16; and effect of lapwings on golden plovers 210-16; random arrival rate 34-44; and sequence of departures 207-8; size distribution of arriving/departing groups of plovers 199-200; and size and species composition 192-216; stability 37-44; statically stable 37-44, 193-5; and time budgeting 192-216; variable flocks 194-9; variation in arrival/departure rates of plovers 200-2; and worm density 195-6
flock positive and negative species 25, 44-5
flock size: and flock dynamics 31, 32; anti-predator costs and benefits 34, 44 (*see also* flocking), feeding costs and benefits 34-44 (*see also* flocking); and food density 73-7, 258; of foraging flocks 81; time limited 38-9; variance 196-8; *see also* time budgeting
flocking, mixed species: aggression and competition 11-13, 165-74; anti-predator effects of 143; attendant species 24; copying 24; early detection of predators 11-13, 27; changes in foraging niche 11-13, 24; changes in foraging strategy 26; food depletion 11-13, 167; foraging niche overlap 155-6; formation of 29, 36; forms of aggression within 165; indicating good feeding areas 76,96-9, 204, 216, 224, 276; increased food availability 11-13; in plovers 53-6; interference within 167, 170, 181; and modulating response to alarm 259-61; more time to feed 11-13; local information about food 29, 30; nucleus species 24; partitioning by kleptoparasites 59; prey disturbance 11-13, 53; reduced risk 11-13
flocking, single species 2; aggression and food-stealing 7-10; attracting predators 7-10; area-copying 7-10; confusion effect 14; and decreased risk of doing badly 4-7, 21; dilution effect 14; and disease 7-10; early detection 14, 15; and exploiting good food supplies 4-7, 16, 204, 216; impaired vigilance 7-10; interference 23; in plovers 53-6; leaders 19, 205; local information about food 4-7, 17, 205; and modulating response to alarm 258-9; and more time to feed 4-7; prey depletion 7-10, 22; prey disturbance 7-10, 22, 53; reduced risk of predation 14
flycatcher, pied (*Ficedula hypoleuca*) 227
following behaviour 18
food availability 25, 28; decrease in and aggression 168-72; effect on feeding efficiency 145-65; and rates of depletion 198; effect on time budgeting 145-65
food chain 2
food demand 85
food distribution; patchiness and capture rate 137; patchiness and defendability 205, 215; patchiness and position in flock 196-203, 215; patchy 16, 18, 21, 77-8, 89, 93, 107, 137, 172, 253; random 18; uniform 18
foot-trembling 86, 172, 173
foraging area 107
foraging circuit: in pied wagtail 108
foraging flock: choice of feeding site 66-92; distribution 66-92, 95; local species composition 96; population size and distribution of 81; posture in 102; size and species composition 143-91; types of 193-9, 263
foraging strategy 34, 93, 279; in black-headed gull 52-3, 252-3; individual variation in 185-9
fox (*Vulpes vulpes*) 84
frost 105
functional response (type 2): and

feeding efficiency 162-5; and pasture age 162-5; and prey density 162-5

games theory 256
giraffe (*Giraffa* spp.) 27
giving up time: and food patch quality 227, 253
goose: barnacle (*Branta leucopsis*) 5, 22, 102, 206, 218; brent (*B. bernicla*) 5, 8, 108; white-fronted (*Anser albifrons*) 5, 14, 15
godwit: bar-tailed (*Limosa lapponica*) 10
goshawk (*Accipiter gentilis*) 3
grackle: boat-tailed (*Cassidix mexicanus*) 8, 172
grasses (*Lolium perenne*) 206; (*Poa pratensis*) 206
grassquit: yellow-faced (*Tiaris olivacea*) 11, 26
grebe 25; Rolland's (*Rollandia rolland*) 13, 268; silver (*Podiceps occipitalis*) 13, 268; western (*Aechmophosus occidentalis*) 268
grosbeak: evening (*Hesperiphona vespertina*) 12, 26
guard 27; *see also* sentinel
gull (excluding black-headed) 13; black-billed (*Larus bulleri*) 204-5; brown-hooded (*L. maculipennis*) 13, 268; common (*L. canus*) 58-9; feeding strategies in 61-2, 273; Franklin's (*L. pipixcan*) 268; kleptoparasitic species 61-2; laughing (*L. atricilla*) 218
guppy 3

habitat change: golden plover 49; improvement and afforestation 49; lapwing 48
Halobates robustus 3, 14
handling: by targets 224; *see also* worm size selection
head turning 121, 256
heron: great blue (*Ardea herodias*) 5, 17
honeycreeper: Hawaiian (*Loxops virens*) 108
hornbill 27; *Tockus deckeni* 26; *T. T. flavirostris* 26

ice age 53-6
impala (*Aepyceros melampus*) 3, 27

information centre: versus assembly point 204-9, 216, 276; selective association with plovers by gulls 206-9; sequence of group departures 207-8; *see also* roost
instantaneous feeding rate 16
intention movement 218, 259
internal flight 112
interscan interval 236-41, 258

jackdaw (*Corvus monedula*) 13, 25, 112
'judge' strategy 257
junco 196, 198; yellow-eyed (*Junco phaeonotus*) 4, 7, 15, 16, 30-3, 256-7

katydid 26
kestrel (*Falco tinnunculus*) 261
kinglets (*Regulus satripa*) 11, 25
kleptoparasitism 23, 26, 52, 58, 273-9; alternative strategies of 20; attack efficiency 20, 149, 217-41; costs of 133, 179; costs to hosts 217-54; ecological and behavioural factors favouring 21; effects of gulls on plover feeding efficiency 146-65; energetic cost 218-19, 224; and flock dynamics 210-16; in gulls 59-62, 217-54; host choice 247-50; host manouverability 159, 190, 224, 248, 253; host responses to 20, 253-4, 273, 279; host vulnerability 218, 227, 231, 243; intraspecific 18; and post-roost aggregations of plovers 206-9; selectivity in attack 224; snatching 18, 19, 20; risk of attack 161; territoriality and 20; *see also* attack
knot (*Calidris canutus*) 53, 56

lipid index 183
loafing 49
local enhancement 17, 19; and worm density 99-102
Lumbricus: *rubellus* 125; *terrestris* 73, 79, 87, 158

magpie (*Pica pica*) 13, 112
maintenance activities 255
merganser; red-breasted (*Mergus serrator*) 6, 22, 25
merlin (*Falco columbarius*) 14
migration 86; black-headed gull 52; golden plover 57, 183-4; lapwing

47, 57, 184; spring in geese 108; and reduced food availability 86
minnow 3, 14
mobbing 3, 14
mongoose: dwarf (*Helogale undulata rufula*) 26, 27
moonphase: and foraging in plovers 67; and night feeding 87-8
musk ox (*Ovibus moschatus*) 3

Nephthys 119
Nereis 119
night feeding 87-8, 182-3, 277; *see also* moonphase
nuthatch (*Sitta* spp.) 11; brown-headed (*S. pusilla*) 25; red-breasted (*S. canadensis*) 24

olfactory prey detection 158
optimal diet selection 117-42, 279; breadth 136-42, 244-7; in gulls 241-7, 252-4; in plovers 120-42; in waders 118-42
optimal flock (group) size 2, 30-45; perception of 42; versus actual 32-6, 40; unstable 34, 40-4
optimal foraging theory 116-42; classical 243
optimal scanning rate 256-7
optimal species (flock) composition 2, 36
ostrich (*Struthio camelus*) 5, 25
oystercatcher (*Haematopus ostralegus*) 9, 58, 59, 82, 120, 134, 165

parasitism 19; astome ciliate 133; monocystid gregarines 133; nematode 133; parasite load 125, 134; *Parvatrema affinis* 134; risk of infection 133; syngamiasis 134
partridge: grey (*Perdix perdix*) 134; red-legged (*Alectoris rufa*) 134
pasture age: and field preference 71-92, 269, 275; and local bird density 102-3; and local soil quality 99; and spatial predictability of prey 78, 93-115; and surface characteristics 80-5, 105, 275; and types of foraging flock 195; and wet weight biomass 74-7; and worm availability 81, 105, 204-5; and worm density 73-7; and worm patchiness 77-8
peregrine falcon (*Falco peregrinus*) 3
pheasant (*Phasianus colchicus*) 134
phenotype-limited characteristics 34
pike (*Esox lucius*) 3, 14
plover: association between species 53-6; evolution of pasture use 53-7; fossil record 53-4; grey (*Pluvialis squatorola*) 85, 144, 177, 183; ringed (*Charadrius hiaticula*) 9, 85, 144
plover's page 259-60
plumage variation: in golden plover 49; in lapwing 185
pollutant concentration: in earthworms 125, 134; heavy metals 134; pesticides 134; radio-active soil contaminants 134
population: closed 257; effect of size on flock distribution 89; extinction 49; golden plover 49; lapwing 47; of plover spp. in study area 64, 85; size 36-44
post-foraging flock 68-70, 87
predation: risk of 256; and edge of flock 202; and field size 39, 215; and flock dynamics 215; of plovers 261; in relation to scanning 121
predator attack strategy 258
predator detection 256
preference index 70, 71
pre-foraging (post-roosting) flock 68-70, 87, 192, 205-9, 275-6; gathering 206; and selectivity in gull departures 206-9
prey depletion 82-4, 105, 278; and aggression 168, 170, 174; and pasture age 105; and renewal 105, 276
prey recognition 117, 118; in gulls 238
prey rejection 133, 141, 220, 243, 277
primate 27
protector species 268
protein reserve 183
puffin (*Fratercula arctica*) 25

range: black-headed gull 51-3; golden plover 48-9, 58; lapwing 47-8
raven (*Corvus corax*) 25
reciprocal altruism 19
redshank (*Tringa totanus*) 9, 23, 58, 59, 119, 171; ingestion rate in 119
redwing (*Turdus iliacus*) 12, 28-30, 84, 269
relocation flight 219, 227, 253; and

capture rate 229
resettlement flight 112-13, 261; and field size 270-2; and local depletion 112; and local worm density 112-13; and worm distribution 112-13
return times 22; foraging groups and 108; optimal 107-9; in plovers 109-11; and prey depletion 103-11
robin, American (*Turdus migratorius*) 188
rook (*Corvus frugilegus*) 6, 10, 13, 84, 112, 261-2
roost 2, 33, 36, 192, 194; communal 18; departure for 201; as information centre 18, 19, 204-5; mutual benefit of 19; roosting flock 72
rule of thumb 95, 157, 277

safety in numbers 2, 15; in mixed flocks 27
sandpiper (Scolopacidae) 6, 9, 14, 53, 86; purple (*Calidris maritima*) 12
scanning *see* time budgeting; vigilance
scavenging 62
sea plaintain (*Plantago maritima*) 22, 108
search path 121; meander ratio of 172
'searcher' strategy 18
searching time: and location of profitable lapwing flocks 208
secretary bird (*Sagittarius serpentarius*) 25
seedeater: black (*Sporophila aurita corvina*) 11, 26; white-collared (*S. torquilla*) 11, 26
selfish herd effect 14
sentinel 259; *see also* guard
sex 47; and aggression 186-7; and competitive ability 191; and feeding territories 186; and foraging behaviour 186-9
shearwater (*Puffinus* spp.) 13
shelduck (*Tadorna tadorna*) 5, 8
shrike (*Lanius* spp.) 26
shrimp (*Corophium volutator*) 23, 171
siskin: pine (*Carduelis pinus*) 12, 26
skua 13
sleep 70, 255
snail (*Hydrobia ulvae*) 23
social facilitation 17, 19
spacing: as anti-kleptoparasitic strategy 224, 236, 279; between species 58; between gulls and plovers 224; clumping of plovers 228; distribution within flocks 59; inter-neighbour distance 167-8, 224
sparrow: *Harris* (*Zonotrichia querula*), 4, 8, 18, 20; house (*Passer domesticus*) 4, 7, 15-19, 22, 26, 37-9, 176-7, 196, 258, 269; vesper (*Pooecetes gramineus*) 15, 269
sparrowhawk (*Accipiter nisus*) 261
spatial memory 115
starling (*Sturnus vulgaris*) 4, 7, 15, 23, 134, 202
stickleback (*Gasterosteus aculeatus*) 3, 14
stork (*Ciconia* spp.) 25
study area 62-4; land use in 63-4
sub-roost 70

temperature: and choice of feeding site 67-92; choice within feeding site 111; and diet selection 121-91; and dispersion from roost 206-7; effect on energy demand 85, 94; effect on flock formation 86, 90-1; effect on foraging behaviour 143; and equilibrium flock size 196-7; and existence energy requirement 176; and feeding priority 272; and fighting 31; and flock size 33; and flock size variance 196-8; effect on time budgeting 145-65; effect on time spent feeding 181-2; and gull's food requirement 207; and prey availability 32, 88, 105; and prey capture rate 85; and reduced prey activity 86, 144
tern (*Sterna* spp.) 22; Forster's (*S. forsteri*) 268
territoriality 107, 108; breeding 108; defence and worm density 187; defence and worm patchiness 187; feeding territories in plovers 187, 190; mobile feeding territories in gulls 59, 254
threat: in black-headed gulls 59
time budgeting 15, 28-33, 120, 121, 217, 258, 275-9; crouching 28, 29, 30, 120, 121, 130-3, 136-42, 144-91, 213, 219, 234, 237-40, 243, 253, 277, 279; definition of 120-1; effects of flock/subflock size on 149-65, 200-2; and feeding efficiency 143-91; and flock dynamics 210-16;

in golden plovers 150-2; in lapwings 146-50; looking 29; and metabolic costs 177-8, 251-2; and responsiveness 269-74; scanning 16, 28-33, 102, 120, 121, 143-91, 202, 221, 255-74
tit 11; coal (*Parus ater*) 11, 24; crested (*P. cristatus*) 11, 24; great (*P.major*) 4, 18, 24; tufted (*P. bicolor*) 12; willow (*P.montanus*) 8, 11, 24
tit-for-tat 257
touch foraging 23, 86
trade-off: between foraging and territorial defence 117; between foraging and vigilance 117; between scanning and crouching 149, 276-7; between worm profitability and cost of attack 232-4
Trafalgar effect 3
turnstone (*Arenaria interpres*) 12, 53, 56

ungulate 27

vigilance: of target 221, 234; in gulls and plovers 255-74; and responsiveness 261-9; social behaviour and vigilance for predators 255-61; subflock 262, *see also* corporate vigilance; *see also* time budgeting
visual foraging 23, 58, 80, 158-9
vulture: black (*Coragyps atratus*) 7, 22

wagtail: pied (*Motacilla alba*) 107; territoriality in 107
warbler (*Dendroica* spp.) 11; pine (*D. pinus*) 25
weasel (*Grison vittatus*) 268
weaverbird (*Quelea quelea*) 4, 8, 18, 258
wheeling 101, 115
whinchat (*Saxicola rubetra*) 4
windspeed and chill factor 179; and energy demand 85; and handling time 144; and heat loss 85
winter feeding ground, lapwing 47
wolf (*Canis lupus*) 14
woodpecker: downy (*Picoides pubescens*) 5, 8, 12
woodpigeon (*Columba palumbus*) 3, 6
worm density 94, 103, 107; and arrival of gulls 209; and capture rate 135, 136; and crouching 149, 155, 277; effects of ploughing on 74-7, 101; and position in flock 203; renewal rate 110-11; vertical distribution of 80
worm size: availability 234; depth and profitability 155; and distance from gulls 231; distribution 94; and energy content 121-6, 224, 231-2, 241; and orientation time 132, *see also* worm size selection; in relation to depth 130-42; range 155; vertical distribution of 155
worm size selection 80, 116-42, 241; assessment and 130-3, 149, 152, 156-9, 277, *see also* crouching; and breakage 79, 125-6, 241, 277; costs of 126-34, 241-3; costs of kleptoparasitism 133, 243; by gulls 231; handling time 117-42, 145, 231, 234, 237-50, 253; host and 241-50; and nutrients 117-18, 140; and profitability 117-91, 241, 243-7; searching time 117-42, 279

zebra (*Equus* spp.) 27

GPSR Compliance

The European Union's (EU) General Product Safety Regulation (GPSR) is a set of rules that requires consumer products to be safe and our obligations to ensure this.

If you have any concerns about our products, you can contact us on ProductSafety@springernature.com

In case Publisher is established outside the EU, the EU authorized representative is:

Springer Nature Customer Service Center GmbH
Europaplatz 3
69115 Heidelberg, Germany

Batch number: 10823775

Printed by Printforce, the Netherlands